Current Topics in Microbiology and Immunology

204

Springer
Berlin
Heidelberg
New York
Barcelona
Budapest
Hong Kong
London
Milan
Paris
Santa Clara
Singapore
Tokyo

Transposable Elements

Edited by H. Saedler and A. Gierl

With 42 Figures

Springer

Professor Dr. Heinz Saedler
Max-Planck-Institut für Züchtungsforschung
Carl-von-Linné-Weg 10
D-50829 Köln, Germany

Professor Dr. Alfons Gierl
Lehrstuhl und Institut für Genetik
Technische Universität München
Lichtenbergstraße 4
D-85747 Garching, Germany

Cover illustration: The phenomenon of variegation in plants has always attracted scientists and artists. The latter is evident from the Japanese wood block prints collected by Lord Yoritaka Matsudaira (1711–1772). Shaseiga-chou represents a collection of Japanese morning glory (Asagao) phenotypes, an old medicinal plant, including variegated flowers. The insert shows a photograph kindly provided by Professor Shigeru Iida, Science University of Tokyo, exhibiting somatic excisions of the Tpnl element from Japanese morning glory (see p. 155, this volume). (The wood block print is reprinted here with kind permission of Yoritaka Matsudaira, The Matsudaira Foundation.)

Cover design: Künkel+Lopka, Ilvesheim

ISSN 0070-217X
ISBN-13: 978-3-642-79797-2 e-ISBN-13: 978-3-642-79795-8
DOI: 10.1007/978-3-642-79795-8

Softcover reprint of the hardcover 1st edition 1996

Library of Congress Catalog Card Number 15-12910

Typesetting: Thomson Press (India) Ltd, Madras
SPIN: 10495150 27/3020/SPS – 5 4 3 2 1 0 – Printed on acid-free paper.

Preface

Most genes are lined up on chromosomes like pearls on a string. However, a certain class of genes differ by being highly mobile; they are termed transposons. Their properties and the mechanism of transposition will be described in this book.

Where uniformity is the rule, irregularities like a spot on a plain-coloured surface strike the eye. Thus the phenomenon of variegation among organisms has long been a source of fascination. In plants, variegation is most easily recognised as irregularities in pigment patterns on leaves, flowers and seeds, but other characteristics such as leaf or flower form might also show variegation. In 1588, such a variegation pattern was described in kernels of *Zea mays* by Jacob Theodor of Bergzabern, a village south of Strasbourg. The report by Theodor is so detailed that if one counts the different kernel phenotypes described it becomes clear that the author is looking at a Mendelian segregation. It goes without saying that Latin-American Indians had already bred such variegated material much earlier, but no descriptions have yet been uncovered. Meanwhile, genetically heritable variegation patterns have been described at many different loci in more than 34 different plant species.

In the late 1940s, Barbara McClintock developed the concept of what she termed "controlling elements" based on variegation patterns. These "controlling elements" were genetic units associated with a gene, for example one involved in kernel pigmentation of *Zea mays,* thus controlling the expression of that gene. This would result in colourless kernels which feature many coloured spots due to the instability of the element, a phenotype already described by Jacob Theodor in 1588. Through her very through genetic analysis, for which McClintock received the Nobel prize in 1983, the current picture of transposable elements ultimately evolved.

The concept is that transposable elements are genetic entities which can migrate through the genome of an organism. However, a molecular understanding of this simple picture required a long journey from early genetic to current molecular studies, which involved many diverse organisms and laboratories. In the mid 1960s, revertible polar mutations in the galactose operon of *Escherichia coli* were under study in Cologne and elsewhere. In 1968 these were shown to be due to DNA insertion (IS) elements. Their relationship to "controlling elements" in corn soon became apparent. At about that time a mutation of the white locus of *Drosophila melanogaster* was described, which had similar properties to an IS1 mutation in the gal operon of *E. coli,* i.e., both induced chromosomal deletions flanking the mutation. However, it was not until 1973, when heteroduplex studies with F and R DNA revealed the presence of IS elements on these plasmids, that a larger group of scientists became interested in DNA insertion elements, thus merging the field of medical bacteriology with molecular studies of DNA insertion elements. It soon became clear that IS elements could border other genes, in particular antibiotic resistance genes, and cause them to transpose from one DNA molecule to another. Subsequently it was found that many antibiotic resistance genes were organized in such structures, thereby allowing transposition.

In the 1980s, the field expanded further due to the isolation of plant transposable elements. Molecular studies revealed that McClintock's "controlling elements" were indeed DNA insertion elements. The analysis of their structure and function prevailed throughout the 1980s and the early 1990s. While in the bacterial and fly worlds researchers had moved to the analysis of the mechanism of transposition, in the plant world emphasis was placed on the use of these elements as tools to isolate other genes of interest. Transposon tagging soon became a powerful technique to isolate fly and plant genes even in heterologous hosts.

Although our knowledge about transposons in various organisms such as bacteria, flies and plants is ever increasing, their biological relevance remains obscure. In the current volume the state of research on transposons in various organisms is reviewed, in the hope of attracting researchers from other fields to enjoy and hopefully contribute to this fascinating field of science.

Köln

H. Saedler
P. Starlinger

List of Contents

List of Contributors

(Their addresses can be found at the beginning of their respective chapters.)

Bacterial Insertion Sequences

E. Ohtsubo and Y. Sekine

1 Introduction

While DNA has a property of being fundamentally stable as invariable genetic information, studies on gene expression and gene organization have revealed that the genome is often subject to dynamic changes. Some of these changes are brought about by mobile genetic elements which have been found in prokaryotic and eukaryotic genomes so far studied. Insertion sequences (ISs) are bacterial mobile DNA elements which cause various kinds of genome rearrangements, such as deletions, inversions, duplications, and replicon fusions, by their ability to transpose. These were discovered during investigation of mutations that are highly polar in the galactose and lactose operons of *Escherichia coli* K-12 (Jordan et al. 1968; Malamy 1966, 1970; Shapiro 1969) and in the early genes of bacteriophage λ (Brachet et al. 1970). Many of these mutations were shown by

Institute of Molecular and Cellular Biosciences, The University of Tokyo, Bunkyo-ku, Yayoi 1-1-1, Tokyo 113, Japan

electron microscope heteroduplex analysis to be insertions of distinct segments of DNA which are hence called insertion sequences (Fiandt et al. 1972; Hirsch et al. 1972; Malamy et al. 1972). It was later shown that the transcription of flanking genes can originate from promoters located within an IS or from hybrid promoters created by the insertion event or by the IS-mediated genome rearrangements. An important note here is that the finding of IS elements as mobile elements to new loci to turn genes either off or on would re-evaluate the controlling elements described in maize by McClintock (1956, 1965) (see Starlinger and Saedler 1976).

In addition to the IS elements found in early investigations (IS1, IS2, IS3, and IS4), numerous others have been identified in the genomes, plasmids, and bacteriophages of a wide range of bacterial genera and species. They range in length from 800 to 2500 base pairs (bp) and can be found in the genomes of many different bacteria at multiplicities between a few and a few hundred per genome. They are often associated with genes responsible for resistance to antibiotics, heavy metal ions, etc. as components of transposons that frequently appear on natural plasmids.

IS elements contain one or more open reading frames encoding an enzyme, transposase, which is required for transposition. The termini of the majority of known IS elements carry inverted repeats (IRs) of about 10–40 bp, which are also required for transposition. These terminal repeats serve as recognition sites for transposase during the transposition process. IS elements can move to new sites by mechanisms largely independent of the homology-dependent recombination pathway. Upon insertion, these elements generate small, directly repeated duplications of the target DNA at the insertion point. This is presumably due to the staggered cleavage of target DNA by transposases. Many elements appear to induce a duplication of a fixed number, ranging from 2 to 13 bp, but some show variations in target duplication length. In a variety of DNA rearrangements, a target sequence at the illegitimate recombination junction is also duplicated.

In this chapter, we describe various IS elements and their characteristics, focusing on two elements, IS1 and IS3, which we have been studying for the past few years. We will present in the first section the finding that the majority of IS elements found in gram-negative and gram-positive bacteria actually belong to several families, which are classified based on structures, mechanisms of transposition and gene expression, and homologous genes encoding transposases. It is remarkable that the range of hosts for some families is found to be extremely broad, even including fungi, plants, invertebrate and vertebrate metazoa, and ciliated protozoa. It is particularly interesting that the most conserved transposase domain of several IS families is shared by retrovirus and retrotransposon integrases. Moreover, both prokaryotic and eukaryotic members of some families require translational frameshifting to produce the active transposase or integrase which apparently promotes the cleavage and integration of the elements. In the second and final sections, therefore, we will present a mechanism of the expression of transposases of several elements by translational frameshifting and the generation of possible transposition intermediates by the action of transposases, to show that some IS elements use mechanisms in both gene expression and transposition similar to those used by retroelements. For the most comprehensive recent review, see GALAS and CHANDLER (1989), who have described the current understanding of the nature, occurrence, and genetic activities of various IS elements in detail.

2 Families of Insertion Sequences

Until recently, IS elements were considered to be a heterogeneous class of bacterial mobile DNA elements, and affiliations were restricted to highly similar ISs originating from related hosts. With the increasing number of known ISs, more distant relationships were pointed out, leading to the emergence of families of elements with conserved transposase domains, common structural features of similar functional properties. The majority of the IS elements so far identified may belong to the following families (see Table 1).

2.1 IS1 Family

IS1 (FIANDT et al. 1972; HIRSCH et al. 1972; MALAMY et al. 1972) is present in various copy numbers in chromosomes as well as in plasmids of bacteria belonging to Enterobacteriaceae (NYMAN et al. 1981; RAMIREZ et al. 1992; LAWRENCE et al. 1992; BISERCIC and OCHMAN 1993; MATSUTANI and OHTSUBO 1993). IS1 is the element that appears as spontaneous insertion mutations in various genes much more frequently than other ISs (for recent examples, see TOBA and HASHIMOTO 1992; RODRIGUEZ et al. 1992; OU et al. 1992; SKALITER et al. 1992). IS1 from resistance plasmid R100 is 768 bp, the smallest known, and has imperfect IRs (IRL and IRR) of about 30 bp at its termini (see Fig. 1) (OHTSUBO and OHTSUBO 1978). The majority of sequenced IS1 insertions generate target duplication of 9 bp, as previously identified (GRINDLEY 1978; CALOS et al. 1978). However, duplications of 7, 8, 10, 11, and 14 bp have been also observed (see GALAS and CHANDLER 1989 for references). IS1 is involved in various kinds of genomic rearrangements including co-integration between two replicons to form a characteristic co-integrate with two copies of IS1 (see Fig. 2) (IIDA and ARBER 1980; E. OHTSUBO et al. 1980). The co-integration event generates target duplication at the recombination junctions (Fig. 2) (E. OHTSUBO et al. 1980). Note that many IS elements including IS10R which transpose in a nonreplicative manner (see N. KLECKNER et al., this volume) do not generate the co-integrate.

IS1 contains two open reading frames, called insA and insB, which are essential for transposition (E. OHTSUBO et al. 1981; Y. MACHIDA et al. 1982, 1984; JAKOWEC et al. 1988). insA and insB code for putative proteins of 91 and 125 amino

Table 1. A list of insertion sequences and related elements identified in each family

Family	ISs and related elements
IS1	IS1 (R100) or IS1R, IS1A~IS1G, IS1(SD), IS1(vξ), IS1(SS), IS1(SF), IS1SFO, IS1s1~3, IS1Efe, IS1Ehe, IS1Sfl, IS1Sso, IS1Sdy
IS3	IS2, IS3, IS21, IS26, IS51, IS120, IS136, IS150, IS240, IS426, IS476, IS481, IS600, IS629, IS861, IS904, IS911, IS986, IS3411, IS6110, ISR1, ISL1, ISS1W, [IS232, IS981, IS987, IS1076L, IS1076R, IS1138, IS1533] (HTLV1, HTLVII, HIV1, Human-D, HumERVKA, Visna, FIV, EIAV, BLV, RSV, Mouse IAP, MMTV, MoMuLV, BaEV, SNV, HSpuENV, *Drosophila* 412, *Drosophila* 17.6, *Drosophila* 1731, Gypsy, Copia, *Bombyx* Mag, Tobacco Tnt, *Arabidopsis* Ta1-3, *Trichplusia ni* Ted, Ty1-17, Ty3-1, Ty3-2, S. *pombe* Tf1)
IS4	Group A: IS4, IS10R, IS50R, IS186A, IS186B, IS231A~IS231F, IS231V, IS231W, IS421, IS701, IS942, IS1151, IS5377, ISH26, ISH27-1~ISH27-3, ISH51-3, [IS231G, IS231H]
	Group B: IS5, IS102, IS112, IS402, IS427, IS493, IS702, IS869, IS903, IS1031A, IS1031C, IS1031D, IS1096, IS1106, ISH1, ISH11, ISH28, ISRm4, ISTUB, ISVM5-2, Tn4811 [IS6501, IS31831]
IS630-Tc1	IS630, IS895, IS1066, [IS870]
	(Tc1, IpTc1, CbTc2, Tc3, CbTc, Tec1, Tec2, TBE1, *Mariner*, CpMar, Tes1, Bari, Minos, RSa, RiATL, DmUhu, DmHB1, DmHB2)
Unusual ISs	IS110, IS116, IS117, IS492, IS900, IS901, IS902, IS1000, IS1111a, IS1533
IS1071 and Tn3	IS1071
	Tn3 family transposons: γδ, Tn3, Tn21, Tn501, Tn917, Tn2501, Tn3926, Tn4430, Tn4556
IS30	IS30, IS1086, IS4351, [ISAS2]
IS15 (or IS6)	IS15Δ, IS15(P-21), IS15R, IS26, IS46, IS176, IS240A, IS240B, IS257L, IS257R1, IS257R2, IS431L, IS431R, IS946, IS6100, ISS1N, ISS1S, ISS1T, ISS1W, [IS904]
IS91	IS91, IS801
IS256	[IS256, IS406, IS905, IS1081, IS6120, IST2, ISRm3]
not classified	[IS53, IS66, IS200, IS298, IS407, IS466, IS511, IS866, IS891, IS892, IS986, IS1016, IS1131, IS1133, IS1136, ISC1217, ISAE1, ISAS1, IS-PA-4, IS-PA-5]

ISs and related elements in eukaryotes (in parentheses) are listed whose transposase/integrase sequence relations have been studied. For IS1 family members, see OHTSUBO et al. (1984), UMEDA and OHTSUBO (1991), MILLS et al. (1992) and LAWRENCE et al. (1992); for IS3 family members, see FAYET et al. (1990), KHAN et al. (1991), KULKOSKY et al. (1992) and REZSÖHAZY et al. (1993a); for IS4 family members, see REZSÖHAZY et al. (1993a); for IS630-Tc family members, see HENIKOFF (1992) and DOAK et al. (1994); for unusual IS members, see LENICH and GLASGOW (1994); for IS1071 and Tn3 family members, see NAKATSU et al. (1991) and MAEKAWA and OHTSUBO (1994); for IS30 family members, see DONG et al. (1992) REZSÖHAZY et al. (1993a); for IS6 (IS15) family members, see KATO et al. (1994) and REZSÖHAZY et al. (1993a). Note that IS4 family elements are separated into two groups, A and B (see text). The other IS elements, including those which may or may not show homology with the family members, are shown in square brackets without references.

acids (aa), respectively, and the two frames are 47 bp apart (see Fig. 1). There is an open reading frame of 126 bp, named B' frame, in the region extending from the 5'-end of insB. This B' frame overlaps insA in the –1 reading frame (Fig. 1). As will be described in the next section in detail, IS1 uses translational frameshifting to produce a transframe protein from two out-of-phase reading frames, insA and B'-insB (SEKINE and OHTSUBO 1989; LUTHI et al. 1990; ESCOUBAS et al. 1991). The frameshifting event in the –1 direction occurs at a run of six adenines present in the

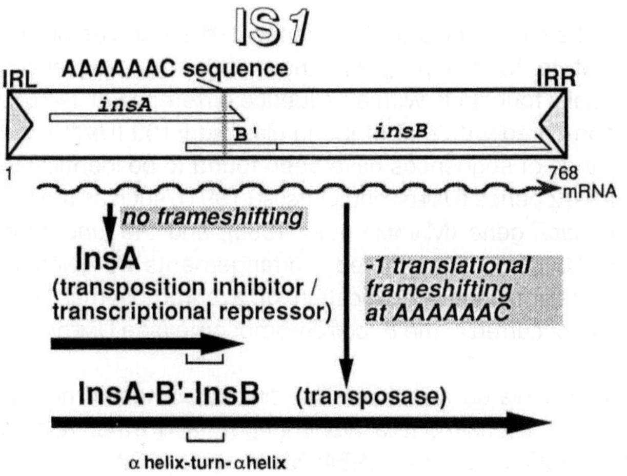

Fig. 1. Structure and gene expression of IS1

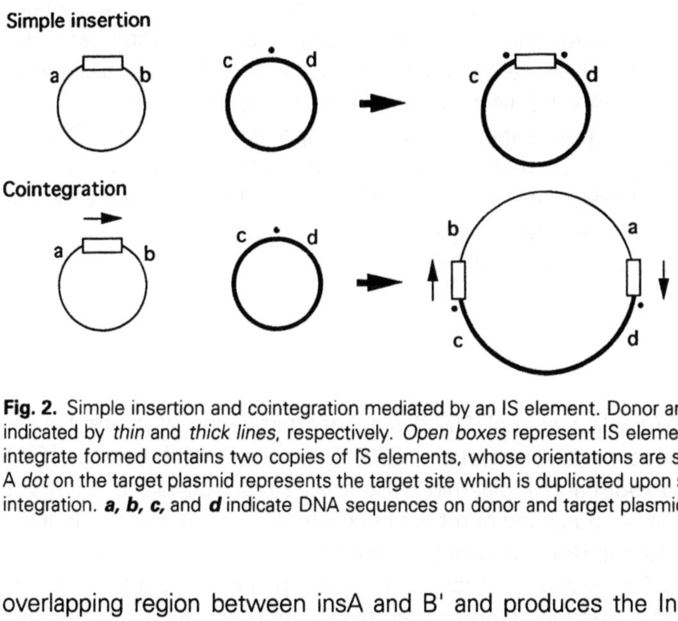

Fig. 2. Simple insertion and cointegration mediated by an IS element. Donor and target plasmids are indicated by *thin* and *thick lines*, respectively. *Open boxes* represent IS elements. Note that the cointegrate formed contains two copies of IS elements, whose orientations are shown by *thin arrows*. A *dot* on the target plasmid represents the target site which is duplicated upon simple insertion or cointegration. ***a, b, c,*** and ***d*** indicate DNA sequences on donor and target plasmid

overlapping region between insA and B' and produces the InsA-B'-InsB fusion protein, i.e., IS1 transposase (SEKINE et al. 1992). Unless frameshifting occurs, only the InsA protein is produced (see Fig. 1).

E. coli K-12 strains contain various number of copies of IS1 in their chromosomes. For W3110, the reference strain of the Kohara phage library (KOHARA et al. 1987), seven copies of IS1 were identified and six were mapped at 0.4, 6.3, 6.5, 22.3, 75.6, and 87.5 min (UMEDA and OHTSUBO 1989) and a possible seventh at 7 min (VAN HOVE et al. 1990) (IS1A-IS1G in Table 1). Recently, the eighth copy of IS1 identified was mapped at 49.6 min (ZUBER and SCHUMANN 1993). BIRKENBIHL and

Vielmetter (1989) also found seven copies of IS1 and located six copies on the physical map (although not to Kohara phages). Their nucleotide sequences revealed that there are actually four kinds with a sequence difference of 1~10% by base substitutions as compared with that of IS1 in plasmid R100 (Umeda and Ohtsubo 1991). Only two kinds of sequences have been found to be identical to the insertions in the *E. coli* K-12 genes (Umeda and Ohtsubo 1991), such as the *lacI* gene (Johnsrud 1979), the *lacZ* gene (Malamy et al. 1985), and the *unc* gene cluster (Kanazawa et al. 1984). The IS1-mediated rearrangements by deletion, tandem duplication, transposition, and amplification of a chromosomal DNA segment, are shown to have occurred in the *E. coli* chromosome (see Umeda and Ohtsubo 1989, 1991).

IS1 elements have been isolated also from the chromosomes of natural strains of *E. coli*, *Shigella* strains, *Salmonella typhimurium*, and other related enteric bacteria (see Table 1) (Ohtsubo et al. 1984; Mills et al. 1992; Lawrence et al. 1992; Bisercic and Ochman 1993). Their nucleotide sequences were very similar to those identified in *E. coli* K-12, except for one in *S. dysenteriae*, called IS1 (νξ) (H. Ohtsubo et al. 1981), which shows sequence difference of about 45% as compared with that of IS1 in R100. The similarity of the IS1 sequences suggests that they have been transferred horizontally among Enterobacteriaceae (H. Ohtsubo et al. 1981; Nyman et al. 1981; Ramirez et al. 1992; Lawrence et al. 1992; Bisercic and Ochman 1993; Matsutani and Ohtsubo 1993). Almost all the IS1 family elements contain open reading frames corresponding to insA and B'-insB, suggesting that they produce their transposases by translational frameshifting (Sekine and Ohtsubo 1989; Lawrence et al. 1992). An IS1-like element of only 116 bp, with IRs and some internal regions of IS1, was found to be widespread in Enterobacteriaceae (Goussad et al. 1991).

2.2 IS3 Family

IS3 (Fiandt et al. 1972; Malamy et al. 1972) is an insertion element present in the *E. coli* K-12 chromosome (Deonier et al. 1979; Birkenbihl and Vielmetter 1989; Umeda and Ohtsubo 1989) and in plasmid F (Hu et al. 1975; Yoshioka et al. 1990). IS3 is also present in various copy numbers in chromosomes of bacteria belonging to Enterobacteriaceae (Matsutani et al. 1987; Lawrence et al. 1992; see Galas and Chandler 1989). This element (1258 bp in length) has imperfect terminal IRs (IRL and IRR) of 39 bp (see Fig. 3) and generates target duplication of 3 bp upon insertion (Sommer et al. 1979; Timmerman and Tu 1985; Yoshioka et al. 1987; Spielmann-Ryser et al. 1991; Lawrence et al. 1992). Unlike IS1, IS3 does not mediate co-integration and is thus supposed to transpose in a nonreplicative manner (Spielmann-Ryser et al. 1991; Sekine et al. 1994). IS3 codes for two open reading frames, orfA and orfB. A reading frame (B') extending upstream from the initiation codon ATG of orfB overlaps orfA in the −1 frame (Fig. 3). Like IS1, a transframe protein of 317 aa, i.e., IS3 transposase, is produced by −1 translational frameshifting at an AAAAG sequence present in the overlapping region between

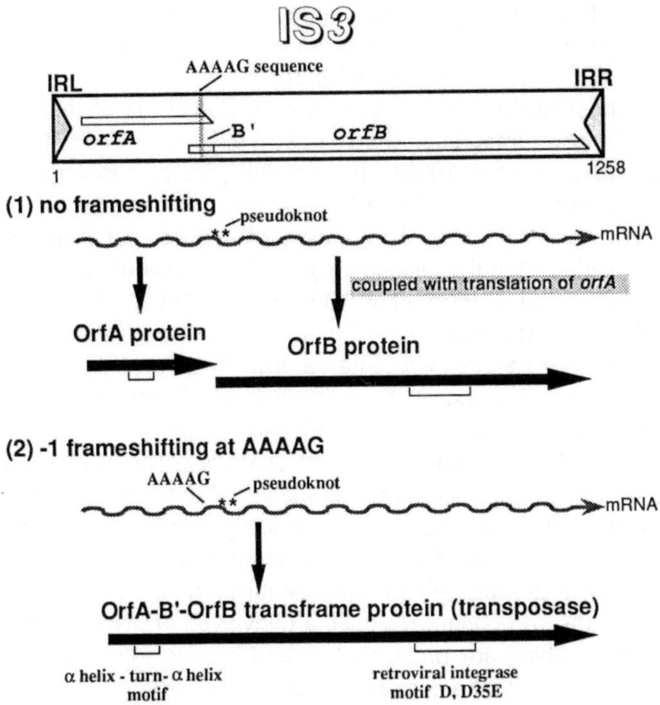

Fig. 3. Structure and gene expression of IS3

the two orfs (Fig. 3; see next section). Unlike IS1, however, both OrfA (99 aa) and OrfB (288 aa) proteins are produced from orfA and orfB, respectively, unless frameshifting occurs (Fig. 3; SEKINE et al. 1994).

There exists a group of IS elements, called the IS3 family (SCHWARTZ et al. 1988), which are structurally related to IS3. This is a large group (FAYET et al. 1990) including IS2, IS150, and IS911 (see Table 1). They are present not only in gram-negative bacteria but also in gram-positive bacteria (see GALAS and CHANDLER 1989). Several copies of IS2 and IS3 (BIRKENBIHL and VIELMETTER 1989; UMEDA and OHTSUBO 1989) and one copy of IS150 (BIRKENBIHL and VIELMETTER 1989) have been mapped on the *E. coli* K-12 chromosomes. IS2 and IS3 are importantly involved in both integration of plasmid F to form Hfr strains and excision from the Hfr strains to form F-prime factors (UMEDA and OHTSUBO 1989). A particular pair of IS3 is involved in inversion of a chromosomal segment (KOMADA et al. 1991; AJDIC et al. 1991). Almost all members of this group have two open reading frames corresponding to those in IS3. Although the predicted amino acid sequences of orfAs show little similarity, those of orfBs are significantly related (SCHWARTZ et al. 1988; FAYET et al. 1990; PRÈRE et al. 1990). Transposases of IS911 and IS150 have been shown to be produced from the two open reading frames by translational frameshifting in the –1 direction (POLARD et al. 1991; VÖGELE et al. 1991), as will be described below in detail. The other IS3 family elements have frameshifting signals similar to those of IS3, IS911, and IS150, suggesting that they also employ frameshifting during expression of their transposase genes (SEKINE and OHTSUBO 1991; LAWRENCE et al. 1992; CHANDLER and FAYET 1993).

Interestingly, the most conserved IS3 transposase domain is shared by retrovirus and retrotransposon integrases (FAYET et al. 1990; KHAN et al. 1991; KULKOSKY et al. 1992; REZSÖHAZY et al. 1993a). In retroviral/retrotransposon integrase proteins of several organisms (see Table 1), sequence comparison of their deduced aa sequences reveals strong conservation of a constellation of aa characterized by two invariant Asp (D) residues and a Glu (E) residue, which is referred to as the D, D(35)E region (KHAN et al. 1991; KULKOSKY et al. 1992). The same constellation is in the transposases of the IS3 family elements. The invariant acidic D and E residues in Rous sarcoma virus integrase (KULKOSKY et al. 1992) and human immunodeficiency virus integrase (KULKOSKY et al. 1992; VAN GENT et al. 1992; ENGELMAN and CRAIGIE 1992; DRELICH et al. 1992) are actually shown to be important in site-specific cleavage and integration of viral DNA. The invariant D and E residues are proposed to participate in coordination of the metal cofactor (Mn^{2+} or Mg^{2+}) required for the catalytic activities of integrases. A metal-DNA complex may be necessary to position both LTR and target DNA substrates for nucleophilic attack during the cleavage and joining reactions. The conservation of this region suggests that the component residues are involved in DNA recognition, cutting, and joining, which are properties shared among integrases of divergent origin. It is interesting that a comparison of terminal IRs of IS3 family elements shows a high frequency at the tips of these elements of the sequence 5'-TG...CA-3'; the same sequence is found at the proviral DNA ends of retroviruses and is part of the *cis*-acting region required for integration (FAYET et al. 1990; KHAN et al. 1991). It should be noted here that transposons such as Tn7, Tn552, Tn5090, and Mu, with the ends beginning by 5'-TG, also produce transposases with the D, D(35)E region (RÅDSTRÖM et al. 1994; see also N. Craig, this volume).

2.3 IS4 Family

IS4 (FIANDT et al. 1972), which is 1426 bp with IRs of 18 bp, generates direct repeats of the target sequence of 11~13 bp upon insertion (KLAER et al. 1981). Unlike IS1 and IS3, IS4 displays one long open reading frame encoding a putative transposase of 442 aa and, in fact, produces it (TRINKS et al. 1981). A unique IS4 copy has been mapped at about 97 min in the *E. coli* K-12 chromosome (KLAER and STARLINGER 1980). MAHILLON et al. (1985) have described that IS231 is similar to IS4 in overall structure and in fact shares homology in transposase of IS4. They further noticed that IS231 is similar to not only IS4, but also other IS elements such as IS10R, IS50R, IS5, IS903, and ISH1, thus defining a new family of IS elements (see Table 1). IS231 itself defines a family of eight ISs, 1.7–2.0 kb in length, originating from the gram-positive entomopathogen *Bacillus thuringiensis* (MAHILLON et al. 1985, 1987; REZSÖHAZY et al. 1992, 1993a). These elements are delimited by 20-bp IRs. Six of them (IS231A~IS231F) display one long open reading frame encoding a 477/478-aa transposase. The other two (IS231V and IS231W) show two slightly overlapping open reading frames (ORFA and ORFB) on the same DNA strand. It was speculated that in these two elements +1 (or +2) translational frameshifting could lead to the synthesis of a single 472 aa transposase (REZSÖHAZY et al. 1993b). Altogether, the eight IS231 transposases share 40% sequence identity, with five regions displaying more than 60% identity. One of these regions, designated C1, corresponds to a conserved 60 aa C-terminal box, which is shared by IS4, IS10R, IS50R, and ISH1 transposases (MAHILLON et al. 1985).

This IS4 family now comprises more than 40 ISs from widely different origins (REZSÖHAZY et al. 1993a). Members of this family all display the conserved C1 region. Moreover, they also share a second region (designated N3) highly conserved in the N-terminal half of the IS231 transposases (REZSÖHAZY et al. 1993a). The pattern of sequence similarities within these conserved regions and their relative location within the transposase define two groups within the IS4 family, the IS4 and IS5 groups (see groups A and B in Table 1). These relationships are reinforced by sequence conservation found within the IR sequences of these elements, in which the three external nucleotides, defining the limits of the elements, are rather group specific, 5'-CAT... for group A and either 5'-GGC... or 5'-GAG... for group B (REZSÖHAZY et al. 1993a). Interestingly, the N3 and C1 regions of the IS4 family may correspond to the integrase domain shared by retroelements and the IS3 family members (REZSÖHAZY et al. 1993a). This relationship might indicate a common step in the transposition mechanism of these otherwise unrelated mobile genetic elements. It is interesting to point out that the IS elements in group A include IS10R and IS50R, which transpose in a nonreplicative manner (see N. Kleckner et al., this volume), whereas those in group B include IS102 and IS903, which can form co-integrates (H. OHTSUBO et al. 1980b; GRINDLEY and JOYCE 1981). IS5 in group B is present in the *E. coli* K-12 chromosome in more copies than the others, and their locations have been determined (BIRKENBIHL and VIELMETTER 1989; UMEDA and OHTSUBO 1990a). They are involved in rearrangements of the chromosome by inversion, tandem duplication, etc. (UMEDA and OHTSUBO 1990a).

2.4 IS630-Tc1 Family

IS630 is a 1153-bp element with terminal IRs of 28 bp and has an open reading frame encoding a putative 343-aa protein (MATSUTANI et al. 1987; TENZEN et al. 1990). IS630, which was identified in *Shigella sonnei*, has been shown to transpose in *E. coli* K-12 specifically to the dinucleotide 5'-TA-3' in the core of at least 4-bp palindromic sequences, such as CTAG, TTAA, and ATAT (TENZEN et al. 1990), in which the CTAG sequences are used as hot spots for transposition (TENZEN and OHTSUBO 1991). IS630, like IS3, does not mediate co-integration, indicating that it transposes in a nonreplicative manner.

In prokaryotes, *Pseudomonas* IS1066 (van der Meer et al. 1991), *Anabaena* IS895 (Alam et al. 1991), and *Agrobacterium* IS870 (Furnier et al. 1993) were identified to be related to IS630. In eukaryotes, however, many transposable elements are known to transpose into the dinucleotide TA. Examples are transposable elements, Tc1 and Tc3 of *Caenorhabditis elegans,* Tc2 and Tc6 of *C. elegans,* Tc1-like elements found in arthropods and vertebrates (see R.H.A. Plasterk, this volume), Tec1 and Tec2 of the ciliate *Euplotes crassus* (Jaraczewski and Jahn 1993), and *pogo* of *Drosophila* (Tudor et al. 1992). Henikoff (1992) used blocks of aligned protein segments derived from the Tc1 family members to search a nucleotide sequence databank and detected the relatives of Tc1, IS630, and Tc1-like elements in arthropods and vertebrates. Doak et al. (1994) have reported that the transposable elements TBE1, Tec1, and Tec2 of hypotrichous ciliated protozoa appear to encode a protein that belongs to the IS630-Tc1 family of transposases (see Table 1). Doak et al. (1994) noted that most family members transpose into the dinucleotide target, TA, and that members with eukaryotic hosts have a tendency for somatic excision that is carried to an extreme by the ciliate elements. Alignments including the additional members, and also *mariner* elements, show that transposases of this family share strongly conserved residues in a large C-terminal portion, including a fully conserved dipeptide, DE, and a block consisting of a fully conserved D residue and highly conserved E residue, separated by 35 (or 34) residues (D35E). This D35E motif likely is homologous to the D35E motif of the family of retroviral-retrotransposon integrases and IS3-like transposases. The homologous relations suggest that the two families share homologous catalytic transposase domains and that members of both families may share a common transposition mechanism.

Recently, an element named Tnr1 (235 bp in length) with terminal inverted repeats of 75 bp has been identified in rice (Umeda et al. 1991; Tenzen et al. 1994). Because of its small size, Tnr1 is supposed to be a defective form of an autonomous element capable of transposing by itself. Comparison of nucleotide sequences of the regions with or without a Tnr1 member revealed that Tnr1 transposes to 5'-PuTAPy-3' duplicating TA (Tenzen et al. 1994), suggesting that Tnr1 may be a member of the IS630-Tc1 family. A family of elements named *stowaway,* homologous to Tnr1, are associated with the genes of both monocotyledonous and dicotyledonous plants (Bureau and Wessler 1994). The finding of these elements suggests that the range of hosts for the aggregate IS630-Tc1 family is extremely broad, including bacteria, fungi, plants, invertebrate and vertebrate metazoa, and ciliated protozoa.

2.5 A Family of Unusual IS Elements

Lenich and Glasgow (1994) reported that the predicted amino acid sequence of Piv, an essential protein in site-specific DNA inversion of the pilin segment in *Moraxella lacunata,* shows significant homology with the transposases of a family of IS elements (see Table 1) which were previously proposed to be a group

(Hoover et al. 1992; Kunze et al. 1991; Leskiw et al. 1990; Moss et al. 1992; Stark et al. 1992). These proteins contain four regions where highly conserved aa are clustered, representing aa motifs or domains that are important for transposition. Many of the IS elements in this family are unusual in structure and behavior, and can be subgrouped on the basis of the absence or presence of terminal IRs and of whether insertion results in target site duplications. One subgroup is composed of IS elements found in the gram-positive species *Streptomyces clavuligerus* (IS116), *Mycobacterium paratuberculosis* (IS900), *M. avium* (IS901), and *M. avium* subsp. *silvaticum* (IS902) (Kunze et al. 1991; Leskiw et al. 1990; Moss et al. 1992; Green et al. 1989). These IS elements do not have terminal IRs and do not generate a short repeat at the site of insertion: IS110 and IS117 from *Streptomyces coelicolor* may be part of this subgroup though they have imperfect terminal IRs, and it is uncertain whether IS110 generates target site duplications (Bruton and Chater 1987; Henderson et al. 1989). IS1000 from *Thermus thermophilis* and IS111 a from *Coxiella burnetii* also give no sequence duplications at the site of insertion, but they do have terminal IRs (Ashby and Berquist 1990; Hoover et al. 1992). IS492 from *Pseudomonas atlantica* generates a 5-bp repeat at its target sites but has no terminal IRs (Bartlett and Silverman 1989). IS1533 has imperfect terminal IRs and may generate a 2-bp duplication upon insertion (Zuerner 1994). No common features in the ends of the IS elements or in their target sites have been noted for this family of transposable elements.

2.6 IS1071, a Member of Tn3 Family

Nakatsu et al. (1991) described the structure of a 17-kb transposon Tn5271, which resides in the plasmids or chromosome of an *Alcaligenes sp.* strain. The transposon is flanked by a directly repeated sequence of 3201 bp, designated IS1071. The IRs of IS1071 and the derived aa sequence of the single open reading frame within IS1071 are related to the IRs and transposase proteins of the Tn3 family transposons (see Table 1) (Nakatsu et al. 1991; Maekawa and Ohtsubo 1994). Tn3 family transposons generally code for two genes (tnpA and tnpR) that are necessary for its transposition: tnpA encodes the transposase, which catalyzes the first step of transposition that is the formation of a co-integrate; tnpR encodes the resolvase/repressor, which acts at the res sites to resolve the co-integrate molecule with two copies of Tn3 into two products, each with a single copy of Tn3 (for a recent review, see Sherratt 1989). The absence of tnpR within IS1071 suggests that this element is capable of determining the first step only. This was confirmed by observations on the IS1071-dependent formation of stable co-integrates in a recombination-deficient *E. coli* K-12 strain (Nakatsu et al. 1991). The existence of IS1071 may support an evolutionary scheme in which the Tn3 family transposons descended from simpler insertion sequences.

2.7 Others

Dong et al. (1992) reported that there is a family of IS elements related to IS30 encoding a single open reading frame for a protein (see Table 1). Locations of IS30 on the *E. coli* K-12 chromosome have been determined (Birkenbihl and Vielmetter 1989; Umeda and Ohtsubo 1990b). Rezsöhazy et al. (1993a) noted that members of this family display the conserved regions, which correspond to the integrase domain shared by retroelements and the IS3 family members. Interestingly, the transposase domain found in these members has both the length feature of amino acid spacer characteristic of the IS3 family and the amino acid conservation specific of the C1 domain of the IS4-related elements, indicating a common step in the transposition mechanism of the IS elements belonging to the IS3, IS4, and IS30 families. There is another family, called the IS15 (or IS6) family (see Table 1) (Martin et al. 1990; Kato et al. 1994). Some members of this family have been identified as those of the IS3 family (see Table 1). Rezsöhazy et al. (1993a) noted that five members (IS15Δ, IS240A, IS431L, IS946, and ISS1S), however, display the conserved regions corresponding to the integrase domain but are distinct from the IS3 family elements. These suggest that the IS15 (IS6) family members may be divided into at least two groups.

There exist IS elements which seem not to belong to any of the families described above. The most interesting of them is *E. coli* IS91, which specifically inserts at CAAG or GAAC of target and does not duplicate any sequence upon insertion (Mendiola and de la Cruz 1989). The related IS element to IS91 is *Pseudomonas* IS801 (Romantschuk et al. 1991; Mendiola et al. 1992). IS91/IS801 transposase is interestingly related to the rolling-circle-type replication proteins of the pUB110 family of plasmids which produce a single-strand nick in a specific site, suggesting that transposition of IS91/IS801 involves single-strand nicking by the transposases (Mendiola and de la Cruz 1992). The other elements, which may or may not show homology with the IS elements described above, are listed without references in Table 1. These include IS256 and its relatives, forming a family (see Table 1). It is not our aim to describe characteristics of all of the unclassified elements in detail here. With the increasing number of new ISs, distant relationships will be further pointed out, leading to more or fewer families of elements.

3 Translational Frameshifting in Production of Transposases Encoded by IS Elements

IS1 and IS3 family elements use translational frameshifting in the –1 direction to produce their transposases. In this section we will describe this event in some detail and compare it with the frameshifting event well known in retroelements as producing a transframe polyprotein from which integrase is derived by processing.

3.1 IS1 Transposase

As described in the previous section, translational frameshifting occurs in the –1 direction from the 3'-end region of insA to B'-insB in IS1 to produce the InsA-B'-InsB transframe protein, i.e., transposase (see Fig. 1) (SEKINE and OHTSUBO 1989; ESCOUBAS et al. 1991). This finding could explain many things that cannot be readily accounted for in the expression of insA and the (B')-insB frame: The protein product from insB or B'-insB was not detected, but the insA product was (ARMSTRONG et al. 1986; ZERBIB et al. 1987; ESCOUBAS et al. 1991); two IS1 mutants defective in insA and insB, respectively, do not complement each other in restoring their co-integration ability; a 5-bp insertion in the B' frame results in loss of the ability of IS1 to mediate co-integration (MACHIDA et al. 1982).

The frameshifting in the –1 direction occurs within the run of six adenines in the sequence 5'-TTAAAAAACTC-3' at nucleotide (nt) position 305~315 in the overlapping region (Fig. 1) and produces transposase with a polypeptide segment Leu-Lys-Lys-Leu at residues 84~87 (SEKINE and OHTSUBO 1989; SEKINE et al. 1992). An IS1 mutant with a single base insertion in the run of adenines which produces the transframe protein with the segment Leu-Lys-Lys(or Arg)-Leu without frameshifting could efficiently mediate co-integration and adjacent deletion (SEKINE and OHTSUBO 1989; SEKINE et al. 1992). Substitution mutations at each of three (2nd, 3rd and 4th) adenine residues, which comprise a codon for Lys in insA, caused serious negative effects in frameshifting, but those introduced in the region flanking the run of adenines did not. These indicate that the AAA codon for Lys is the site of frameshifting and that $tRNA^{Lys}$ thus plays an important role in frameshifting (SEKINE and OHTSUBO 1992).

In many genetic systems to be described below, secondary structures downstream of the frameshift site are supposed to elevate opportunities for a change of reading frames. Several possible secondary structures in the region downstream of the run of adenines are, however, not required for frameshifting in IS1, but the termination codon of insA located at 17 bp downstream of the run of adenines plays an important role in enhancement of frameshifting (SEKINE et al. 1992; SEKINE and OHTSUBO 1992). Enhancement of –1 frameshifting by a termination codon immediately downstream of the frameshift site has been reported also in an artificial context (WEISS et al. 1987). The efficiency of frameshifting in IS1 is only 0.2~0.3%, however, which is very low compared with that of other genetic systems (see below). IS1 may adopt the low level of frameshifting, which results in production of a low amount of transposase, to avoid deleterious rearrangement of the host chromosome containing IS1 (SEKINE and OHTSUBO 1989; ESCOUBAS et al. 1991; SEKINE et al. 1992).

The InsA protein has the carboxyl-terminal region containing an α-helix-turn-α-helix motif (see Fig. 1) (ZERBIB et al. 1987; SEKINE and OHTSUBO 1991), which is observed in many DNA binding proteins (PABO and SAUER 1984). It has in fact been shown that InsA specifically binds to both IRL and IRR (ZERBIB et al. 1987, 1990b; SEKINO et al. 1995). IRL contains a promoter for the transcript coding for InsA and transposase, whereas IRR contains a promoter used for synthesizing RNA being oriented opposite to insA and insB (CHAN and LEBOWITZ 1982; C. MACHIDA et al. 1984). Since the InsA-binding regions overlap the –35 region of the promoters, InsA could inhibit the transcription by interacting with RNA polymerase at IRL and IRR. Inhibition of transcription from IRL by InsA has actually been shown in vivo (MACHIDA and MACHIDA 1989; ZERBIB et al. 1990a; ESCOUBAS et al. 1991). The transposase (InsA-B'-InsB) protein, which includes the same DNA-binding motif, binds to the DNA segment with or without the IR sequence but preferentially to that with IR (SEKINO et al. 1995). The nonspecific DNA-binding ability of transposase may be involved in recognition of the target DNA, an important process of transposition of IS1. It is speculated that IS1 transposase consists of at least two domains, the N-terminal half, which almost entirely overlaps InsA, and the C-terminal half, which almost entirely overlaps B'-InsB. The frameshifting event adds the latter domain to the former to give the transposase activity recognizing IRs and the target sequence to initiate the transposition reaction.

3.2 IS3 Transposase

As described also in the previous section, IS3 produces its transposase by –1 frameshifting at an AAAAG sequence present in the overlapping region between orfA and B'-orfB (see Fig. 3). Amino acid sequencing analysis has shown that the sequence 5'-CAAAAGGC-3' at nt position 327–334 in the overlapping region encodes the transframe polypeptide segment Gln-Lys-Gly, suggesting that frameshifting occurs at either codon CAA or codon AAG (SEKINE et al. 1994). Mutational analysis indicates that AAG is the site of frameshifting (SEKINE et al. 1994). This suggests that $tRNA^{Lys}$ recognizing this codon plays an important role in –1 frameshifting. This frameshifting requires a pseudoknot structure in the region downstream of the AAAAG sequence (see Fig. 3) (SEKINE et al.1994). A mutant IS3 with a single base insertion in the sequence, which results in in-frame alignment of orfA and B'-orfB, mediated adjacent deletion to produce various miniplasmids at a very high frequency (SEKINE and OHTSUBO 1991).

The IS3 family members IS150 and IS911 also produce their transposases by –1 frameshifting at a motif AAAAAAG present in the overlapping region between two open reading frames. In IS150 (VÖGELE et al. 1991), the sequence 5'-CUAAAAAAGCU-3' at nt position 528–538 encodes Leu-Lys-Lys-Ala, indicating that a frameshifting occurs at either codon CUA for Leu or one of the consecutive codons, AAA and AAG, for Lys in the 0-frame. This frameshifting requires a single stem-loop structure in the region downstream of the frameshifting site (VÖGELE et al. 1991). In IS911 (POLARD et al. 1991), the sequence 5'-TTAAAAAAGGC-3' at nt position 322–332 was preliminarily shown to encode Leu-Lys-Lys-Gly, indicating that a frameshifting occurs at either codon TTA for Leu or one of the consecutive codons, AAA and AAG, for Lys in the 0-frame.

Unlike IS1, however, IS3, IS150, and IS911 encode two proteins, OrfA and OrfB, in addition to the transposase protein, from two open reading frames unless frameshifting occurs. In IS3, OrfB is produced in a manner dependent on translation of orfA, whose termination codon overlaps the ATG codon of orfB. In other words, OrfB is produced due to translational coupling between orfA and orfB (see Fig. 3). The pseudoknot structure required for frameshifting is important also for translational coupling in production of OrfB. Presumably, the initiation codon for orfB and its upstream sequence within the pseudoknot structure are occluded, unless ribosomes proceed toward the termination codon of orfA, allowing exposure of the region essential for translation of orfB (SEKINE et al. 1994). IS150 and IS911, however, produce OrfB using the mechanism different from that in IS3: In IS150, the OrfB protein, whose coding region begins from an ATG codon which is in phase with orfA and is located upstream of the AAAAAAG motif, is produced in a manner absolutely dependent on frameshifting at the motif (VÖGELE et al. 1991); in IS911, OrfB is produced by utilizing an unusual initiation codon AUU in phase with orfB located upstream of the AAAAAAG motif (POLARD et al. 1991).

Both OrfA and OrfB proteins are not required for transposition of IS3 (SEKINE et al. 1994). The OrfA protein contains an α-helix-turn-α-helix DNA-binding motif in the middle (PRÈRE et al. 1990; SEKINE and OHTSUBO 1991). This motif is also present in the transposase protein (see Fig. 3). It is therefore possible that OrfA competes with transposase to bind to the terminal IRs (IRL and IRR) and could thus become a transposition inhibitor. Since IRL contains a possible promoter sequence for transcription of the IS3-coded genes, OrfA could also inhibit transcription from this promoter, as has been shown in IS1. In fact, our recent results indicate that OrfA inhibits IS3-mediated deletion and that OrfB enhances the inhibitory effect by OrfA, while OrfB alone has no effect (Y. Sekine, K. Izumi, E. Ohtsubo, unpublished results). It has been reported that OrfA encoded by IS911 is, however, not a transposition inhibitor but stimulates intermolecular transposition (POLARD et al. 1992).

As described in the first section, the orfB frame of IS3 family members contains a conserved aa sequence motif found in retroviral integrase. The actual function of the OrfB proteins is not known at present. In the transframe proteins, the conserved region may function in catalysis of the transposition reaction and be directed to its site of action at the IS ends using the N-terminal portion encoded by orfA (SCHWARTZ et al. 1988; PRÈRE et al. 1990; SEKINE and OHTSUBO 1991). The occurrence of frameshifting suggests the two-domain structure in the transposase proteins.

3.3 Other Transframe Proteins Including Retroviral Polyproteins

In prokaryote, a chromosomal gene *dnaX* encoding the τ subunit of DNA polymerase III holoenzyme uses the −1 frameshifting to produce the transframe protein that is γ subunit (BLINKOWA and WALKER 1990; FLOWER and MCHENRY 1990; TSUCHIHASHI and KORNBERG 1990). The sequence 5'-GCAAAAAAGAG-3' in *dnaX* encodes Ala-Lys-Lys-Glu (TSUCHIHASHI and KORNBERG 1990), indicating that a frameshifting occurs at GCA for Ala or one of the consecutive codons, AAA and AAG, for Lys in the 0-frame. Mutational analysis of the region containing the AAAAAAG motif indicates that codon AAG is the site of frameshifting (TSUCHIHASHI and BROWN 1992). It is therefore likely that in IS150 and IS911, which use the same AAAAAAG motif for frameshifting, codon AAG is the site of frameshifting.

Translational frameshifting in the –1 direction has been reported in retroviruses (for a review, see VARMUS and BROWN 1989) and other eukaryotic viruses (BRIERLEY et al. 1989; DINMAN et al. 1991). These viruses have a conserved heptanucleotide motif at which frameshifting occurs as proposed in the simultaneous slippage model (JACKS et al. 1988). In mouse mammary tumor virus (MMTV), a –1 frameshifting occurs at the gag-pro overlapping region with the sequence AAAAAAC (JACKS et al. 1987; MOORE et al. 1987) identical to that of IS1 to produce its transposase, such that the last codon in gag (0 frame) is codon AAC (HIZI et al. 1987), which is downstream by one codon compared with the site in IS1. This difference between IS1 and MMTV [or other retroviruses which are suggested to use the AAAAAAC motif for frameshifting (RICE et al. 1985; SHIMOTOHNO et al. 1985)] might be due to the structural or functional differences between prokaryotic molecules participating in the translational process and those of eukaryotes.

The efficiency of frameshifting is 5~25% in retroviruses (for a review, see VARMUS and BROWN 1989), 40~50% in *dnaX* (FLOWER and MCHENRY 1990; TSUCHIHASHI and KORNBERG 1990), 30~40% in IS150 (VÖGELE et al. 1991), about 15% in IS911 (POLARD et al. 1991), and about 6% in IS3 (SEKINE et al. 1994). Pseudoknot structures have been shown to have an important role in frameshifting in IBV (BRIERLEY et al. 1989, 1991), MMTV (CHAMORRO et al.1992), the yeast double-stranded RNA virus L-A (DINMAN et al. 1991), and IS3 (SEKINE et al. 1994). A single stem-loop structure(s) downstream of the frameshift site has been demonstrated to stimulate frameshifting in *dnaX* (FLOWER and MCHENRY 1990; TSUCHIHASHI and KORNBERG 1990) and IS150 (VÖGELE et al. 1991). Such structures in mRNA are supposed to cause translating ribosomes to pause at the frameshift site, thereby providing tRNA on the ribosomes with elevated opportunities for a change of reading frames (JACKS et al. 1988; BRIERLEY et al. 1989).

As described here, frameshifting is the event that allows synthesis of several different proteins from a single template, and this leads to the merit of storing as much information as possible in a limited amount of DNA and RNA. It is also conceivable that frameshifting may be an exquisite strategy by which some fine regulations are carried out. Some IS elements, including IS231V and IS231W in the IS4 family, which have a set of smaller ORFs assumed to encode transposase may also employ frameshifting as translation trickery during expression of transposase genes.

4 Possible Intermediate Molecules of Transposition of IS Elements

The transposition reaction has not been reproduced in vitro in any IS elements except IS10R (see N. KLECKNER et al., this volume). Transposases from only a few IS elements have been isolated but not characterized for their enzymatic properties other than DNA binding to the terminal IRs. Therefore, the molecular mechanism of transposition of the IS element is not yet well understood. Some IS elements or their related elements in eukaryotes, however, have been observed to generate characteristic circular and linear molecules, whose structures are similar to those identified as intermediates of transposition in the transposons, including those described in the chapters by N. KLECKNER et al. and N. CRAIG, this volume, and retroviruses. Here, we will describe IS elements which produce possible transposition intermediates and discuss the possibility that the IS3 family elements and retroelements may share a common transposition mechanism.

4.1 IS Circles

Circular forms of excised DNA have been found in several sequence-specific DNA rearrangement processes as either an intermediate or an end product. In transposons, circular products have been observed for Tn10 (IS10R) and conjugative transposon Tn916. IS10R generates two types of circular molecules upon the action of transposase (see N. Kleckner et al., this volume). One is a circle, a possible end product, formed by transposition of the element into itself. The second structure is a noncovalently closed structure that consists of a protein-DNA complex formed between the transposase and the ends of the element and can thus be converted to a linear form. Tn916 generates a covalently closed circular intermediate (CAPARON and SCOTT 1989). The junction of the transposon termini in the Tn916 circular intermediates is a heteroduplex that contains extra nucleotides derived from adjacent chromosomal sequences from each end of the integration site. This circular product is capable of reintegration (SCOTT et al. 1988).

POLARD et al. (1992) have reported that IS911, an IS3 family element, generates minicircles consisting of the entire sequence of IS911 and a 3-bp sequence intervening between the IRs. The 3-bp sequence is the same as the direct repeats of a target sequence flanking IS911. The minicircles are assumed to be formed by joining one end of IS911 with the overhanging 3-bp sequence on the other side of IS911. SEKINE et al. (1994) have reported that the IS3 mutant, which overproduces transposase without frameshifting, efficiently generates IS3 circles similar to the IS911 circles, in addition to miniplasmids formed by the IS3-mediated deletion from the parental plasmid. The IS3 circles have the intervening 3-bp sequence identical to either one of the sequences flanking IS3 in the parental plasmid or its miniplasmid derivatives (see Fig. 4) (SEKINE et al. 1994). Though it is not clear at present whether the IS3 circles participate as substrates in transposition or not, POLARD et al. (1992) have suggested that the IS911 circles are not the obligatory transposition intermediates. It has been reported, however, that two copies of IS elements separated by a few base pairs are active in transposition of IS21 (REIMMANN et al. 1989), IS3 (SPIELMANN-RYSER et al. 1991), and IS30 (OLASZ et al. 1993). It should be noted that the active junction composed of two IRs in the tandem repeats of IS elements resembles the IS3 (IS911) circle junction. As will

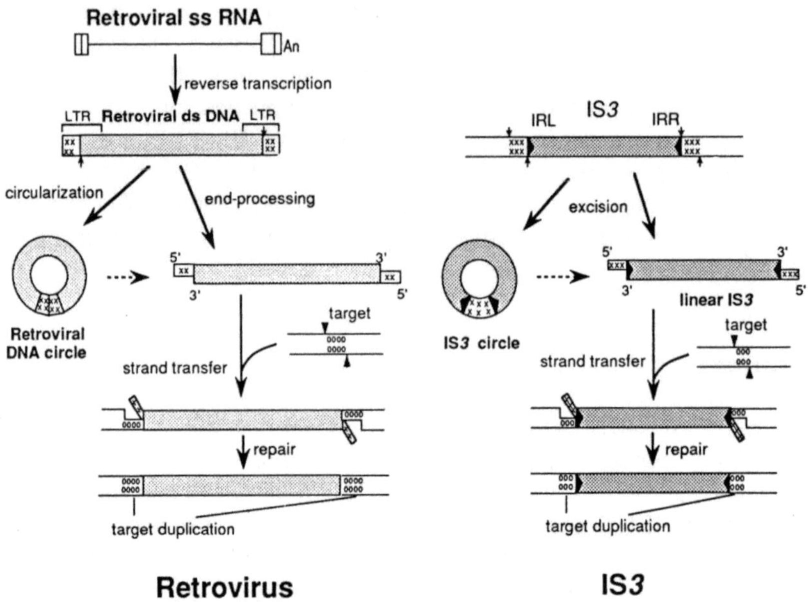

Fig. 4. Retrovirus integration and IS3 transposition

be described below, the IS3 transposase generates linear molecules with 3-nt overhangs at the 5'ends of IS3 which are possible intermediates of IS3 transposition. It is possible that the IS3 circles are converted to such linear molecules (see Fig. 4).

Similarly to IS911 and IS3, an IS1 mutant overproducing transposase generates characteristic circles consisting of the entire IS1 sequence and an intervening sequence, mostly 6–9 bp in length, between the IS1 ends (Y. Sekine, N. Eisaki, K. Kobayashi, E. Ohtsubo, unpublished results). The intervening sequences are derived from either one of the sequences flanking IS1 in the parental plasmid. Such IS1 circles may be formed during a process(es) of transposition or co-integration mediated by IS1.

Retroviruses are known to generate circular DNAs with two LTR sequences after reverse transcription from viral RNA (see Fig. 4). The circular molecules of retroviruses are considered not to be the transposition intermediates during integration of viral DNA (see Brown et al. 1987; Fujiwara and Mizuuchi 1988), but can be converted to linear molecules by integrase, which introduces site-specific staggered breaks at the junction between the two LTRs (see Fig. 4) (Grandgenett et al. 1986; Terry et al. 1988).

In the eukaryotic transposable elements related to IS630, circular forms of the *C. elegans* Tc1 element and *E. crassus* Tec elements have been detected (Rose and Snutch 1984; Ruan and Emmons 1984; Jaraczewski and Jahn 1993). In these circles, the inverted repeats of each element are joined in head-to-head orientation (see R.H.A. Plasterk, this volume). In Tec circles, the inverted repeat junctions consist of both copies of the target site duplication surrounding ten additional bases. An unusual nuclease-sensitive conformation exists at a circular junction that may be the result of the heteroduplex DNA like that seen in Tn916 circles (JARACZEWSKI and JAHN 1993). To date, there is no direct evidence for reinsertion of the circular molecules, although the frequency of transposition correlates with the presence of circular forms.

4.2 IS Linears

In transposons Tn10 (IS10R) and Tn7, which transpose in a nonreplicative manner, double-strand breaks occur at both end regions of the elements to given an intermediate linear molecule, which is subseqûently inserted into the target site (see N. Kleckner et al., this volume). IS3 does not mediate co-integration and is thus supposed to transpose in a nonreplicative manner (SEKINE et al. 1994). This leads us to propose that the IS3 transposase may excise the IS3 sequence from the donor molecules in the IS3 transposition reaction. We have, in fact, found linear molecules of IS3 with 3-nt overhangs at its 5'ends which were generated in addition to the IS3 circles from the plasmid carrying the IS3 mutant overproducing transposase (SEKINE et al. 1995). The nucleotide sequences of the overhangs are the same as those flanking IS3 in the parental plasmid, implying that the linear IS3 molecules are excised from the parental plasmid DNA or from IS3 circles by staggered double-strand breaks at the end regions of IS3 (Fig. 4).

IS3 generates both circular and linear molecules, while IS10R (Tn10) and Tn7 generate linear molecules but not circles. Retroviruses are, however, known to generate double-stranded linear DNA molecules in addition to circular DNAs (see Fig. 4). In this respect, IS3 resembles retroviruses. Note here that, as described in the earlier section, the most conserved IS3 transposase domain is shared by retrotransposon and retrovirus integrases. The conservation of this region suggests that the component residues are involved in DNA recognition, cutting, and joining, since these properties are shared among these proteins of divergent origin. In retroviruses, 2 nt from each 3' end of the linear viral DNA are removed by integrase to produce 5'-protruding ends (CRAIGIE et al. 1990; KATZ et al. 1990; KATZMAN et al. 1989; SHERMAN and FYFE 1990), and the 3' ends of the linear molecules are subsequently joined to the 5' ends generated at a target site (Fig. 4) (FUJIWARA and MIZUUCHI 1988; BROWN et al. 1989). Thus in IS3, the 3'-OH of the linear IS3 molecule is likely to be joined to 5'-P of the target DNA, which is supposed to be exposed by 3-bp staggered breaks, and subsequently the 3-nt gap on the opposite strand is repaired to convert the gap to a duplex form and to remove the 3-nt donor sequence attached to the 5' end of the linear IS3 molecule (see Fig. 4).

As described in this section, the molecular mechanism of transposition of the IS elements including IS3 is poorly understood compared with those of some transposons and retroviruses. Further investigation by reproducing in vitro the transposition reaction of a representative IS element(s) in each family and by purification and characterization of its transposase is needed.

Acknowledgments. We would like to thank Dr. H. Ohtsubo for stimulating discussions, critical comments on the manuscript, and encouragement in preparing the manuscript.

This work was supported by a Grant-in-Aid for Scientific Research from the Ministry of Education, Science and Culture of Japan.

References

Ajdic D, Jovanovic G, Glisin V, Hejna J, Savic DJ (1991) Nucleotide sequence analysis of the inversion termini located within IS3 element α3β3 and β5α5 of *Escherichia coli.* J Bacteriol 173: 906–909

Alam J, Vrba JM, Cai Y, Martin JA, Weislo LJ, Curtis SE (1991) Characterization of the IS895 family of insertion sequences from the Cyanobacterium *Anabaena* sp. strain PCC 7120. J Bacteriol 173: 5778–5783

Armstrong KA, Ohtsubo H, Bauer WR, Yoshioka Y, Miyazaki C, Maeda Y, Ohtsubo E (1986) Characterization of the gene products produced in minicells by pSM1, a derivative of R100. Mol Gen Genet 205:56–65

Ashby MK, Berquist PL (1990) Cloning and sequence of IS1000, a putative insertion from *Thermus thermophilus* HB8. Plasmid 24: 1–11

Bartlett DH, Silverman M (1989) Nucleotide sequence of IS492, a novel insertion sequence causing variation in extracellular polysaccharide production in the marine bacterium, *Pseudomonas atlantica.* J Bacteriol 171: 1763–1766

Birkenbihl RP, Vielmetter W (1989) Complete maps of IS1, IS2, IS3, IS4, IS5, IS30 and IS150 locations in *Escherichia coli* K12. Mol Gen Genet 220: 147–153

Bisercic M, Ochman H (1993) The ancestry of insertion sequences common to *Escherichia coli* and *Salmonella typhimurium.* J. Bacteriol 175: 7863–7868

Blinkowa AL, Walker JR (1990) Programmed ribosomal frameshifting generates the *Escherichia coli* DNA polymerase III γ subunit from within the τ subunit reading frame. Nucleic Acids Res 18: 1725–1729

Brachet P, Eisen H, Rambach A (1970) Mutations of coliphage lambda affecting the expression of replicative functions O and P. Mol Gen Genet 108: 266–276

Brierley I, Digard P, Inglis SC (1989) Characterization of an efficient coronavirus ribosomal frameshifting signal: requirement for an RNA pseudoknot. Cell 57: 537–547

Brierley I, Rolley NJ, Jenner AJ, Inglis SC (1991) Mutational analysis of the RNA pseudoknot component of a coronavirus ribosomal frameshifting signal. J Mol Biol 220: 889–902

Brown PO, Bowerman B, Varmus HE, Bishop JM (1987) Correct integration of retroviral DNA in vitro. Cell 49: 347–356

Brown PO, Bowerman B, Varmus HE, Bishop JM (1989) Retroviral integration: structure of the initial covalent product and its presursor, and a role for the viral IN protein. Proc Natl Acad Sci USA 86: 2525–2529

Bruton CJ, Chater KF (1987) Nucleotide sequence of IS110, an insertion sequence of *Streptomyces coelicolor* A3(2). Nucleic Acids Res 15: 7053–7065

Bureau TE, Wessler SR (1994) *Stowaway:* a new family of inverted repeat elements associated with the genes of both monocotyledonous and dicotyledonous plants. Plant Cell 6: 907–916

Calos MP, Johnsrud L, Miller JH (1978) DNA sequence at the integration sites of the insertion element IS1. Cell 13: 411–418

Caparon MG, Scott JR (1989) Excision and insertion of the conjugative transposon Tn916 involves a novel mechanism. Cell 59: 1027–1934

Chamorro M, Parkin N, and Varmus HE (1992) An RNA pseudoknot and an optimal heptameric shift site are required for highly efficient ribosomal frameshifting on a retroviral messenger RNA. Proc Natl Acad Sci USA 89: 713–717

Chan PT, Lebowitz J (1982) Mapping RNA polymerase binding sites in R12 derived plasmids carrying the replication-incompatibility region and the insertion element IS1. Nucleic Acids Res 10: 7295–7311

Chandler M, Fayet O (1993) Translational frameshifting in the control of transposition in bacteria. Mol Microbiol 7: 497–503
Craigie R, Fujiwara T, Bushman F (1990) The IN protein of Moloney murine leukemia virus processes the viral DNA ends and accomplishes their integration in vitro. Cell 62: 829–837
Deonier RC, Hadley RG, Hu M (1979) Enumeration and identification of IS3 elements in *Escherichia coli* strains. J Bacteriol 137: 1421–1424
Dinman JD, Icho T, Wickner RB (1991) A −1 ribosomal frameshift in a double-stranded RNA virus of yeast forms a gag-pol fusion protein. Proc Natl Acad Sci USA 88: 174–178
Doak TG, Doerder FP, Jahn CL, Herrick G (1994) A proposed superfamily of transposase genes: transposon-like elements in ciliated protozoa and a common "D35E" motif. Proc Natl Acad Sci USA 91: 942–946
Dong Q, Sadouk A, van der Lelie D, Taghavi S, Ferhat A, Nuyten JM, Borremans B, Mergeay M, Toussaint A (1992) Cloning and sequencing of IS1086, an *Alcaligenes eutrophus* insertion element related to IS30 and IS4351. J Bacteriol 174: 8133–8138
Drelich M, Wilhelm R, Mous J (1992) Identification of amino acid residues critical for endonuclease and integration activities of HIV-1 IN protein in vitro. Virology 188: 459–468
Engelman A, Craigie R (1992) Identification of conserved amino acid residues critical for human immunodeficiency virus type 1 integrase function in vitro. J Virol 66: 6361–6369
Escoubas JM, Prère MF, Fayet O, Salvignol I, Galas D, Zerbib D, Chandler M (1991) Translational control of transposition activity of the bacterial insertion sequence IS1. EMBO J 10: 705–712
Fayet O, Ramond P, Polard P, Prère MF, Chandler M (1990) Functional similarities between retroviruses and the IS3 family of bacterial insertion sequences? Mol Microbiol 4: 1771–1777
Fiandt M, Szybalski W, Malamy MH (1972) Polar mutations in lac, gal, and phage λ consist of a few IS-DNA sequences inserted with either orientation. Mol Gen Genet 119: 223–231
Flower AM, McHenry CS (1990) The γ subunit of DNA polymerase III holoenzyme of *Escherichia coli* is produced by ribosomal frameshifting. Proc Natl Acad Sci USA 87: 3713–3717
Fournier P, Paulus F, Otten L (1993) IS870 requires a 5'-CTAG-3' target sequence to generate the stop codon for its large ORF1. J Bacteriol 175: 3151–3160
Fujiwara T, Mizuuchi K (1988) Retroviral DNA integration: structure of an integration intermediate. Cell 54: 497–504
Galas DJ, Chandler M (1989) Bacterial insertion sequences. In: Berg DE, Howe MM (eds) Mobile DNA. American Society for Microbiology, Washington DC, pp 109–162
Goussard S, Sougakoff W, Mabilat C, Bauernfeind A, Courvalin (1991) An IS1-like element is responsible for high-level synthesis of extended-spectrum β-lactamase TEM-6 in Enterobacteriaceae. J Gen Microbiol 137: 2681–2687
Grandgenett DP, Vora AC, Swanstrom R, Olsen JC (1986) Nuclease mechanism of the avian retrovirus pp32 endonuclease. J Virol 58: 970–974
Green EP, Tizard MLV, Moss MT, Thompson J, Winterborne DJ, McFadden JJ, Hermon-Taylor J (1989) Sequence and characteristics of IS900, an insertion element identified in human Crohn's disease isolate of *Mycobacterium paratuberculosis*. Nucleic Acids Res 17: 9063–9073
Grindley NDF (1978) IS1 insertion generates duplication of a nine base pair sequence at its target site. Cell 13: 419–426
Grindley NDF, Joyce CM (1981) Genetic DNA sequence analysis of the kanamycin resistance transposon Tn903. Proc Natl Acad Sci USA 77: 7176–7180
Henderson DJ, Lydiate DJ, Hopwood DA (1989) Structural and functional analysis of the minicircle, a transposable element of *Streptomyces coelicolor* A3(2). Mol Microbiol 3: 1307–1318
Henikoff S (1992) Detection of *Caenorhabditis* transposon homologs in diverse organisms. New Biol 4: 382–388
Hirsch H-J, Starlinger P, Brachet P (1972) Two kinds of insertions in bacterial genes. Mol Gen Genet 119: 191–206
Hizi A, Henderson LE, Copeland TD, Sowder RC, Hixson CV, Oroszlan S (1987) Characterization of mouse mammary tumor virus gag-pro gene products and the ribosomal frameshift site by protein sequencing. Proc Natl Acad Sci USA 84: 7041–7045
Hoover TA, Vodkin MH, Williams JC (1992) A *Coxiella burnetii* repeated DNA element resembling a bacterial insertion sequence. J Bacteriol 174: 5540–5548
Hu S, Ptashne K, Cohen SN, Davidson N (1975) αβ sequence of F is IS3. J Bacteriol 123: 687–692
Iida S, Arber W (1980) On the role of IS1 in the formation of hybrids between the bacteriophage P1 and the R plasmid NR1. Mol Gen Genet 177: 261–270

Jacks T, Townsley K, Varmus HE, Majors J (1987) Two efficient ribosomal frameshifting events are required for synthesis of mouse mammary tumor virus gag-related polyproteins. Proc Natl Acad Sci USA 84: 4298–4302

Jacks T, Madhani HD, Masiarz FR, Varmus HE (1988) Signals for ribosomal frameshifting in the Rous sarcoma virus gag-pol region. Cell 55: 447–458

Jakowec M, Prentki P, Chandler M, Galas DJ (1988) Mutational analysis of the open reading frames in the transposable element IS1. Genetics 120: 47–55

Jaraczewski JW, Jahn CL (1993) Elimination of Tec elements involves a novel excision process. Genes Dev 7: 95–105

Johnsrud L (1979) DNA sequence of the transposable element IS1. Mol Gen Genet 169: 213–218

Jordan E, Saedler H, Starlinger P (1968) 0-zero and strong polar mutations in the gal operon are insertions. Mol Gen Genet 102: 353–363

Kanazawa H, Kiyasu T, Noumi T, Futai M, Yamaguchi K (1984) Insertion of transposable elements in the promoter proximal region of the gene cluster for *Escherichia coli* H^+-ATPase: 8 base pair repeat generated by insertion of IS1. Mol Gen Genet 194: 179–187

Kato K, Ohtsuki K, Mitsuda H, Yomo T, Negoro S (1994) Insertion sequence IS6100 on plasmid pOAD2, which degrades nylon oligomers. J Bacteriol 176: 1197–1200

Katz RA, Merkel G, Kulkosky J, Leis J, Skalka AM (1990) The avian retroviral IN protein is both necessary and sufficient for integrative recombination in vitro. Cell 63: 87–95

Katzman M, Katz RA, Skalka AM, Leis J (1989) The avain retroviral integration protein cleaves the terminal sequences of linear viral DNA at the in vivo sites of integration. J Virol 63: 5319–5327

Khan E, Mack JPG, Katz RA, Kulkosky J, Skalka AM (1991) Retroviral integrase domains: DNA binding and the recognition of LTR sequences. Nucleic Acids Res 19: 851–860

Klaer R, Starlinger P (1980) IS4 at its chromosomal site in *E. coli* K-12. Mol Gen Genet 178: 285–291

Klaer R, Kühn S, Tillman E, Fritz H-J, Starlinger P (1981) The sequence of IS4. Mol Gen Genet 181: 169–175

Kohara Y, Akiyama K, Isono K (1987) The physical map of the whole *E. coli* chromosome: application of a new strategy for rapid analysis and sorting of a large genomic library. Cell 50: 495–508

Komoda Y, Enomoto M, Tominaga A (1991) Large inversion in *Escherichia coli* K-12 1485IN between inversely oriented IS3 elements near lac and cdd. Genetics 129: 639–645

Kulkosky J, Jones KS, Katz RA, Mack JPG, Skalka AM (1992) Residues critical for retroviral integrative recombination in a region that is highly conserved among retroviral/retrotransposon integrases and bacterial insertion sequence transposases. Mol Cell Biol 12: 2331–2338

Kunze ZM, Wall S, Appelberg R, Silva MT, Portaels F, McFadden JJ (1991) IS901, a new member of a widespread class of atypical insertion sequences is associated with pathogenicity in *Mycobacterium avium.* Mol Microbiol 5: 2265–2272

Lawrence JG, Ochman H, Hartl DL (1992) The evolution of insertion sequences within enteric bacteria. Genetics 131: 9–20

Lenich AG, Glasgow AC (1994) Amino acid sequence homology between Piv, an essential protein in site-specific DNA inversion in *Moraxella lacunata,* and transposases of an unusual family of insertion elements. J Bacteriol 176: 4160–4164

Leskiw BK, Mevarech M, Barritt LS, Jensen SE, Henderson DJ, Hopwood DA, Bruton CJ, Chater KF (1990) Discovery of an insertion sequence, IS116, from *Streptomyces clavuligerus* and its relatedness to other transposable elements from actinomycetes. J Gen Microbiol 136: 1251–1258

Luthi K, Moser M, Ryser J, Weber H (1990) Evidence for a role of translational frameshifting in the expression of transposition activity of the bacterial insertion element IS1. Gene 88: 15–20

Machida C, Machida Y (1989) Regulation of IS1 transposition by the insA gene product. J Mol Biol 208: 567–574

Machida C, Machida Y, Ohtsubo E (1984) Both inverted repeat sequences located at the ends of IS1 provide promoter functions. J Mol Biol 177: 247–267

Machida Y, Machida C, Ohtsubo H, Ohtsubo E (1982) Factors determining frequency of plasmid cointegration mediated by insertion sequence IS1. Proc Natl Acad Sci USA 79: 277–281

Machida Y, Machida C, Ohtsubo E (1984) Insertion element IS1 encodes two structural genes required for its transposition. J Mol Biol 177: 229–245

Maekawa T, Ohtsubo E (1994) Identification of the region that determines the specificity of binding of the transposases encoded by Tn3 and $\gamma\delta$ to the terminal inverted repeat sequences. Jpn J Genet 69: 269–285

Mahillon J, Seurinck J, van Rompuy L, Delcour J, Zabeau M (1985) Nucleotide sequences and structural organization of an insertion sequence element (IS231) from *Bacillus thuringiensis* strain berlıner 1715. EMBO J 4: 3985–3899

Mahillon J, Seurinck J, Delcour J, Zabeau M (1987) Cloning and nucleotide sequence of different iso-IS231 elements and their structural association with the Tn4430 transposon in *Bacillus thuringiensis*. Gene 51: 187–196

Malamy MH (1966) Frameshift mutations in the lactose operon of *E. coli*. Cold Spring Harb Symp Quant Biol 31: 189–201

Malamy MH (1970) Some properties of insertion mutations in the lac operon. In: Beckwith JR, Zipser D (eds) The loctose operon. Cold Spring Harbor Laboratory, Cold Spring Harbor, New York, pp 359–373

Malamy MH, Fiandt M, Szybalski W (1972) Electron microscopy of polar insertions in the lac operon of *Escherichia coli*. Mol Gen Genet 119: 207–222

Malamy MH, Rahaim PT, Hoffman CS, Baghdoyan D, O'Connor MB, Miller JF (1985) A frameshift mutation at the junction of an IS1 insertion within lacZ restores β-galactosidase activity via formation of an active lacZ-IS1 fusion protein. J Mol Biol 181: 551–555

Martin C, Timm J, Rauzier J, Gomez-Lus R, Davies J, Gicquel B (1990) Transposition of an antibiotics resistance element in mycobacteria. Nature 345: 739–743

Matsutani S, Ohtsubo E (1993) Distribution of the *Shigella sonnei* insertion elements in Enterobacteriaceae. Gene 127: 111–115

Matsutani S, Ohtsubo H, Maeda Y, Ohtsubo E (1987) Isolation and characterization of IS elements repeated in the bacterial chromosome. J Mol Biol 196: 445–455

McClintock B (1956) Controlling elements and the gene. Cold Spring Harb Symp Quant Biol 21: 197–216

McClintock B (1965) The control of gene action in maize. Brookhaven Simp Biol 18: 162–184

Mendiola MV, de la Cruz F (1989) Specificity of insertion of IS91, an insertion sequence present in alpha-hemolysin plasmids of *Escherichia coli*. Mol Microbiol 3: 979–984

Mendiola MV, de la Cruz F (1992) IS91 transposase is related to the rolling-circle-type replication protein of the pUB110 family of plasmids. Nucleic Acids Res 20: 3521

Mendiola MV, Jubete Y, de la Cruz F (1992) DNA sequence of IS91 and identification of the transposase gene. J Bacteriol 174: 1345–1351

Mills JA, Venkatesan MM, Baron LS, Buysse JM (1992) Spontaneous insertion of an IS1-like element into the *virF* gene is responsible for avirulence in opaque colonial variants of *Shigella flexneri* 2a. Infect Immun 60: 175–182

Moore R, Dixon M, Smith R, Peters G, Dickson C (1987) Complete nucleotide sequence of a milk-transmitted mouse mammary tumor virus: two frameshift suppression events are required for translation of gag and pol. J Virol 61: 480–490

Moss MT, Malik ZP, Tizard MLV, Green EP, Sanderson JD, Hermon-Taylor J (1992) IS902, an insertion element of the chronic-enteritis-causing *Mycobacterium avium* subsp. *silvaticum*. J Gen Microbiol 138: 139–145

Nakatsu C, Ng J, Singh R, Straus N, Wyndham C (1991) Chlorobenzoate catabolic transposon Tn5271 is a composite class I element with flanking class II insertion sequences. Proc Natl Acad Sci USA 88: 8312–8316

Nyman K, Nakamura K, Ohtsubo H, Ohtsubo E (1981) Distribution of the insertion sequence IS1 in gram-negative bacteria. Nature 289: 609–612

Ohtsubo E, Zenilman M, Ohtsubo H (1980) Plasmids containing insertion elements are potential transposons. Proc Natl Acad Sci USA 77: 750–754

Ohtsubo E, Zenilman M, Ohtsubo H, McCormick M, Machida C, Machida Y (1981) Mechanism of insertion and cointegration mediated by IS1 and Tn3. Cold Spring Harb Symp Quant Biol 45: 283–295

Ohtsubo E, Ohtsubo H, Doroszkiewicz W, Nyman K, Allen D, Davison D (1984) An evolutionary analysis of iso-IS1 elements from *Escherichia coli* and *Shigella* strains. J Gen Appl Microbiol 30: 359–376

Ohtsubo H, Ohtsubo E (1978) Nucleotide sequence of an insertion element, IS1. Proc Natl Acad Sci USA 75: 615–619

Ohtsubo H, Nyman K, Doroszkiewicz W, Ohtsubo E (1981) Multiple copies of iso-insertion sequences of IS1 in *Shigella dysenteriae* chromosome. Nature 292: 640–643

Ohtsubo H, Zenilman M, Ohtsubo E (1980) Insertion element IS102 resides in plasmid pSC101. J Bacteriol 144: 131–140

Olasz F, Stalder R, Arber W (1993) Formation of the tandem repeat (IS30)2 and its role in IS30-mediated transpositional DNA rearrangements. Mol Gen Genet 239: 177–187

Ou JT, Huang CJ, Houng HS, Baron LS (1992) Role of IS1 in the conversion of virulence (Vi) antigen expression in Enterobacteriaceae. Mol Gen Genet 234: 228–232

Pabo C, Sauer R (1984) Protein-DNA recognition. Annu Rev Biochem 53: 293–321

Polard P, Prère MF, Chandler M, Fayet O (1991) Programmed translational frameshifting and initiation at an AUU codon in gene expression of bacterial insertion sequence IS911. J Mol Biol 222: 465–477

Polard P, Prère MF, Fayet O, Chandler M (1992) Transposase-induced excision and circularization of the bacterial insertion sequence IS911. EMBO J 11: 5079–5090
Prère MF, Chandler M, Fayet O (1990) Transposition in *Shigella dysenteriae*: isolation and analysis of IS911, a new member of the IS3 group of insertion sequences. J Bacteriol 172: 4090–4099
Rådström P, Sköld O, Swedberg G, Flensburg J, Roy PH, Sundström (1994) Transposon Tn5090 of plasmid R751, which carries an integron, is related to Tn7, Mu and retroelements. J Bacteriol 176: 3257–3268
Ramirez SJ, Alvarez G, Cisneros E, Gomez EM (1992) Distribution of insertion sequence IS1 in multiple-antibiotic resistant clinical Enterobacteriaceae strains. FEMS Microbiol Lett 72: 189–193
Reimmann C, More R, Little S, Savioz A, Willetts NS, Haas D (1989) Genetic structure, function and regulation of the transposable element IS21. Mol Gen Genet 215: 416–424
Rezsöhazy R, Hallet B, Delcour J (1992) IS231D, E, and F, three new insertion sequences in *Bacillus thuringiensis:* extension of the IS231 family. Mol Microbiol 6: 1959–1967
Rezsöhazy R, Hallet B, Delcour J, Mahillon J (1993a) The IS4 family of insertion sequences: evidence for a conserved transposase motif. Mol Microbiol 9: 1283–1295
Rezsöhazy R, Hallet B, Mahillon J, Delcour J (1993b) IS231V and W from *Bacillus thuringiensis* subsp. *israelensis*, two distant members of the IS231 family of insertion sequences. Plasmid 30: 141–149
Rice NR, Stephens RM, Burny A, Gilden RV (1985) The gag and pol genes of bovine leukemia virus: nucleotide sequence and analysis. Virology 142: 357–377
Rodriguez H, Snow ET, Bhat U, Loechler (1992) An *Escherichia coli* plasmid-based, mutational system in which supF mutants are selectable: insertion elements dominate the spontaneous spectra. Mutat Res 270: 219–231
Romantschuk M, Richter GY, Mukhopadhyay P, Mills (1991) IS801, an insertion sequence element isolated from *Pseudomonas syringae phaseolicola.* Mol Microbiol 5: 617–622
Rose AM, Snutch TP (1984) Isolation of the closed circular form of the transposable element Tc1 in *Caenorhabditis elegans.* Nature 311: 485–486
Ruan KS, Emmons SW (1984) Extrachromosomal copies of transposon Tc1 in the nematode *Caenorhabditis elegans.* Proc Natl Acad Sci USA 81: 4018–4022
Schwartz E, Kröger M, Rak B (1988) IS150: distribution, nucleotide sequence and phylogenetic relationships of a new *E. coli* insertion element. Nucleic Acids Res 16: 6789–6802
Scott JR, Kirchman PA, Caparon MG (1988) An intermediate in transposition of the conjugative transposon Tn916. Proc Natl Acad Sci USA 85: 4809–4813
Sekine Y, Ohtsubo E (1989) Frameshifting is required for expression of IS1 transposase. Proc Natl Acad Sci USA 86: 4609–4613
Sekine Y, Ohtsubo E (1991) Translational frameshifting in IS elements and other genetic systems. In: Kimura M, Takahata N (eds) New aspects of the genetics of molecular evolution. Japan Sci Soc Press, Tokyo/Springer, Berlin Heidelberg New York, pp 243–261
Sekine Y, Ohtsubo E (1992) DNA sequences required for translational frameshifting in production of the transposase encoded by IS1. Mol Gen Genet 235: 325–332
Sekine Y, Nagasawa H, Ohtsubo E (1992) Identification of the site of translational frameshifting required for production of the transposase encoded by insertion sequence IS1. Mol Gen Genet 235: 317–324
Sekine Y, Eisaki N, Ohtsubo E (1994) Translational control in production of transposase and in transposition of insertion sequence IS3. J Mol Biol 235: 1406–1420
Sekine Y, Eisaki N, Ohtsubo E (1995) Identification and characterization of the linear IS3 molecules generated by staggered breaks. J Biol Chem (in press)
Sekino N, Sekine Y, Ohtsubo E (1995) IS1-encoded proteins, InsA and the InsA-B'-InsB transframe protein (transposase): functions deduced from their DNA-binding ability. Adv Biophys 31: 209–222
Shapiro JA (1969) Mutations caused by insertion of genetic material into the galactose operon of *Escherichia coli.* J Mol Biol 40: 93–105
Sherman PA, Fyfe JA (1990) Human immunodeficiency virus integration protein expressed in *Escherichia coli* possesses selectine DNA cleaving activity. Proc Natl Acad Sci USA 87: 5119–5123
Sherratt D (1989) Tn3 and related transposable elements: site-specific recombination and transposition. In: Berg DE, Howe MM (eds) Mobile DNA. American Society for Microbiology, Washington DC, pp163–184
Shimotohno K, Takahashi Y, Shimizu N, Gojobori T, Golde DW, Chen IS, Miwa YM, and Sugimura T (1985) Complete nucleotide sequence of an infectious clone of human T-cell leukemia virus type II: an open reading frame for the protease gene. Proc Natl Acad Sci USA 82: 3101–3105
Skaliter R, Eichenbaum Z, Shwartz H, Ascarelli GR, Livneh Z (1992) Spontaneous transposition in bacteriophage lambda *cro* gene residing on a plasmid. Mutat Res 267: 139–151

Sommer H, Cullum J, Saedler H (1979) Integration of IS3 into IS2 generates a short sequence duplication. Mol Gen Genet 177: 85–89

Spielmann-Ryser J, Moser M, Kast P, Weber H (1991) Factors determining the frequency of plasmid cointegrate formation mediated by insertion sequence IS3 from *Escherichia coli*. Mol Gen Genet 226: 441–448

Stark WM, Boocock MR, Sherratt DJ (1992) Catalysis by site-specific recombinases. Trends Genet 8: 432–439

Starlinger P, Saedler H (1976) IS-elements in microorganisms. In: Compans RW, Cooper M, Koprowski H et al. (eds) Current topics in microbiology and immunology, vol 75. Springer, Berlin Heidelberg New York, pp 111–152

Tenzen T, Ohtsubo E (1991) Preferential transposition of an IS630-associated composite transposon to TA in the 5'-CTAG-3' sequence. J Bacteriol 173: 6207–6212

Tenzen T, Matsutani S, Ohtsubo E (1990) Site-specific transposition of insertion sequence IS630. J Bacteriol 172: 3830–3836

Tenzen T, Matsuda Y, Ohtsubo H, Ohtsubo E (1994) Transposition of Tnr1 in rice genomes to 5'-PuTAPy-3' duplicating the TA sequence. Mol Gen Genet 245: 441–448

Terry R, Soltis DA, Katzman M, Cobrinik D, Leis J, Skalka AM (1988) Properties of avain sarcoma leukosis virus pp32–related pol-endonuclease produced in *Escherichia coli*. J Virol 62: 2358–2365

Timmerman KP, Tu CD (1985) Complete sequence of IS3. Nucleic Acids Res 13: 2127–2139

Toba MM, Hashimoto GT (1992) Characterization of the spontaneous elimination of streptomycin sensitivity (SmS) on high-copy-number plasmids: SmS-enforcement cloning vectors with a synthetic *rpsL* gene. Gene 121: 25–33

Trinks K, Habermann P, Beyreuther K, Starlinger P, Ehring R (1981) An IS4-encoded protein is synthesized in minicells. Mol Gen Genet 182: 183–188

Tsuchihashi Z, Brown PO (1992) Sequence requirements for efficient translational frameshifting in the *Escherichia coli dnaX* gene and the role of an unstable interaction between $tRNA^{Lys}$ and an AAG lysine codon. Genes Dev 6: 511–519

Tsuchihashi Z, Kornberg A (1990) Translational frameshifting generates the γ subunit of DNA polymerase III holoenzyme. Proc Natl Acad Sci USA 87: 2516–2520

Tudor M, Lobocka M, Goodell M, Pettitt J, O'Hare K (1992) The pogo transposable element family of *Drosophila melanogaster*. Mol Gen Genet 231: 126–134

Umeda M, Ohtsubo E (1989) Mapping of insertion elements IS1, IS2 and IS3 on the *E. coli* K-12 chromosome: role of insertion elements in formation of Hfrs and F-prime factors and in rearrangement of bacterial chromosomes. J Mol Biol 208: 601–614

Umeda M, Ohtsubo E (1990a) Mapping of insertion element IS5 in the *Escherichia coli* K-12 chromosome: chromosomal rearrangement mediated by IS5. J Mol Biol 213: 229–237

Umeda M, Ohtsubo E (1990b) Mapping of insertion element IS30 on the *Escherichia coli* K-12 chromosome. Mol Gen Genet 222: 317–322

Umeda M, Ohtsubo E (1991) Four types of IS1 with difference in nucleotide sequences residue in the *Escherichia coli* K-12 chromosome. Gene 98: 1–5

Umeda M, Ohtsubo H, Ohtsubo E (1991) Diversification of the rice *wx* gene by insertion of mobile DNA elements into introns. Jpn J Genet 66: 569–586

van der Meer JR, Zehnder AJB, de Vos WM (1991) Identification of a novel composite transposable element, Tn5280, carrying chlorobenzene dioxygenase genes of *Pseudomonas* sp. strain P51. J Bacteriol 173: 7077–7083

van Gent DC, Oude Groeneger AAM, Plasterk RHA (1992) Mutational analysis of the integrase protein of human immunodeficiency virus type 2. Proc Natl Acad Sci USA 89: 9598–9602

van Hove B, Staudenmaier H, Braun V (1990) Novel two-component transmembrane transcriptional control: regulation of iron dicitrate transport in *Escherichia coli* K-12. J Bacteriol 172: 6749–6758

Varmus H, Brown P (1989) Retroviruses. In: Berg DE, Howe MM (eds) Mobile DNA. American Society for Microbiology, Washington DC, pp 53–108

Vögele K, Schwartz E, Welz C, Schiltz E, Rak B (1991) High-level ribosomal frameshifting directs the synthesis of IS150 gene products. Nucleic Acids Res 19: 4377–4385

Weiss RB, Dunn DM, Atkins JFM, Gesteland RF (1987) Slippery runs, shifty stops, backward steps, and forward hops: –2, –1, +1, +2, +5, and +6 ribosomal frameshifting. Cold Spring Harb Symp Quant Biol 52: 687–693

Yoshioka Y, Ohtsubo H, Ohtsubo E (1987) Repressor gene *finO* in plasmids R100 and F: constitutive transfer of plasmid F is caused by insertion of IS3 into *finO*. J Bacteriol 169: 619–623

Yoshioka Y, Fujita Y, Ohtsubo E (1990) Nucleotide sequence of the promoter distal region of the tra operon including traI (DNA helicase I) and *traD* genes of plasmid R100. J Mol Biol 214: 39–53
Zerbib D, Jakowec M, Prentki P, Galas DJ, Chandler M (1987) Expression of proteins essential for IS1 transposition: specific binding of InsA to the ends of IS1. EMBO J 6: 3163–3169
Zerbib D, Polard P, Escoubas JM, Galas D, Chandler M (1990a) The regulatory role of the IS1-encoded InsA protein in transposition. Mol Microbiol 4: 471–477
Zerbib D, Prentki P, Gamas P, Freund E, Galas DJ, Chandler M (1990b) Functional organization of the ends of IS1: specific binding for an IS1-encoded protein. Mol Microbiol 4: 1477–1486
Zuber U, Schumann W (1993) The eighth copy of IS1 in *Escherichia coli* W3110 maps at 49.6 min. J Bacteriol 175: 1552
Zuerner RL (1994) Nucleotide sequence analysis of IS1533 from *Leptospira borgpetersenii*: identification and expression of two IS-encoded proteins. Plasmid 31: 1–11

Transposon Tn7

N.L. Craig

1 Introduction

Most transposable elements move at low frequencies and insert into many different target sites, displaying only modest or little target site-selectivity. The bacterial transposon Tn7 (Barth et al. 1976) is distinguished by its ability to insert at high frequency into a specific site called an "attachment" site, or *attTn7*, in the chromosomes of many bacteria, including *Escherichia coli* (reviewed by Craig 1989, 1991). In the absence of *attTn7*, for example when Tn7 inserts into

---

Howard Hughes Medical Institute, Department of Molecular Biology and Genetics, 615 PCTB, 725 North Wolfe Street, Johns Hopkins School of Medicine, Baltimore, MD 21205, USA

plasmids, Tn7 resembles other elements and transposes to many sites at low frequency. Through both genetic and biochemical studies, it is now known that Tn7 can recognize and transpose to two different types of target sites through the action of two distinct but overlapping sets of Tn7-encoded transposition proteins.

Dissection in vitro of Tn7 insertion into *attTn7* has led to reconstitution of this reaction with purified proteins and allowed molecular analysis of the reaction mechanism. Tn7 transposition appears to occur within a multiprotein nucleoprotein complex containing the transposon ends and the target DNA and uses a nonreplicative, cut-and-paste mechanism in which the transposon is first excised from the donor site and then inserted into the target site. There is considerable mechanistic similarity between the transposition of Tn7 and that of other mobile elements, including bacteriophage Mu and retroviruses.

2 The Tn7 Transposition Machinery and Antibiotic Resistance Genes

Tn7 is a relatively large bacterial transposon, some 14kb in length. This element encodes an elaborate array of transposition genes (*tns ABCDE*), *cis*-acting sites at the transposon termini which are the substrates for recombination, and antibiotic resistance genes.

2.1 The *tns* Genes

The *tns* genes were identified by analysis of Tn7 mutants (HAUER and SHAPIRO 1984; ROGERS et al . 1986; WADDELL and CRAIG 1988). The sequences of *tns* genes have been determined (SMITH and JONES 1986; FLORES et al. 1990; ORLE and CRAIG 1991) and their protein products identified (SMITH and JONES 1986; ORLE and CRAIG 1991; Fig. 1). It is notable that the sum of the molecular weights of the Tns proteins approaches 300 kD; thus, considerable information is devoted to Tn7 transposition. Overlapping sets of these genes mediate two distinct transposition pathways (see below).

The *tns* genes are all oriented in the same direction in Tn7, with their N-termini closest to the right end of Tn7, Tn7R (SMITH and JONES 1986; WADDELL and CRAIG 1988; FLORES et al. 1990; ORLE and CRAIG 1991). Little is known about how expression of these genes is controlled or how their level of expression may affect transposition. It is known that *tnsA* and *tnsB* form an operon whose promoter is located about 100 bp from the tip of Tn7R (GAY et al. 1986; ROGERS et al. 1986; MCKOWN et al. 1987; WADDELL and CRAIG 1988). Transcription from this *tnsAB* promoter appears to be repressed by TnsB, so that these *tns* genes are under autoregulation (ROGERS et al. 1986; MCKOWN et al. 1987); ORLE and CRAIG 1991; ARCISZEWSKA et al. 1991; TANG et al. 1991). Genetic analysis of the effects on

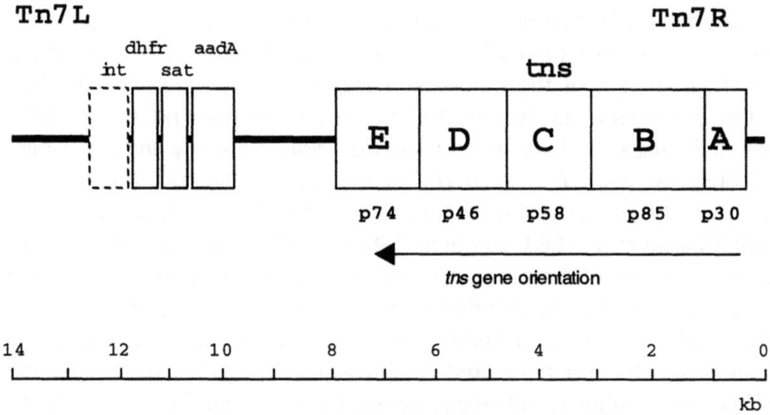

Fig. 1. Map of Tn7. The Tn7-encoded transposition genes, *tnsABCDE*, and the sizes of the Tns proteins they encode are shown; all the *tns* genes are oriented in the same direction with their amino termini adjacent to Tn7R. The Tn7-encoded antibiotic resistance genes *dhfr* (trimethoprim resistance), *sat* (streptothricin resistance) and *aadA* (streptomycin/spectinomycin resistance) are shown. Also indicated is putative integrase ORF *(int)* which, when functional, may mediate rearrangements of the drug resistance cassettes

transposition of polar insertions in the *tns* genes has revealed that, although *tnsA* and *tnsB* do form an operon, expression of the other *tns* genes is not entirely dependent on the *tnsAB* promoter (Waddell and Craig 1988), suggesting that the *tns* genes are not contained within a single operon. Also, other promoters within the *tns* genes have been detected by transcriptional fusion (Smith and Jones, 1986; Rogers et al. 1986). A more direct analysis of *tns* gene expression is now possible using direct mRNA characterization and anti-Tns antibodies; it seems likely that control of the expression of these genes will occur at many levels (Kleckner 1990).

2.2 Different Sets of *tns* Genes Direct Tn7 to Different Types of Target Sites

Analysis of Tn7 transposition in the presence of the various combinations of *tns* genes revealed that different subsets of the *tns* genes promote recombination to different classes of target sites (Hauer and Shapiro 1984; Rogers et al. 1986; Waddell and Craig 1988; Kubo and Craig 1990). All Tn7 recombination reactions require *tnsABC,* but these genes alone are insufficient to promote transposition to any target site; no recombination is observed in the presence of only *tnsABC*. In the presence of *tnsABC+D,* transposition to *attTn7* is observed, whereas in the presence of *tnsABC+E* transposition to many non-*attTn7* sites is observed. The frequency of *tnsD*-dependent insertion into *attTn7* is high, whereas the frequency of *tnsE*-dependent insertion is 100- to 1000-fold less.

The high frequency of Tn7 insertion into the *E. coli* chromosome at *attTn7* is notable (Barth et al. 1976; Barth and Datta 1977). This high frequency is revealed by the fact that when a plasmid containing Tn7 is introduced into cells lacking Tn7, the bacterial chromosome acquires Tn7 in *att Tn7* at high frequency in the absence of selection. Examination of the chromosomal DNA in such cells reveals that 1–10% of the chromosomes contain Tn7 in *attTn7* (Barth and Datta 1977; Lichtenstein and Brenner 1981; Hauer and Shapiro 1984; R. DeBoy and N.L. Craig, unpublished observations). *E. coli attTn7* is located at min. 84, about 15 kb leftwards (counterclockwise) of the bacterial origin of replication (Lichtenstein and Brenner 1982; Walker et al. 1984; Gay et al. 1986). In *attTn7,* Tn7 insertion occurs about 20 bp downstream of the bacterial *glmS* gene, an essential gene involved in cell wall metabolism, in the transcription terminator of this gene (Gay et al. 1986; Gringauz et al. 1988; Fig. 2). Tn7 insertion into *attTn7* is not accompanied by any obvious deleterious effect upon the host. An interesting but unanswered question is whether Tn7 transposition is influenced by expression of *glmS* (Gay et al. 1986; Gringauz et al. 1988; McKown et al. 1988); such a connection could provide a coupling between transposition and cell growth.

When *attTn7* is unavailable, *tnsABC+D* can also promote low-frequency insertion into a limited set of other sites in the *E. coli genome* (Kubo and Craig 1990). The frequency of insertion into these other *tnsD* sites is 100- to 1000-fold less than that into *attTn7* and is thus about the same as the frequency of *tnsABC+E* insertion. Sequence analysis of these secondary insertion sites reveals that they have significant homology with *attTn7*; they are termed pseudo-*attTn7* sites. Thus *tnsD* activates the *tnsABC machinery* to mediate target sequence-selective transposition, i.e., all sites have a common sequence determinant. As described below, TnsD is a sequence-specific DNA-binding protein that interacts specifically with *attTn7.*

Site-specific insertion of Tn7 has also been observed in many different bacteria (reviewed in Craig 1989, 1991), but only a few of these insertion events have been examined at the nucleotide sequence level. The site-specific insertion of Tn7 in many bacteria likely reflects broad conservation of *glmS* and Tn7 insertion downstream of this gene in other bacteria.

Tn7 insertion via *tnsABC+E* occurs at a frequency 100- to 1000-fold lower than insertion into *attTn7* and thus is comparable to the frequencies observed

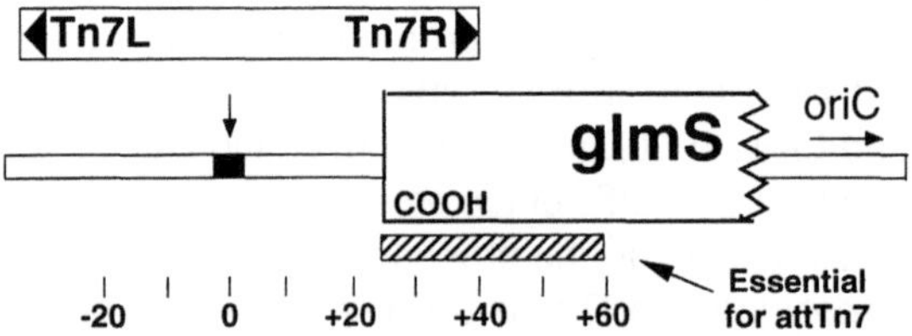

Fig. 2. *E. coli attTn7.* The *attTn7* region of the *E. coli* chromosome is shown. Tn7 (not to scale) inserts into the chromosome in site- and orientation-specific fashion downstream of bacterial *glmS* (glucosamine synthetase). Duplication of 5 bp of chromosomal sequence (*black box*) accompanies Tn7 insertion. The centre base pair of the duplication is designated "0"; base pairs to the right (towards *glmS*) are "+", those leftwards are "–". The nucleotides required for *attTn7* target activity, i.e., to promote high-frequency site- and orientation-specific insertion, are indicated by the *cross-hatched box*

with other well-studied transposons. Examination of *tnsABC+E* target sites reveals that this pathway has little target site-selectivity: insertion can occur into many different sites, and multiple insertions into a single site have not been observed (Rogers et at. 1986; Waddell and Craig 1988; Kubo and Craig 1990). There is no obvious sequence similarity among *tnsE* sites or between *tnsD* and *tnsE* sites other than that a 5-bp target duplication occurs in each pathway. It is notable that *attTn7* is ignored by *tnsABC+E*; thus, targeting specificity does not lie in the common *tnsABC* machinery.

2.3 Why Does Tn7 Encode Two Distinct Transposition Pathways?

Tn7 is unusual in its ability to transpose to two classes of target sites, one class of particular sequence and the other class of many different sequences. Insertion into *attTn7* is not deleterious to the bacterial host, so that even high frequency insertion can be tolerated. Thus insertion into *attTn7* provides Tn7 with a "safe haven" in the bacterial genome, much as does site-specific insertion of bacteriophages into particular sites during lysogeny, and provides a stable association between the transposon and its host (Campbell 1992).

An attractive idea is that Tn7 has an additional pathway that can use many different target sites to facilitate its horizontal transmission between bacteria via plasmids. The low target site-selectivity of the *tnsE* pathway allows Tn7 ready access to plasmids, which can then be transmitted between bacteria. The target site-selectivity of the *tnsD* pathway is considerable, and the number of pseudo-*attTn7* sites that can be efficiently recognized are also limited; for example, no *tnsD*-dependent transposition to endogenous pseudo-*attTn7* sites on the F plasmid can be detected (Waddell and Craig 1988). Thus, the *tnsE* pathway allows Tn7 access to a wide variety of target DNAs.

2.4 Transposition Substrates: the Ends of Tn7

The ends of Tn7 provide the *cis*-acting DNA sequences that are the substrates for transposition; these are the cognate sites recognized and acted upon by the Tn7 recombination enzymes (Fig. 3). Analysis in vivo of mini-Tn7 elements containing various end segments revealed that recombination requires considerable information at each end of Tn7: about 150 bp at the end of Tn7L and about 100 bp at the end of Tn7R (Hauer and Shapiro 1984; Arciszewska et al. 1989; R. DeBoy and N.L. Craig unpublished observations). These end segments each contain multiple binding sites for TnsB, a protein required for all Tn7 transposition reactions (McKown et al. 1987; Arciszewska et al. 1991; Arciszewska and Craig 1991; Tang et al. 1991); thus TnsB plays a key role in recognizing the ends of Tn7.

The arrays of TnsB sites are different in each end; in Tn7R there are four contiguous TnsB sites and three separated sites in Tn7L. Within each end, the

TnsB sites are oriented as direct repeats, but the Tn7L and Tn7R sites are inverted with respect to each other. It seems likely that the differential array of TnsB sites in the ends underlies the functional asymmetry of Tn7L and Tn7R. This asymmetry is revealed by the fact that while mini-Tn7 elements containing two TnR ends can transpose, elements containing two Tn7L ends cannot (ARCISZEWSKA et al. 1989). Also, insertion into *attTn7* is orientation-specific as well as site-specific, with Tn7R always inserting adjacent to *glmS* (LICHTENSTEIN and BRENNER 1982; McKOWN et al. 1988).

Although each end of Tn7 contains multiple TnsB binding sites, DNA breakage and joining occurs only at the extreme termini of Tn7, adjacent to the terminal TnsB binding site. What distinguishes these terminal TnsB sites from the internal TnsB sites that are not the sites of DNA breakage and joining? Inspection of the sequences of the ends reveals that the very tips of Tn7 are markedly different from the similarly positioned nucleotides in other TnsB sites (Fig. 3); it is tempting to speculate that at least some of these terminal nucleotides are critical in defining the active termini at the ends of Tn7. Analysis of the ends of other elements has revealed that they are bipartite, with one region providing information for trans-

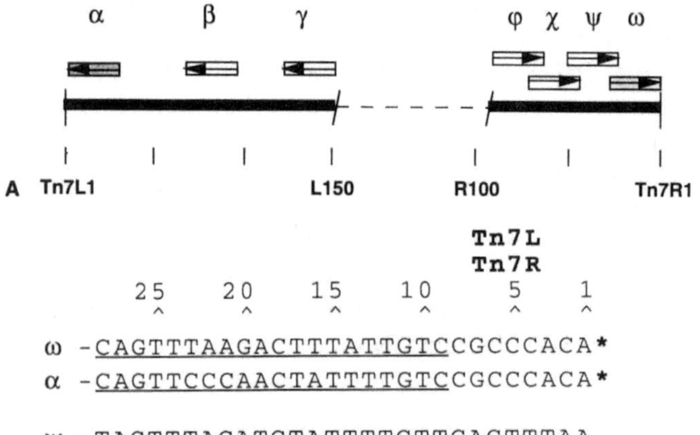


```
                    Tn7L
                    Tn7R
  25   20   15   10   5   1
  ^    ^    ^    ^    ^   ^
ω -CAGTTTAAGACTTTATTGTCCGCCCACA*
α -CAGTTCCCAACTATTTTGTCCGCCCACA*

ψ -TAGTTTAGATCTATTTTGTTCAGTTTAA
χ -TAGTTTAAGACTTTATTGTCATAGTTTA
φ -CAGTTTACAACTATTCTGTCTAGTTTAA
β -CAGTTCCCATCTATTTTGTCACTCTTCC
γ -TAGTTCCAAACTATTTTGTCATTTTTAA
```

B →

Fig. 3A, B. The ends of Tn7. **A** The Tn7 end regions that contain the *cis*-acting sequences essential for transposition are shown; the recombination sequences in Tn7L are included in a DNA segment about 150 bp long; the recombination sequences in Tn7R are about 100 bp. Both Tn7L and Tn7R contain specific binding sites for TnsB, whose position and orientation are indicated by the *boxed arrows*. **B** The sequences of the TnsB binding sites in each end of Tn7 are shown. The top two sites are those at the very tips of Tn7; the 3' terminal *A*s, where breakage and joining occur, are marked by *asterisks*. The most conserved sequences are *underlined*; the sequences of the 3' 8 bp are divergent and distinct between the terminal α and ω sites, where breakage and joining occur, and the other internal sites

posase binding and another involved in a subsequent step, likely breakage and joining (DERBYSHIRE et al. 1987; HUISMAN et al. 1989; JILK et al. 1993). The 3' terminal nucleotides of Tn7L and Tn7R are -CA3'; this 3' terminal A is of particular interest, because during transposition this nucleotide is exposed by DNA breakage and is the site at which the initial covalent linkage of Tn7 to the target DNA occurs (BAINTON et al. 1991). The identity of these terminal nucleotides is important to Tn7 recombination: changing the terminal -CA3' blocks both breakage and joining at the 3'end (BAINTON 1992; P. Gary and N.L. Craig, unpublished observations).

2.5 Transposition Substrates: Target Selection and Transposition Immunity

As described above, Tn7 can recognize and insert into either of two types of targets site: TnsD directs Tn7 to insertion sites of particular sequence, and TnsE can direct Tn7 to many non-*attTn7* insertion sites. There is, however, an additional feature of a potential target DNA which can strongly influence its susceptibility to Tn7 insertion: Tn7 exhibits transposition or target immunity; that is, the presence of Tn7 in a target molecule much reduces the frequency of insertion of a second copy of Tn7 into that target (HAUER and SHAPIRO 1984; ARCISZEWSKA et al. 1989). Transposition immunity is a *cis*-acting inhibition acting only on the DNA which contains the Tn7 copy, not a global inhibition of transposition. Although the presence of Tn7 in a particular potential target DNA greatly reduces the frequency of Tn7 insertion into that DNA, transposition to other targets lacking Tn7 is not inhibited. Transposition immunity is effective in both the *tnsD*- and *tnsE*-dependent transposition reactions. Transposition immunity is fundamentally related to transposition itself: the ends of Tn7, i.e., the DNA segments that are the transposition substrates, are the features of Tn7 which provide immunity (ARCISZEWSKA et al. 1989). The bacterial transposons Tn3 (LEE et al. 1983) and bacteriophage Mu (REYES et al. 1987; DARZINS et al. 1988) also display transposition immunity (reviewed MIZUUCHI 1992).

An especially intriguing aspect of transposition immunity is its ability to be effective over large distances. For example, the presence of a copy of Tn7 in a large (>50 kb) derivative of the conjugable F plasmid can reduce the frequency of Tn7 insertion at any position in that plasmid (ARCISZEWSKA et al. 1989). Thus, transposition immunity does not reflect prior occupancy of a preferred target site; rather, it is the ability of the ends of Tn7 at one position to influence insertion into adjacent DNA, where "adjacent" means anywhere on the target plasmid. More recent experiments have revealed that Tn7 target immunity can also be effective over more than 150 kb in the *E. coli* chromosome (R. DeBoy and N.L. Craig, unpublished observations).

The ability of a particular DNA site (for example, a transposon end) to influence a reaction (for example, transposon insertion) on DNA at another position is a fundamental problem common to many DNA transactions. Thus understanding the molecular basis of transposition immunity is of general interest. Transposition immunity is effective in in vitro transposition systems for Mu (ADZUMA and MIZUUCHI 1988) and Tn7 (BAINTON et al. 1993); dissection of the Mu system has provided a molecular model for this interesting process (see below).

How is target immunity useful to the transposon? Although transposition immunity is described above as a phenomenon affecting intermolecular transposition, its most important role may be to decrease the possibility of intramolecular transposition, a potentially lethal event. Especially for elements which can insert into many target sequences, the potential target DNA closest to the transposon ends is actually the donor backbone. Transposition immunity, i.e., inhibition of insertion into a DNA containing the transposon, may thus discourage intramolecular recombination, promoting intermolecular events which have less possibility of lethality.

2.6 Tn7 Antibiotic Resistance Genes

Like many bacterial transposons, Tn7 encodes antibiotic resistance determinants in addition to its transposition functions: cells containing Tn7 are resistant to antifolates such as trimethoprim and to the aminoglycosides streptothricin, streptromycin, and spectinomycin. Trimethoprim resistance is provided by the Tn7 *dhfr* gene, which encodes a novel dihydrofolate reductase resistant to this inhibitor (FLING and RICHARDS 1983; SIMONSEN et al. 1983). Streptothricin resistance is provided by the *sat* gene, which encodes a transacetylase that inactivates this aminoglycoside (SUNDSTROM et al. 1991; TIETZE and BREVET 1991). Resistance to streptomycin and spectinomycin is provided by the *aadA* gene, which encodes an adenylyl transferase that inactivates these compounds (FLING et al. 1985).

The antibiotic resistance genes of Tn7 are clustered towards the left end of Tn7, Tn7L (Fig. 1). The drug resistance genes appear to be encoded on gene cassettes which can undergo rearrangement through excision and reintegration (SUNDSTROM et al. 1991; HALL et al. 1991). Some other transposons such as Tn1825 and Tn1826 (TIETZE et al. 1987), which are similar to Tn7 in their transposition functions but have distinct antibiotic resistances, may reflect such rearrangements. To the left of the *dhfr* gene in Tn7 there is an ORF which has homology to a lambdoid-like integrase; this is the presumed recombinase of such resistance cassette rearrangements (SUNDSTROM et al. 1991; HALL et al. 1991). This integrase region is not required for Tn7 transposition (WADDELL and CRAIG 1988).

3 The Mechanism of Tn7 Transposition

A cell-free system for *tnsD*-dependent transposition has been developed (BAINTON et al. 1991). The site-specific insertion of Tn7 into *attTn7* has been reconstituted with purified proteins: recombination requires four Tn7-encoded proteins, TnsA, TnsB, TnsC, and TnsD, and two essential cofactors, Mg^{2+} and ATP (BAINTON et al. 1993). Recombination is efficient, and insertion into *attTn7* in vitro recapitulates insertion in vivo; i.e., insertion into *attTn7* is site and orientation specific. The requirement for these four Tns proteins for transposition in vitro accounts for the genetic requirement for their respective genes for transposition in vivo (ROGERS et al. 1986; WADDELL and CRAIG 1988). Dissection of this in vitro system has allowed analysis of Tn7 transposition at the molecular level (BAINTON et al. 1991, 1993).

3.1 Tn7 Is Excised from the Donor Backbone by Double-strand Breaks

When the recombination substrates are a donor plasmid containing a Tn7 element and a target plasmid containing *attTn7*, the reaction products are a simple insertion of Tn7 into *attTn7* and the donor backbone from which Tn7 has been excised (Fig. 4A). Therefore, Tn7 transposition occurs by a nonreplicative cut-and-paste mechanism; i.e., Tn7 is not copied by DNA replication during the act of transposition. Thus replication is limited to repair of the joints between the newly inserted transposon and the target DNA.

The transposon is broken away from the donor backbone by double-strand breaks; donor molecules with a double-strand break at either Tn7L or Tn7R can also be readily observed early in the reaction. It is notable that, although these double-strand break species are transposition intermediates and are generated prior to the covalent involvement of *attTn7*, the presence of *attTn7* is required to provoke the double-strand breaks. Another feature of the reaction is that such intermediates also are not observed when any protein is omitted. An attractive explanation for the highly concerted nature of Tn7 transposition is that recombination actually occurs in a multiprotein complex that also contains the recombination substrates, i.e., the two transposon ends and *attTn7*. It seems likely that the correct assembly of this elaborate nucleoprotein complex plays a key role in regulating the initiation of recombination (see below).

3.2 Insertion of Excised Tn7 into the Target DNA

The excised transposon is then joined to the target DNA. The predominate products of in vitro transposition are simple insertions; i.e., both transposon ends of a single element are joined to the target DNA. Other species can also be observed, however, in particular, joint molecules with only one end of the transposon joined to the target DNA (BAINTON et al. 1993; P. Gary and N.L. Craig, unpublished observations). Such single-end join species are observed with both the double-strand break intermediates and excised linear transposons. Once formed, however, these single-end join species appear to be only inefficiently converted to simple insertions, suggesting that they are aberrant by-products and

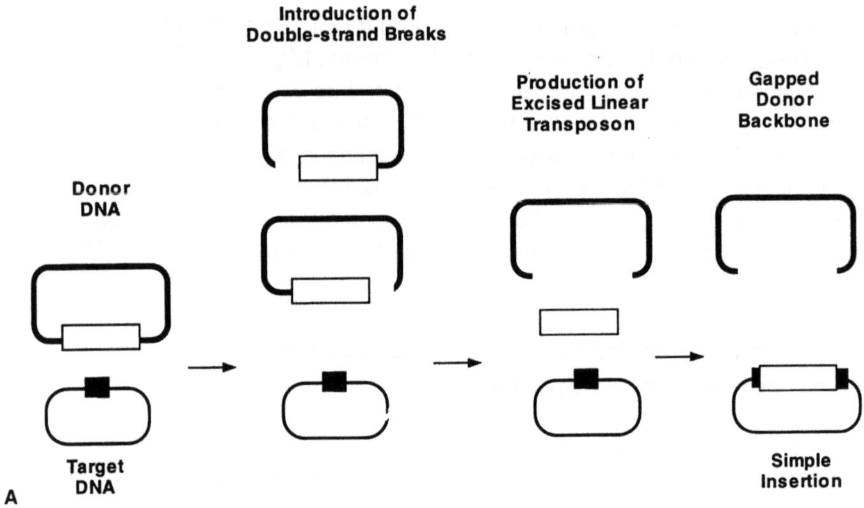

Introduction of Double-strand Breaks
Production of Excised Linear Transposon
Gapped Donor Backbone
Donor DNA
Target DNA
Simple Insertion
A

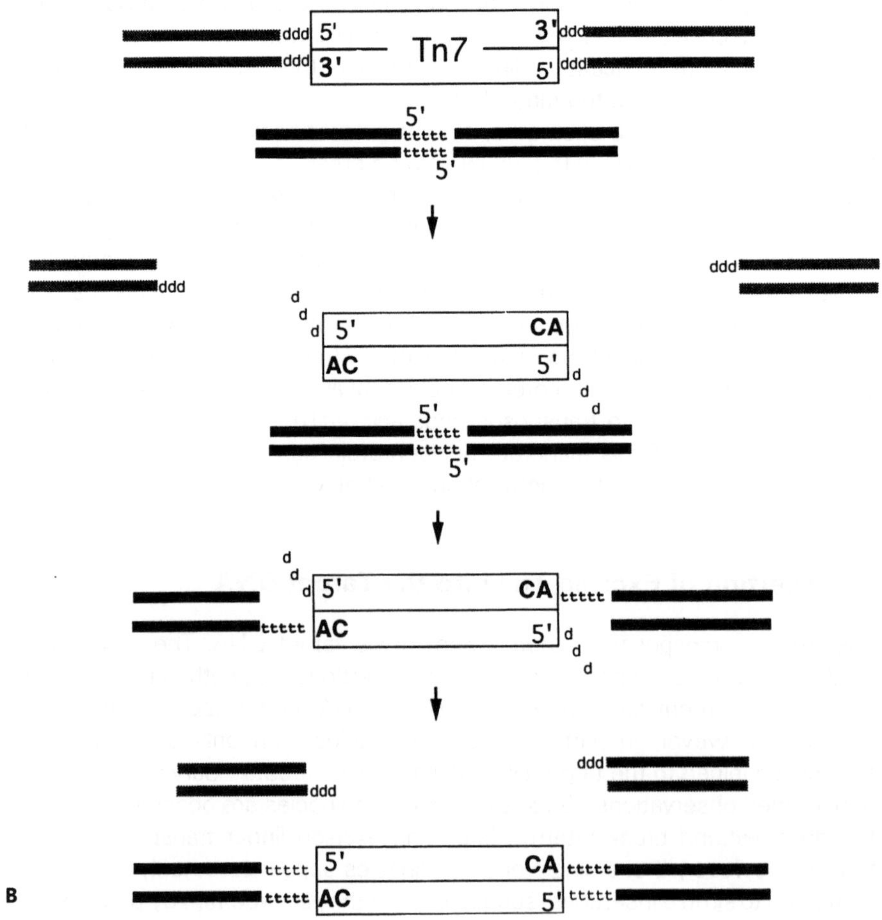

5'
3'
Tn7
ddd
ttttt
CA
AC
d
B

that the reaction proceeds in a standard manner, through concerted joining of the two ends of an excised linear transposon to the target DNA.

3.3 The Chemistry of DNA Breakage and Joining

The double-strand breaks that excise Tn7 from the donor backbone cleanly expose the 3' ends of the transposon but are staggered, such that 3 nucleotides of flanking donor DNA remain attached to the 5' transposon ends (BAINTON et al. 1991; P. Gary and N.L. Craig, unpublished; Fig. 4B). This contrasts with the excision of the bacterial transposon Tn10, which also transposes via an excised linear transposon where the 5' strands are also cut at the transposon termini (BENJAMIN and KLECKNER 1989, 1992). Tn7 is inserted into the target DNA by the joining of the exposed 3' transposon ends to staggered positions on the top and bottom strands of *attTn7*; these positions are separated by 5 bp (BAINTON et al. 1991). The joining of the 3' transposon ends to 5' positions in the target DNA results in the concomitant generation of exposed 3' ends in the target DNA. Thus the newly inserted Tn7 transposon is covalently linked to the target DNA through its 3' ends and is flanked by short gaps at its 5' ends, reflecting the staggered positions of end joining; the intact strands of the flanking gaps are the top and bottom strand of the target DNA between the positions of end joining. The repair of these gaps by the host DNA repair machinery results in the characteristic 5-bp duplications that flank Tn7 insertions; this repair process is also presumed to remove the few nucleotides of donor DNA attached to the 5' transposon ends in the initial recombination product.

The same chemistry of DNA breakage and joining, i.e., breakage to expose the 3' ends of the transposon and subsequent covalent linkage of these ends to 5' ends of target DNA, has been determined for all other elements that have been investigated at the biochemical level including Mu, retroviruses and retrotransposons, and other bacterial elements such as Tn10 (reviewed in MIZUUCHI 1992).

It should be noted that although it appears as if the excised transposon has joined to the target DNA by joining of the transposon ends of a staggered double-strand break at the insertion site, there is no evidence to support the view that Tn7 transposition proceeds through such a mechanism. No such double-strand

Fig. 4A, B. The pathway of transposition. **A** Transposition substrates, intermediates, and products are shown. On the *left* are the substrates, a donor plasmid containing Tn7 (white box) and a target plasmid containing *attTn7* (*black box*). Recombination initiates with double-strand breaks at either transposon end; pairs of breaks on the same substrate generate an excised linear transposon which inserts into the target DNA. The transposition products are a simple insertion and a gapped donor backbone. **B** The chemistry of breakage and joining during Tn7 transposition is shown. Recombination begins with staggered double-strand breaks at the ends of Tn7 which cleanly expose the 3' terminal-*CA*s and leave several nucleotides of donor DNA (*d*) attached to the 5' ends of the transposon. Tn7 joins to the displaced 5' positions on the target DNA though its terminal 3' *A*s. The simple insertion transposition product has covalent linkages between the 3' ends of the transposon and 5' ends of target DNA. The 5' transposon ends are flanked by short gaps; repair of these gaps by the host repair machinery generates 5-bp duplications of target sequences

breaks have been observed at the target site; moreover, in single-end join species, one transposon end is joined to one strand of the target DNA and the other target strand is intact (P. Gary and N.L. Craig, unpublished). These observations support the view that each Tn7 end joins independently to a single target DNA strand. Investigation of the mechanism of DNA strand breakage and joining of the bacteriophage Mu and of retroviruses has revealed that both DNA strand breakage and strand joining appear to occur through a one-step transesterification (MIZUUCHI and ADZUMA 1991; ENGLEMANN et al. 1991; reviewed in MIZUUCHI 1992). In the first stage of transposition, DNA cleavage (hydrolysis) at the 3' end occurs using H_2O as a nucleophile; in the second step, the exposed 3' OH ends of the transposon can be viewed as the attacking nucleophiles in a reaction fundamentally similar to end cleavage. Notably, this transposition reaction mechanism does not require the formation of a protein-DNA intermediate to preserve phosphodiester bond energy. This proposed mechanism of DNA breakage and joining during transposition is in many ways similar to another nucleic acid processing reaction, RNA splicing (CECH 1990), and is distinct from other site-specific recombination systems that have been analyzed, such as the integration/excision cycle of bacteriophage lambda, which proceeds through a topoisomerase-like mechanism involving covalent protein-DNA intermediates (CRAIG 1988). The detailed mechanism of Tn7 strand breakage and joining has not yet been investigated; however, since Tn7, like retroviruses and bacteriophage Mu, executes breakage and joining through its terminal -CA3' and has protein sequence similarity with retroviral integrases (see below), it is not unreasonable to suspect that Tn7 will also use a one-step transesterification mechanism.

4 The Fate of the Donor Site

Tn7 transposes in vitro through a cut-and-paste mechanism in which the transposon is excised from the donor backbone by double-strand breaks, and is inserted into the target DNA (BAINTON et al. 1991, 1993). Several in vivo observations support the view that Tn7 transposition in vivo also occurs through a cut-and-paste mechanism. Both strands of the transposon can be recovered in the same insertion product, indicating that the element is not copied by semi-conservative DNA replication during translocation (M. Lopata, K. Orle, R Gallagher, and N.L. Craig, unpublished observations). Also, induction of the SOS system which can result from double-strand breaks in DNA is observed to accompany Tn7 transposition in vivo (A. Stellwagen and N.L. Craig, unpublished observat.ons). Such a cut-and-paste transposition mechanism generates a broken donor backbone with a gap resulting from transposon excision. What is the in vivo fate of this broken donor molecule?

It has been observed that Tn7 transposition stimulates homologous recombination at the transposon donor site, an observation consistent with the hypothesis that the broken donor backbone undergoes recombinational repair (HAGEMANN and CRAIG 1993). This stimulation of recombination was evaluated with *lacZ* heteroalleles, one copy containing a mobile Tn7 element and the other a non-overlapping deletion. A stimulation of *lacZ* recombination was observed when Tn7 was able to undergo high-frequency transposition to *attTn7*; moreover, there was preferential conversion of the Tn7 donor site, consistent with transposition introducing a double-strand break to stimulate homologous recombination.

These observations support the view that the act of transposition creates a double-strand break in the donor DNA that creates a hot spot for homologous recombination (SZOSTAK et al. 1983), using an intact, homologous template to repair the broken donor (Fig. 5). In the above experiment, recombination was detected through the use of a heteroallele; the usual (and likely preferred) template for such repair is the homologous sister chromosome. When the sister chromosome is used as the template, the repair will be genetically invisible; i.e., after repair, the donor broken and gapped by the transposition reaction will appear identical to the donor site before transposon excision. A similar association between transposition and homologous recombination has been observed with the P element of *Drosophila* (ENGELS et al. 1990) and with Tc1 of *C. elegans* (PLASTERK and GROENEN 1992).

A consequence of repairing the donor site through such double-strand gap repair is to make Tn7 transposition, itself a cut-and-paste, nonreplicative reaction, effectively replicative within the cell. That is, Tn7 appears at a new insertion site through the act of transposition, and another copy of the transposon is generated at the donor site through the stimulation of double-strand gap repair by the break in the donor DNA generated by transposition. It is unknown what role, if any, Tn7-encoded functions may play in stimulating repair at the donor site.

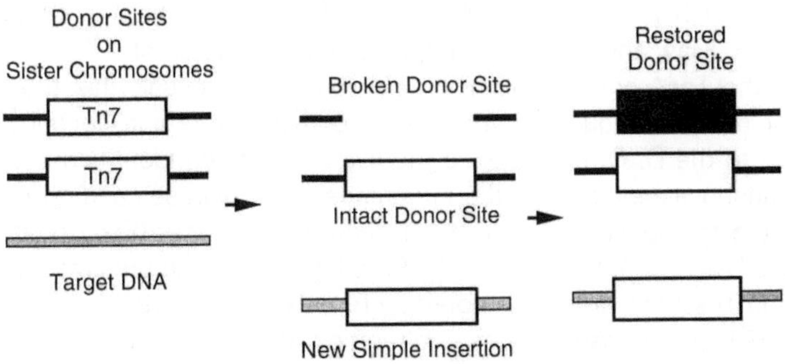

Fig. 5. Repair of the donor site from which Tn7 transposes. Transposition of Tn7 from the donor site results in a broken and gapped DNA. If transposition of Tn7 follows replication, a sister chromosome will contain an intact, unbroken donot site. This intact site can serve as a template for the repair of the gapped donor site through recombinational double-strand gap repair. Note that although Tn7 transposition per se is nonreplicative, i.e., Tn7 is excised from the donor site and inserted into the target site, this overall scheme is effectively replicative because the repair process generates another copy of the transposon

5 Roles of the Tns Proteins in Transposition

Reconstitution of Tn7 insertion into *attTn7* in vitro with purified proteins TnsABC+D revealed that these proteins participate directly in recombination (Bainton et al. 1993). Analysis of the partial activities of these proteins individually and in combination has revealed the principal activities of each of these proteins and has provided some insight into their roles in recombination. Briefly, TnsB specifically recognizes the ends of Tn7, TnsB + TnsA appear to execute DNA breakage and joining, TnsD identifies and selects *attTn7*, and TnsC is a regulator that mediates communication between the transposon ends and the target DNA.

5.1 TnsB + TnsA Mediate DNA Breakage and Joining

Recognition of the transposon ends is a key step in transposition, as it identifies the mobile DNA segment. This function is provided in Tn7 transposition by TnsB, a sequence-specific DNA-binding protein which binds to multiple positions within the regions of Tn7L and Tn7R required for recombination (McKown et al. 1987; Arciszewska et al. 1991; Arciszewska and Craig 1991; Tang et al. 1991).

TnsB and TnsA in combination form the heart of the recombination machinery which executes DNA strand breakage and joining. In support of this view, TnsB+A can, under variant recombination conditions, mediate distinctive intramolecular breakage and joining events (M. Biery, M. Lopata, and N.L. Craig, unpublished observations). Notably, these events occur in the absence of ATP, indicating that ATP does not play a direct role in recombination chemistry, but rather a regulatory role. Examination of the behavior of certain TnsB and TnsA mutant proteins in vitro has revealed that TnsB mediates the DNA breakage and joining events involving the 3' ends of Tn7, whereas TnsA mediates breakage of the 5' ends of Tn7 (R. Sarnovsky, E. May, and N.L. Craig, unpublished observations).

Inspection of the TnsA and TnsB amino acid sequences reveals that they share a region of modest sequence similarity, and that this region contains sequences related to the D, $D_{35}E$ sequence motif that has been identified in a wide variety of transposases including the integrases of retroviruses and retrotransposons and the transposases some other bacterial elements (Fayet et al. 1990; Rowland and Dyke 1990; Kulkosky et al. 1992; Figs. 6, 7). The presence of the D, $D_{35}E$ motif in TnsB has also been noted by Rezsohazy et al. (1993) and by Radstrom et al. (1994). The consequences of alteration of many of these amino acids in TnsA (E. May and N.L. Craig, unpublished observations) and TnsB (R. Sarnovsky and N.L. Craig, unpublished observations) is consistent with the hypothesis that these motifs play a key role in Tn7 transposition. Another observation supporting a fundamental relationship between Tn7 transposition and that of retroviruses is that both reactions proceed with the same chemistry; that is, the critical event is the joining of the 3' end of the transposon to target DNA (Bainton et al. 1991; reviewed in Mizuuchi 1992). Also, the extreme termini of all of these elements are highly conserved, ending in -CA3'.

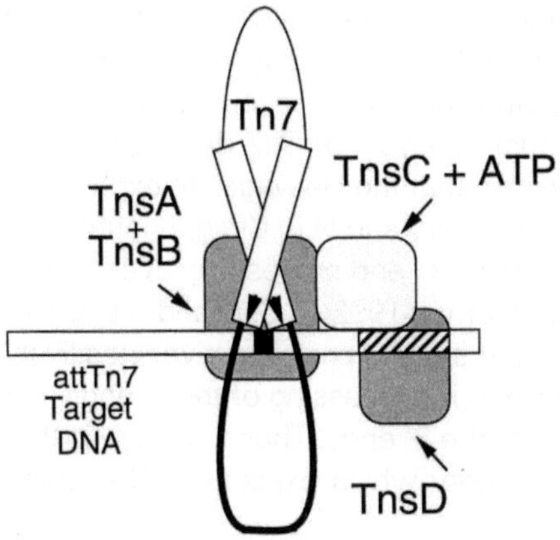

Fig. 6. The Tn7 transposition machinery. The nucleoprotein complex which executes Tn7 transposition is shown. The complex contains the DNA substrates for transposition: a donor DNA containing Tn7 with its ends *arrowheads* closely apposed and an *attTn7* target containing the sequences required for *attTn7* target activity *(cross-hatched box)* and the point of Tn7 insertion (*black box*). Note that the point of insertion is not included in the sequences essential for *attTn7* target activity. Multiple Tns proteins are involved in transposition; the stoichiometry of the proteins in the complex is not known. TnsB mediates end recognition by binding specifically to multiple sites in the ends of Tn7; TnsB + TnsA likely mediate synapsis of the ends and also likely interact with the target DNA to carry out DNA breakage and joining. TnsC is the site of action of the essential ATP cofactor in transposition and is a linker between TnsA+B bound to the ends of Tn7 and TnsD bound specifically to *attn7*. We have proposed that assembly of this complex is required for the initiation of recombination (BAINTON et al. 1993)

```
TnsA   ghgkdYipwlTvqE---vpssgrshriyshktgrvhhllSdlelAvflslewessvldireqfpllpsdtrq
TnsB   gpgsrYeiDaTiadiylvdhhdrqkiigrptLyIvIdvfSrmItGfyigfenpsyvvamqafvnacsdktai
p480   rpneiWqaDhTlldiyildqkgn---inrpwLtIiMddySraIaGyfisfdapnaqntaltlhqaiwnknnt
HIV1   cspgiWqlDcThlegk---------------vIlVaVhvaSgyIeAevipaetgqetayfllklagrwpvk--
RSV    gplqiWqtDfTleprmaprsw----------LaVtVdtaSsaIvvtqhgrvtsvavqhhwataiavlgrpk-
MoMLV  rpgthWeiDfTeikpgl---------ygykyLlVfIdtfSgwIeAfptkketakvvtkklleei--------
IS3    gpnqkWagDiTylrtpe----------gwlyLaVvIdlwSraViGwsmsprmtaqlacdalemal-------

       --------D-----------------------------------------------------------


TnsA iaidsgikh----pvrgvdqvMsTDflvdckdgpfe-qfaiqvkpaaalqdertlekleLErrywqq
TnsB caqhdieisssdwpcvglpdvLlaDrG-elmshqvealvssfnVrvesapprrgdakGiVEstfrtL
p480 ------------nwpvcgipekfyTDhGsdftshhmeqvaidlkInlmfskvgvprgrGkIErffqtV
HIV1 ---------------------tIhTDnGsnftsttvkaacwwagIkqefgipynpqsqGvVEsmnkeL
RSV  ---------------------aIkTDnGscftskstrewlarwgIahttgipgnsqgqAmVEranrlL
MoMLV---------------------vLgTDnGpafvskvsqtvadllgIdwklhcayrpqssGqVErmnrtI
IS3  ---------------------iVhTDrGgqycsadyqaqlkrhnLrgsmsakgccydnAcVEsffhsI

     ------------------------D----------------------------------E------
```

Fig. 7. Comparison of the amino acid sequences of TnsA and TnsB with transposases from retroviruses and other bacterial elements. Amino acids in particular regions of *TnsA, TnsB,* the integrases of retroviruses (*HIV1, RSV, MoMLV*), and the transposases of bacterial elements (*p480 of* Tn552, *IS3* transposase); highly conserved amino acids are in *boldface*. The D, $D_{35}E$ motifs in the retroviruses IS3 and p480 were originally identified by others (FAYET et al. 1990; ROWLAND and DYKE 1990; KULKOSKY et al. 1992). The D, $D_{35}E$ motif in TnsB has been noted by REZSOHAZY et al. (1993) and by RADSTROM et al. (1994)

Although donor processing in Tn7 transposition and that of retroviruses and bacteriophage Mu may seem at first to be fundamentally distinct (i.e., Tn7 end processing involves double-strand breaks and the others involve single-strand breaks to expose only the 3' element ends), analysis of Tn7 derivatives with mutations at their extreme termini has revealed that the cleavages to expose the 3' and 5' ends of Tn7 are separable events (P. Gary and N.L. Craig, unpublished observations). Moreover, mutations that block 3' end-processing events also block the joining of such ends to target DNA (Bainton 1992; P. Gary and N.L. Craig, unpublished observations). This suggests, that several protein-active sites likely mediate Tn7 breakage and joining, one involved in processing of the 5' ends and one involved in both processing and joining of the 3' ends. Thus, our view is that TnsB+A form the core recombination machinery which executes DNA strand breakage and joining.

With TnsA+B as the machinery that does breakage and joining, the other proteins required for transposition, TnsC and TnsD (and likely TnsE), are regulators of transposition that modulate the activity of TnsA+B and also mediate target site selectivity.

5.2 TnsD Mediates Selection of *attTn7* as a Target Site

During insertion into *attTn7*, TnsD, the protein uniquely and specifically required for insertion into *attTn7*, identifies and selects this insertion site as a target. TnsD is a sequence-specific DNA-binding protein that interacts with the nucleotide sequences in *attTn7* essential for its activity (Waddell and Craig 1989; Bainton et al. 1993).

Interestingly, TnsD does not bind to the point of Tn7 insertion in *attTn7*. Rather, the sequences required for *attTn7* activity and TnsD binding are displaced asymmetrically from the insertion site and span a segment between about 25 and 60 bp to the right of the insertion site (McKown et al. 1988; Gringauz et al. 1988; Waddell and Craig 1989; Qadri et al. 1989; Bainton et al. 1993). Thus the sequence information at the point of insertion is not apparently involved in target selection. It is interesting to note that the sequence information required for *attTn7* activity is actually in the COOH terminal coding sequence of the *glmS* gene and thus is likely to be highly conserved among bacteria. Thus Tn7 cleverly exploits information within a gene to direct its insertion but does not destroy that gene upon insertion; rather, it inserts downstream of the gene.

5.3 TnsC Mediates Communication Between the Target DNA and the Transposon Ends

TnsC can bind and hydrolyze ATP and is also an ATP-dependent non-sequence-specific DNA-binding protein (Gamas and Craig 1992; A. Stellwagen and

N.L. Craig, unpublished observations). It is the site of action of the ATP cofactor in transposition (Bainton et al. 1993). ATP appears to play a regulatory role in Tn7 transposition and has not been directly implicated in DNA breakage and joining. We have proposed that TnsC acts as a linker protein that mediates communication between TnsB+A at the transposon ends and TnsD at *attTn7* (Bainton et al. 1993).

A key step during transposition in vitro is the formation of an ATP-dependent TnsC-TnsD complex at *attTn7* (Bainton et al. 1993). We imagine that TnsC is recruited to *attTn7* by a specific interaction with TnsD, although this point has not been directly established. Thus the interaction of TnsC with TnsD positions TnsC at *attTn7*; interactions between TnsA+B and TnsC can then juxtapose the transposon ends to *attTn7* and initiate strand breakage and joining. The observation that, in the presence of the alternative ATP cofactor AMP-PNP, TnsC can stimulate both DNA strand breakage and joining provides evidence that TnsC interacts with TnsA+B (Bainton et al. 1993).

Because TnsC can both stimulate DNA breakage at the transposon ends and provoke joining of the transposon ends to the target DNA, it is suspected that this protein has multiple activities, including the ability to interact with and activate TnsA+B and the ability to interact with the target DNA. TnsC appears to act as a focal point for the regulation of transposition, in particular in evaluating potential target DNAs (see below).

5.4 Possible Roles of TnsE

In vivo analysis revealed that *tnsABC* can also be activated by *tnsE* and directed to random target sites (Rogers et al. 1986; Waddell and Craig 1988; Kubo and Craig 1990). However, no biochemical information is yet available about TnsABC+E transposition, as a cell-free system for this reaction has not yet been established. It is reasonable to expect that TnsABC will have similar roles in TnsE-dependent transposition as in TnsD-dependent recombination, i.e., that TnsA+B mediate the DNA breakage and joining reactions and that TnsC mediates communication between the transposon ends and the target DNA. Several roles can be imagined for TnsE. One possibility is that TnsE, like TnsD, is a DNA-binding protein, likely with low sequence specificity, that interacts with the target DNA and TnsC, thereby recruiting the ends and TnsABC to an insertion site. Another interesting possibility is that TnsE does not interact directly with the target DNA but rather modulates transposition through protein-protein interactions with another components. Perhaps TnsE interacts with a host protein which interacts with target DNA; alternatively, TnsE may be an allosteric effector of the DNA-binding activity of TnsC and thus modulate TnsC binding to a target DNA and its recruitment of the ends of Tn7 bound by TnsA+B.

6 Regulation of Transposition and Transposition Immunity

Tn7 uses an elaborate ensemble of proteins to mediate insertion into *attTn7*: TnsA+B are likely to provide the DNA breakage and joining activities, TnsC is apparently required to activate these functions, and TnsC and TnsD are required to efficiently engage the target DNA. By analogy, it is reasonable to imagine that TnsE is also involved in promoting target acquisition. This Tn7 system is in some ways reminiscent of bacteriophage Mu recombination, in which one protein, MuA, recognizes the transposon ends and executes breakage and joining and another protein, MuB, which, like TnsC, is an ATP-utilizing protein, modulates the activity of MuA and mediates target selection (reviewed in MIZUUCHI 1992). The ability of Tn7 to use two different kinds of target sites can account for the Tn7 system having multiple specialized targeting proteins (TnsD and TnsE); information from these target selectors is likely transduced through another protein used in both pathways (TnsC). The fact that Tn7 transposition proceeds through double-strand breaks in the donor DNA rather than through single-strand breaks, as does Mu, seems to be reflected in the involvement of two Tn7 proteins, TnsA+B, in breakage and joining. Moreover, the central chemical events for both elements is the joining of the 3' transposon ends to the target DNA. Elegant biochemical studies have revealed that Mu recombination involves a variety of specialized nucleoprotein complexes, and that transitions between these complexes are critical control points in transposition which can evaluate and respond to a variety of regulatory signals (reviewed in MIZUUCHI 1992). We imagine the choreography of Tn7 transposition will be similarly elaborate.

A key regulator of both Tn7 and Mu transposition is the nature of the target DNA. One reflection of this in Tn7 transposition is that in TnsD-dependent transposition, the presence of the *attTn7* target is required to initiate recombination, i.e., to provoke the double-strand breaks in the donor DNA which expose the transposon ends. How can the nature of the target DNA be communicated to the transposon ends? As described above, we have suggested that a key step in Tn7 transposition is the assembly of a nucleoprotein complex containing the substrate DNAs and multiple recombination proteins (BAINTON et al. 1993). The initiation of recombination can be coupled to target inspection through the strategy of requiring the assembly of this complex with the target to initiate recombination.

A notable feature of both Tn7 and Mu transposition is that they employ an ATP-utilizing protein to regulate the activity of other proteins which mediate breakage and joining of DNA. This ATP cofactor plays a key role in target evaluation. Biochemical dissection of Mu transposition has revealed that MuB, the ATP protein in Mu transposition, plays a central role in mediating transposition immunity, that is, in discouraging Mu insertion into target DNAs that already contain Mu (ADZUMA and MIZUUCHI 1988, 1989; reviewed in MIZUUCHI 1992). A target DNA already containing the ends of Mu discourages the interaction of that target DNA with MuB by promoting MuB dissociation from that target DNA; MuB dissociation is facilitated by MuB ATP hydrolysis, which is provoked by the interaction of MuB with MuA bound to the Mu ends. Thus ATP provides a critical regulatory switch.

TnsC, the ATP-utilizing protein of Tn7, also plays a central role in controlling Tn7 target immunity, as in Mu transposition; for example, when a nonhydrolyzable ATP analog is used in Tn7 transposition in vitro, Tn7 target immunity is no longer effective (BAINTON et al. 1993). Further biochemical dissection of Tn7 target immunity is required to reveal this process in molecular detail.

7 Tricks with Tn7

Transposable elements provide powerful tools for genetic analysis. Like many other elements, Tn7 has been used as an insertional mutagen; attractive features of Tn7 are its broad host range and it distinctive array of antibiotic resistances. However, the site-specific insertion of Tn7 observed in many organisms had limited its utility as an insertional mutagen (reviewed in CRAIG 1989, 1991). As it is now known that TnsD directs Tn7 to specific insertion sites and TnsE can direct Tn7 to many apparently random sites, it would now be appropriate to use *tnsD*– derivatives for successful random insertional mutagenesis.

The ability of Tn7 to insert into a specific site can also be exploited for genetic analysis. Insertion of a Tn7 derivative lacking its own *tns* functions by complementation with these functions in *trans* is an effective way of stably derivatizing a target DNA at a defined location (GRINTER 1983; BARRY 1986; BAO et al. 1991). The Tn7 element could encode a selectable marker, a gene whose presence in high copy is deleterious, or gene fusions which could then be analyzed in single copy in the chromosome (SHEN et al. 1992). An interesting variation on this theme is to use Tn7 to site-specifically introduce cloned genes into an expression vector (LUCKNOW et al. 1993).

As the components of Tn7 transposition have been identified and defined, i.e., the *tns* genes and the *cis*-acting recombination sequences at the ends of Tn7 and at *attTn7*, and are becoming increasingly well understood, there are many tools available for the construction of Tn7 systems for specialized applications.

8 Concluding Remarks

The bacterial transposon Tn7 provides a rich arena for the study of many aspects of transposition, including the modulation of transposition frequency, target site selection, and the involvement of host-encoded functions in recombination. The dissection of Tn7 transposition at the molecular level has revealed that this reaction is executed by an elaborate multiprotein nucleoprotein complex.

Elucidation of the macromolecular interactions that underlie the structure and function of this complex not only will contribute to a deeper understanding of transposition, but will also provide insight into other protein-nucleic acid transactions which occur in such complexes. The finding that the fundamental chemistry of Tn7 transposition is similar to that of retroviruses, and that recombination of these elements may be executed by related proteins, offers the possibility that dissection of Tn7 transposition will provide insight into the control of recombination by such related elements.

Acknowledgments. Work in my laboratory has been supported by the Howard Hughes Medical Institute and the National Institutes of Health. I am grateful to Anne Stellwagen for her careful reading of the manuscript. I also thank Dr.T. Kusano for providing me with the unpublished sequence of an element related to Tn7 which was useful in sequence alignments.

References

Adzuma K, Mizuuchi K (1988) Target immunity of Mu transposition reflects a differential distribution of MuB protein. Cell 53: 2577–266

Adzuma K, Mizuuchi K (1989) Interaction of proteins located at a distance along DNA: mechanism of target immunity in the Mu DNA strand-transfer reaction. Cell 57: 41–47

Arciszewska LK, Craig NL (1991) Interaction of the Tn7-encoded transposition protein TnsB with the ends of the transposon. Nucleic Acids Res 19: 5021–5029

Arciszewska LK, Drake D, Craig NL (1989) Transposon Tn7 *cis*-acting sequences in transposition and transposition immunity. J Mol Biol 297: 25–52

Arciszewska LK, McKown RL, Craig NL (1991) Purification of TnsB, a transposition protein that binds to the ends of Tn7. J Biol Chem 266: 21736–21744

Bainton RJ (1992) Tn7 transposition in vitro. University of California, San Francisco

Bainton R, Gamas P, Craig NL (1991) Tn7 transposition in vitro proceeds through an excised transposon intermediate generated by staggered breaks in DNA. Cell 65: 805–816

Bainton RJ, Kubo KM, Feng J-N, Craig NL (1993) Tn7 transposition: target DNA recognition is mediated by multiple Tn7-encoded proteins in a purified in vitro system. Cell 72: 931–943

Bao Y, Lies DP, Fu H, Roberts GP (1991) An improved Tn7-based system for the single-copy insertion of cloned genes into chromosomes of gram-negative bacteria. Gene 109: 167–168

Barry G (1986) Permanent insertion of foreign genes into the chromosome of soil bacteria. Bio/Technology 4: 446–449

Barth P, Datta N (1977) Two naturally occurring transposons indistinguishable from Tn7. J Gen Bacteriol 102: 129–134

Barth PT, Datta N, Hedges RW, Grinter NJ (1976) Transposition of a deoxyribonucleic acid sequence encoding trimethoprim and streptomycin resistances from R483 to other replicons. J Bacteriol 125: 800–810

Benjamin HW, Kleckner N (1989) Intramolecular transposition by Tn10. Cell 59: 373–383

Benjamin HW, Kleckner N (1992) Excision of Tn10 from the donor site during transposition occurs by flush double-strand cleavages at the transposon termini. Proc Natl Acad Sci USA 89: 4648–4652

Campbell AM (1992) Chromosomal insertion sites for phages and plasmids. J Bacteriol 174: 7495–7499

Cech TR (1990) Self-splicing of group I introns. Annu Rev Biochem 59: 543–568

Craig NL (1988) The mechanism of conservative site-specific recombination. Annu Rev Genet 22: 77–105

Craig NL (1989) Transposon Tn7. In Berg DE, Howe MM, (eds) Mobile DNA. American Society for Microbiology, Washington DC, pp 211–225

Craig NL (1991) Tn7: a target site-specific transposon. Mol Microbiol 5: 2569–2573

Darzins A, Kent NE, Buckwalter MS, Casadaban MJ (1988) Bacteriophage Mu sites required for transposition immunity. Proc Natl Acad Sci USA 85: 6826–2630

Derbyshire KM, Hwang L, Grindley NDF (1987) Genetic analysis of the interaction of the insertion sequence IS903 transposase with its terminal inverted repeats. Proc Natl Acad Sci USA 84: 8049–8053

Engels WR, Johnson-Schlitz DM, Eggleston WB, Sved J (1990) High-frequency P element loss in *Drosophila* is homolog dependent. Cell 62: 515–525

Engleman A, Mizuuchi K, Craigie R (1991) HIV-1 DNA integration: mechanism of viral DNA cleavage and DNA strand transfer. Cell 67: 1211–1221

Fayet O, Ramond P, Polard P, Prere MF, Chandler M (1990) Functional similarities between the IS3 family of bacterial insertion elements? Mol Microbiol 4: 1771–1777

Fling M, Richards C (1983) The nucleotide sequence of the trimethoprim-resistant dihydrofolate reductase gene harbored by Tn7. Nucleic Acids Res 11: 5147–5158

Fling M, Kopf J, Richards C (1985) Nucleotide sequence of the transposon Tn7 gene encoding an aminoglycoside-modifying enzyme, 3"(9)-0-nucleotidyltransferase. Nucleic Acids Res 13: 7095–7106

Flores C, Qadri MI, Lichtenstein C (1990) DNA sequence analysis of five genes; *tnsA, B, C, D,* and *E,* required for Tn7 transposition. Nucleic Acids Res 18: 901–911

Gamas P, Craig NL (1992) Purification and characterization of TnsC, a Tn7 transposition protein that binds ATP and DNA. Nucleic Acids Res 20: 2525–2532

Gay NJ, Tybulewica VLJ, Walker JE (1986) Insertion of transposon Tn7 into the *Escherichia coli glmS* transcription terminator. Biochem J 234: 111–117

Gringauz E, Orle K, Orle A, Waddell CS, Craig NL (1988) Recognition of *Escherichia coli attTn7* by transposon Tn7: lack of specific sequence requirements at the point of Tn7 insertion. J Bacteriol 170: 2832–2840

Grinter NJ (1983) A broad host-range cloning vector transposable to various replicons. Gene 21: 133–143

Hagemann AT, Craig NL (1993) Tn7 transposition creates a hotspot for homologous recombination at the transposon donor site. Genetics 133: 9–16

Hall RM, Brookes DE, Stokes HW (1991) Site-specific insertion of genes into integrons: role of the 59-base element and determination of the recombination cross-over point. Mol Microbiol 5: 1941–1959

Hauer B, Shapiro JA (1984) Control of Tn7 transposition. Mol Gen Genet 194: 149–158

Huisman O, Errada PR, Signon L, Kleckner N (1989) Mutational analysis of IS10's outside end. EMBO J 8: 2101–2109

Jilk RA, Makris JC, Borchard L, Reznikoff WS (1993) Implications of Tn5-associated adjacent deletions. J Bacteriol 175: 1264–1271

Kleckner N (1990) Regulation of transposition in bacteria. Annu Rev Cell Biol 6: 197–327

Kubo KM, Craig NL (1990) Bacterial transposon Tn7 utilizes two classes of target sites. J Bacteriol 172: 2774–2778

Kulkosky J, Jones KS, Katz RA, Mack JPG, Skalka AM (1992) Residues critical for retroviral integrative recombination in a region that is highly conserved among retroviral/retrotransposon integrases and bacterial inseriton sequence transposases. Mol Cell Biol 12: 2331–2338

Lee C-H, Bhagwhat A, Heffron F (1983) Identification of a transposon Tn3 sequence required for transposition immunity. Proc Natl Acad Sci USA 80: 6765–6769

Lichtenstein C, Brenner S (1981) Site-specific properties of Tn7 transposition into the *E. coli* chromosome. Mol Gen Genet 183: 380–387

Lichtenstein C, Brenner S (1982) Unique insertion site of Tn7 in the *E. coli* chromosome. Nature 297: 601–603

Lucknow VA, Lee SC, Barry GF, Olins PO (1993) Efficient generation of infectious recombinant beculoviruses by site-specific transposon-mediated insertion of foreign genes into a beculovirus genome propagated in *Escherichia coli.* J Virol 67: 4566–4579

McKown RL, Waddell CS, Arciszewska LA, Craig NL (1987) Identification of a transposon Tn7-dependent DNA-binding activity that recognizes the ends of Tn7. Proc Natl Acad Sci USA 84: 7807–7811

MuKown RL, Orle KA, Chen T, Craig NL (1988) Sequence requirements of *Escherichia coli attTn7,* a specific site of transposon Tn7 insertion. J Bacteriol 170: 352–358

Mizuuchi K (1992) Transpositional recombination mechanistic insights from studies of Mu and other elements. Annu Rev Biochem 61: 1011–1051

Mizuuchi K, Adzuma K (1991) Inversion of the phosphate chirality at the target site of Mu DNA strand transfer: evidence for a one-step transesterification mechanism. Cell 66: 129–140

Orle KA, Craig NL (1991) Identification of transposition proteins encoded by the bacterial transposon Tn7. Gene 104: 125–131

Plasterk RHA, Groenen JTM (1992) Targeted alterations of the *Caenorhabditis elegans* genome by transgene instructed DNA double-strand break repair following Tc1 excision. EMBO J 11: 287–290

Qadri MI, Flores CC, Davis AJ, Lichtenstein CP (1989) Genetic analysis of *attTn7*, the transposon Tn7 attachment site in *Escherichia coli*, using a novel M 13-based transduction assay. J Mol Biol 207: 85–98

Radstrom P, Skold O, Swedberg G, Flensburg F, Roy PH, Sundstrom L (1994) Transposon Tn5090 of plasmid R751, which carries as integron, is related to Tn7, Mu, and the retroelements. J Bacteriol 176: 3257–3268

Reyes O, Beyou A, Mignotte-Vieux C, Richaud F (1987) Mini-Mu transduction: *cis*-inhibition of the insertion of Mud transposons. Plasmid 18: 183–192

Rezsohazy R, Hallet B, Delcour J, Mahillon J (1993) The IS4 family of insertion sequences—evidence for a conserved transposase motif. Mol Micro Biol 9: 1283–1295

Rogers M, Ekaterinaki N, Nimmo E, Sherratt D (1986) Analysis of Tn7 transposition. Mol Gen Genet 205: 550–556

Rowland S-J, Dyke KGH (1990) Tn552, a novel transposable element from *Staphylococcus aureus*. Mol Microbiol 4: 961–975

Shen H, Gold SE, Tamaki SJ, Keen NT (1992) Construction of a Tn7-*lux* system for gene expression studies in gram-negative bacteria. Gene 122: 27–34

Simonsen C, Chen E, Levionson A (1983) Identification of the type I trimethoprim-resistant dihydrofolate reductase specified by the *Escherichia coli* R-plasmid 483: comparison with procaryotic and eucaryotic dihydrofolate reductases. J Bacteriol 155: 1001–1008

Smith G, Jones P (1986)Tn7 transposition: a multigene process. Identification of a regulatory gene product. Nucleic Acids Res 14: 7915–7927

Sundstrom L, Roy PH, Skold O (1991) Site-specific insertion of three structural gene cassettes in transposon Tn7. J Bacteriol 173: 3025–3028

Szostak JW, Orr-Weaver T, Rothstein RJ, Stahl FW (1983) The double-strand-break repair model for recombination. Cell 33: 25–35

Tang Y, Lichtenstein C, Cotterill S (1991) Purification and characterization of the TnsB protein of Tn7, a transposition protein that binds to the ends of Tn7. Nucleic Acids Res 19: 3395–3402

Tietze E, Brevet J (1991) The trimethoprim resistance transposon Tn7 contains a cryptic streptothricin resistance gene. Plasmid 25: 217–220

Tietze E, Brevet J, Tschape H (1987) Relationships among the streptothricin resistance transposons in Tn1825 and Tn1826 and the trimethoprim resistance transposon Tn7. Plasmid 18: 246–249

Waddell CS, Craig NL (1988) Tn7 transposition, two transposition pathways directed by five Tn7-encoded genes. Genes Dev 2: 137–149

Waddell CS, Craig NL (1989) Tn7 transposition, recognition of the *attTn7* target sequence. Proc Natl Acad Sci USA 86: 3958–3962

Walker JE, Gay NJ, Saraste M, Eberle AN (1984) DNA sequence around the *Escherichia coli unc* operon. Completion of the sequence of a 17 kilobase segment containing *asnA, oriC, unc, glmS* and *phoS*. Biochem J 224: 799–815

Tn10 and IS10 Transposition and Chromosome Rearrangements: Mechanism and Regulation In Vivo and In Vitro

N. Kleckner, R.M. Chalmers, D. Kwon, J. Sakai, and S. Bolland

---

Department of Biochemistry and Molecular Biology, Harvard University, 7 Divinity Avenue, Cambridge, MA 02138, USA

1 Introduction and Biology

Tn10 is a composite transposon. It comprises a pair of IS10 insertion sequences located in opposite orientation flanking ~6.7 kb of unique sequences; these unique sequences encode a tetracycline resistance determinant and other determinants whose functions remain to be identified (Fig. 1A; KLECKNER 1989). One of Tn10's two IS10 elements, IS10-Right, is structurally and functionally intact and is considered to be the "wild type" IS10. IS10 encodes a single transposase protein which mediates transposition by interacting with specific sequences at two oppositely oriented IS10 (or Tn10) termini. The termini of IS10 are subtly different and are referred to as the "outside" and "inside" end, respectively, by virtue of their position in Tn10. IS10-Left is structurally intact but encodes a substantially defective transposase.

Tn10 is found in a number of enterobacteria, usually on a conjugative resistance transfer factor; it is also common in *Haemophilus* (KLECKNER 1989). A

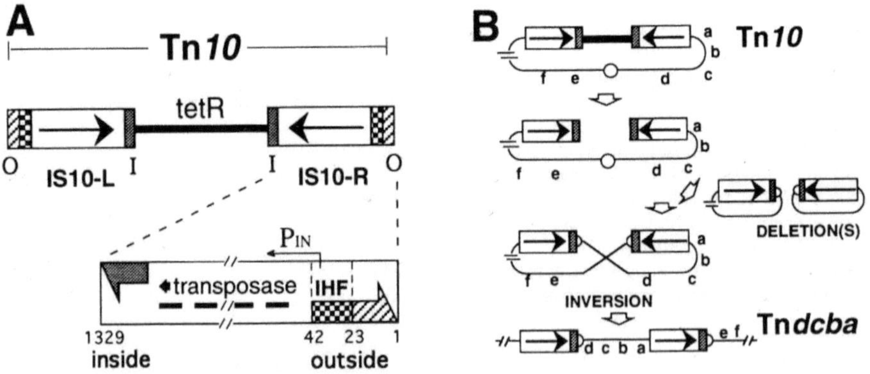

Fig. 1. A Structures of Tn10 and IS10. **B** Chromosome rearrangements promoted by Tn10. Any composite IS-based transposon will generate both deletions and inversions as alternative reaction products. Each inversion event generates a new composite transposon, e.g., Tn-*dcba*

second IS10-based composite transposon has also been described: Tn2921 comprises IS10 elements in direct repeat flanking a fosfomycin resistance determinant (NAVAS et al. 1985).

Transposition has been observed, not only of Tn10, but also of genetically derived composite elements that contain the IS10 elements either in direct repeat (TnGal; RALEIGH and KLECKNER 1984) or in the inverted orientation opposite to that found in Tn10 (FOSTER et al. 1981a). Transposition has also been observed for both IS10-Left and IS10-Right individually in a cell containing Tn10 and for an isolated IS10 (-Right) element acting entirely on its own (CHAO et al. 1983; ROBERTS et al. 1985). Tn10 transposition is rare, about 10^{-6} cell per generation for a chromosomal element (FOSTER et al. 1981b). Transposition of IS10 is much more frequent, 10^{-3}–10^{-4} cell per generation (SHEN et al. 1987). The lower frequency of Tn10 transposition may be attributable largely to "transposition length dependence" (MORISATO et al. 1983; see below).

Tn10 and IS10 are also capable of promoting chromosome rearrangements. Tn10 (and the other IS10-based composite transposons) can promote either inversion or a deletion of a segment that includes chromosomal sequences adjacent to one end of the transposon plus the proximal IS element (Fig. 1B). Inversion events are of particular interest since they yield new composite transposable elements that contain the inverted segment (KLECKNER et al. 1979a; ROSS et al. 1979; Fig. 1B). Tn10-promoted rearrangements occur at a frequency of about 10^{-5}/element per cell per generation (KLECKNER et al. 1979a).

A single IS10 element existing by itself in the genome is also capable of promoting the deletion of adjacent material in which the element remains intact and a continuous set of sequences extending from one end of the element to an adjacent target site is removed (ROBERTS et al. 1991; WEINERT et al. 1984). IS10-promoted adjacent deletions are about 2% as frequent as simple IS10 transposition (ROBERTS et al. 1991). IS-promoted adjacent deletions were one of the original genetic hallmarks of an IS element (STARLINGER and SAEDLER 1976; REIF and SAEDLER 1977).

IS10 does not promote co-integrate formation in a $RecA^-$ host (HARAYAMA et al. 1984; WEINERT et al. 1984). Co-integrates may be formed at a low frequency in a $RecA^+$ host (HARAYAMA et al. 1984; see below).

For previous reviews of Tn10 and IS10, see KLECKNER (1989, 1990 a, b).

2 Tn10 Transposition In Vivo

2.1 Transposition per se Is Nonreplicative

The vast majority of Tn10 transposition events occur without replication of the element as an integral part of the process (BENDER and KLECKNER 1986). A Tn10/IS10 element is excised from its donor site by a pair of double-strand cleavages at the ends of the element and subsequently inserts itself into a new target DNA site; a double-strand gap is left behind at the donor site (Morisato and Kleckner 1984; Roberts and Kleckner 1988; Benjamin and Kleckner 1989; Haniford et al. 1991; Chalmers and Kleckner 1994).

Insertion of Tn10 at a new target site is accompanied by duplication of 9 bp (Kleckner 1979). From the structure of the primary strand transfer product it can be inferred that the 3' transposon ends have become joined to target DNA sequences that are separated by 9 bp on the two strands; host DNA repair functions presumably fill in the resultant 9-bp single-stranded gaps (Fig. 2).

2.2 Biological Features Ensure Retention of the Transposon at the Donor Site

Although Tn10 transposition is mechanistically nonreplicative, the process as a whole, as it occurs in vivo, is essentially replicative in nature: because of the way in which the host deals with the gapped donor molecule, every transposition event results in an increase in the frequency of transposons relative to host chromosomes. Genetic observations demonstrate that every cell in which IS10 has become inserted at a new site retains, in addition, a genetically unaltered copy of the transposon at the donor site (Bender, et al. 1991).

The precise fate of the donor site remains to be established. The double-strand gap resulting from excision of the transposon elicits induction of a cellular SOS response. Probably, in some events, the transposon has inserted into a second chromosome in the cell and the donor chromosome has been completely degraded, and in other events, the donor chromosomes have undergone double-strand gap repair, which reconstitutes the original transposon-containing segment by repair off of a sibling chromosome. The relative contributions of these two processes are not established. Double-strand gap repair may be relatively rare in a wild type *E. coli*, strain, however (Thaler et al. 1987). Using a permissive

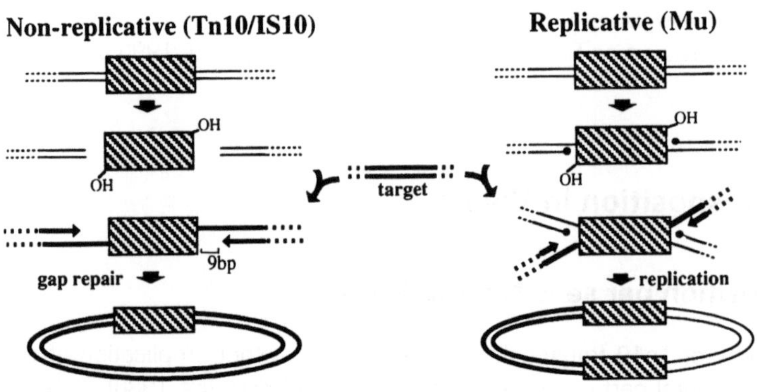

Fig. 2. Nonreplicative vs. replicative transposition

(recD-) mutant host, double-strand gap repair has been observed after Tn7 excision (HAGEMANN and CRAIG 1993).

3 Biochemistry of Transposition

3.1 Chemical Events

The double strand cleavages that occur at each end of the transposon yield flush ends that correspond precisely to the terminal base pairs of the transposon with 3'OH and 5'PO4 strand termini (BENJAMIN and KLECKNER 1992). Strand transfer is presumed to occur via a single-step transesterification reaction in which the 3'OH groups at the two ends of the element directly attack the target DNA (Fig. 2), as is known to be the case for Mu and retroviruses (MIZUUCHI and ADZUMA 1991; ENGELMAN et al. 1991); strong analogies between Tn10 and Mu support this view (HANIFORD et al. 1991; BENJAMIN and KLECKNER 1992; see below).

Transposase has recently been purified to homogeneity in a highly active form. Transposition with this preparation requires only transposase protein, an appropriate buffer, and Mg^{++} (CHALMERS and KLECKNER 1994). IS10 thus resembles other site-specific recombination and transposition reactions, none of which require a high energy cofactor for the chemical steps of the reaction.

Incubation of such transposase with a supercoiled plasmid substrate containing an appropriately oriented pair of transposon ends yields the excision and strand-transfer products expected from in vivo analysis. Excision yields the "excised transposon fragment (ETF)" and "backbone" fragments. Strand-transfer products comprise a complex array that includes both intermolecular events (i.e., insertion of the transposon into a second molecule) and intratransposon events (i.e., insertion into a target site that is located within the transposon segment itself) (Figs. 3, 4). The latter products are topologically complex and comprise a mixture of knotted and unknotted inversion circles and catenated and free deletion circles. These are all products expected if a pair of transposon ends encounters its target DNA by a random collision process (Fig. 4). These efficient in vitro reactions also reveal a complicated array of interesting secondary products (see below).

3.2 Synaptic Complex(es)

For Tn10, as for other well-studied transposons, the chemical steps in transposition all occur within the context of a stable transposase-mediated synaptic complex between the transposon ends. In vivo, interaction of the transposon ends is required prior to and prerequisite to cleavage at either end (HANIFORD and KLECKNER 1994), and a corresponding stable precleavage synaptic complex has been identified in vitro (SAKAI et al. 1995). The stability of the synaptic complex

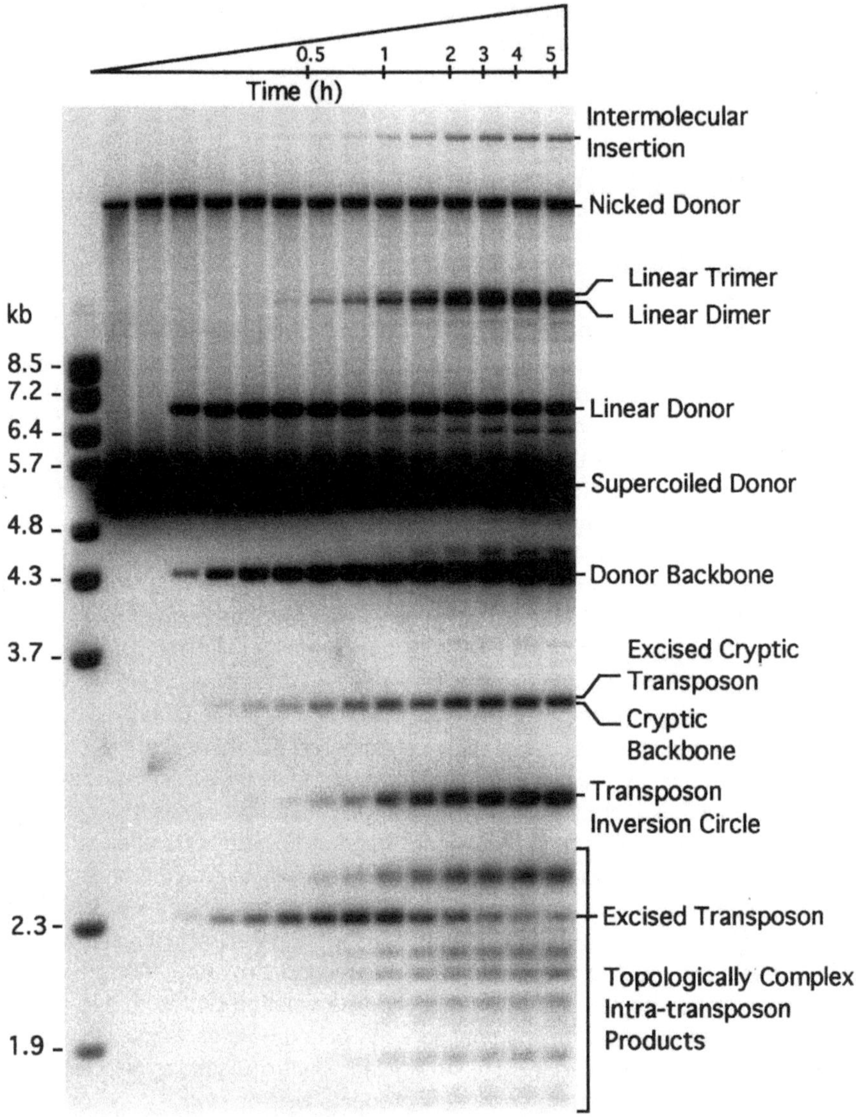

Fig. 3. Transposition reaction in vitro. Reactions were carried out as described in CHALMERS and KLECKNER (1994). Substrate was supercoiled plasmid containing two "inside" ends in appropriate inverted orientation. Cleavage yields the *excised transposon* and the *donor backbone*. Intratransposon strand transfer produces *inversion circles* and the *topologically complex products* (see Fig. 4). Intermolecular strand transfer produces a gapped circle (*intermolecular insertion*) (see Fig. 2). A wild-type transposon end may interact with a pseudo-end that fortuitously exists on the substrate. Products of this reaction are the *excised cryptic transposon* and the *cryptic backbone*. *Linear dimer* and *linear trimer* result from bimolecular synapses (text)

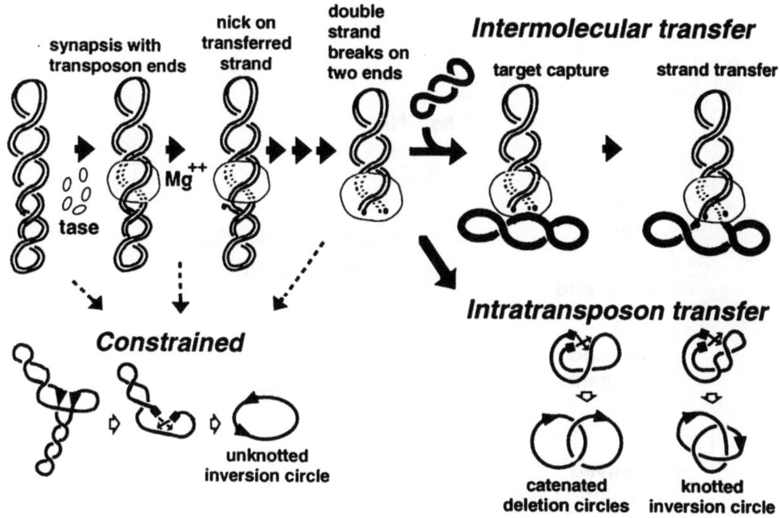

Fig. 4. Progression of the transposition reaction

increases as the reaction progresses. The complex containing the excised transposon is more stable than the precleavage complex, and the strand-transfer complex is the most stable of all (HANIFORD et al. 1991; SAKAI et al. 1995). Thus for Tn10 as for Mu, the transposition reaction is thermodynamically downhill, and progression through the various chemical steps is limited only by a series of kinetic barriers (SURETTE et al. 1987; MIZUUCHI et al. 1992).

In vitro, the formation and progression of synaptic complexes, as well as certain particular kinetic features of the reaction, have been elucidated using an "abbreviated" substrate, a short linear fragment of 136 bp encoding the outside end of IS10 plus some additional DNA on either side. Incubation of such a fragment with both transposase and the host factor IHF results in the formation of protein/DNA complexes that are detectable by gel retardation analysis. Precleavage complexes are formed and subsequently undergo all of the chemical steps of transposition (Fig. 5). IHF is probably required specifically in this assay to compensate for the absence of substrate DNA supercoiling (below).

3.3 Target Capture

Analysis of complexes formed on short linear fragments has revealed an additional step in the transposition reaction, the occurrence of a stable, noncovalent interaction between the "double end cleaved" complex and target DNA (J. Sakai and N. Kleckner, manuscript in preparation). This interaction is not only subsequent to but is also dependent upon double-strand cleavage at both transposon ends within the complex. For intermolecular transposition, this is the first point at

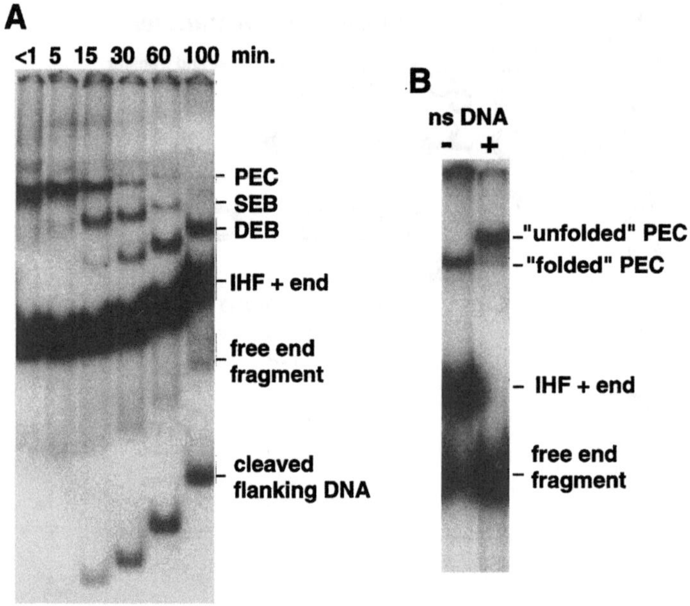

Fig. 5. Transposase-mediated synaptic complexes with transposon ends. **A** A radiolabeled short linear fragment containing one transposon end was incubated with transposase and IHF in the presence of Mg^{++} and electrophoresed in a native acrylamide gel. Free fragment is completely bound by IHF (*IHF + end*). IS10 transposase promotes the formation of a precleavage complex containing two paired ends (*PEC*). During the course of the reaction, sequences included in the complex are cleaved, giving rise to a single-end break complex (*SEB*) and a double-end break complex (*DEB*). Cleaved flanking sequences get separated from the complex. **B** The presence of nonspecific (*ns*) DNA titrates off IHF and promotes the formation of the "*unfolded*" synaptic complex

which target DNA enters the reaction in a meaningful way; it is this interaction which commits the synaptic complex to undergoing strand transfer with a particular target DNA; furthermore, this is the first step at which the presence of target DNA is observed to affect the rate of strand transfer.

3.4 Timing

Completion of the entire sequence of events involved in transposition requires about 1 h. Both in vivo and in vitro, in synchronously initiated reactions, ~50% of the final level of strand-transfer product has been generated after 60–90 min (Figs. 3–5; Haniford et al. 1991; Haniford and Kleckner 1994; Sakai et al. 1995; R. Chalmers and N. Kleckner, manuscript in preparation). The interval between stages increases as the reaction progresses. Assembly of the synaptic complex is very fast, requiring less than 2 min; the excision step is completed about 15 min later and strand transfer is not complete until another 45 min have passed.

3.5 Coordination Between the Two Transposon Ends

Transposon ends and target DNA sequences are all tightly held within the synaptic complex. Most or all of the supercoils present within the transposon segment prior to cleavage are retained following excision (inferred from BENJAMIN and KLECKNER 1989). Similarly, supercoils present within the target DNA segment are retained even after formation, at the DNA level, of the gapped strand-transfer product; in contrast, donor sequences flanking the transposon are released immediately upon cleavage (SAKAI et al. 1995).

Although the two transposon ends are tightly held within the synaptic complex, events at the two ends are closely coordinated but not absolutely interdependent. Cleavages at the first and second end are easily resolved temporally (Fig. 5A); the same is true for strand transfers at the two ends, although the temporal separation is less than at the cleavage stage. It is not yet clear, however, whether events at the " first" end influence the rate or nature of events at the "second" end, or whether events at the two ends are effectively independent.

3.6 Two Interconvertible Types of Synaptic Complexes Reflect the Pathway for Synaptic Complex Assembly

The precleavage synaptic complex formed in the short linear fragment assay can exist in two different forms, "folded" and "unfolded" (J. Sakai and N. Kleckner, manuscript in preparation; Fig. 5B). The form that predominates depends upon the concentration of IHF. At high IHF levels, only the folded form occurs; at low IHF levels, both forms are present.

Initial DNase protection analysis suggests that the folded form of the synaptic complex involves not only specific contacts between transposase and sequences within the inverted repeat, but also an IHF-induced bend adjacent to the transposase interaction site and nonspecific contacts with transposase internal to the bend. The open form, instead, involves contacts between transposase and DNA that are limited to the terminal inverted repeat region.

Moreover, once formed, the unfolded form can be converted to the folded form by addition of either IHF or HU, and the folded form can be converted to the unfolded form by titration of IHF or certain other manipulations. Thus, the difference between the two forms may simply be the presence or absence of wrapping of internal transposon sequences around a core of transposase protein, as determined primarily by the presence or absence of IHF (or HU) within the complex.

Several lines of evidence suggest that the more compact form is likely to be precursor to the more open form. Moreover, while both forms can undergo cleavage, only the open form is capable of intermolecular target capture and strand transfer. It is attractive to suppose that a constellation of weak interactions is used to bring the relevant components together, with subsequent release of the more accessory interactions once the complex has formed. In this sense, formation of the folded Tn10/IS10 complex is functionally analogous to formation of the IAS-mediated complex that precedes stable synaptic complex formation during Mu transposition (LEUNG et al. 1989; MIZUUCHI and MIZUUCHI 1989; SURETTE and CHACONAS 1992).

3.7 The Two Strands at a Single Transposon End Are Cleaved in a Specific Order

The two single-strand cleavages at each end of Tn10/IS10 occur in a specific order, with cleavage of the transferred strand followed by cleavage of the nontransferred strand (BOLLAND and KLECKNER 1995). (a) In a wild-type reaction, nicking of the transferred strand occurs at observable levels, while nicking of the nontransferred strand does not. Moreover, the level of nicking on the transferred strand can be accentuated either by substitution of Mn^{++} for Mg^{++} in an otherwise wild-type reaction, or by certain mutant transposases, or by a change in the terminal nucleotide of the transposon. The last two effects both appear to mimic the Mn^{++} effect exactly, as addition of Mn^{++} to the mutant reactions has no further effect. (b) The nicking in the wild-type reaction exhibits the kinetics expected for a reaction intermediate. The level of such nicking is quite low, however, which suggests that cleavage of the second strand follows rapidly upon cleavage of the first. The functional interdependency between the two cleavages is not known; for example, cleavage of the first strand may be required for cleavage of the second strand.

3.8 One Monomer Cleaves Both Strands

During cleavage at one end of the element, how is the DNA recognized in a specific but asymmetric way so as to mediate cleavage of two strands of opposite polarity? A simple possibility would be that a pair of symmetrically disposed monomers cleaves the two strands in identical fashion. Recent observations suggest, however, that a single monomer of transposase is responsible for cleaving both strands (S. Bolland and N. Kleckner, manuscript in preparation). In reactions carried out by mixtures of wild-type transposase and a catalytically inert transposase, nicked termini are never observed. If two monomers were required to cleave two strands, a significant fraction of synaptic complexes should have exhibited such a nick, because they would contain at least one end loaded with one mutant and one wild-type monomer suitably positioned such that the wild-type monomer would nick the transferred strand. Furthermore, the proportions of complexes that undergo double-strand cleavage at 0, 1, or both ends are incompatible with the possibility that four identical monomers are required and correspond closely to those expected if only two are required; that is, one monomer for each end.

Three models can account for these results. (a) Transposase may contain two different and independent active sites, one for the cleavage of each strand at the end of the element. This is probably not the case: single mutations (catalysis-minus phenotype, see below) abolish all chemical activity of transposase; no cleavage of any of the strands or strand transfer using previously cleaved ends is shown. Such mutations would not occur if the two active sites were independent. (b) Transposase may contain two active sites whose actions are highly coordinated, with cleavage of the second strand dependent upon cleavage of the transferred strand. (c) Transposase may contain only a single active site which mediates cleavage first of the transferred strand and then, after some type of structural rearrangement, of the nontransferred strand. In support of this possibility, two mutations that confer higher than normal levels of nicking (i.e., mutations largely defective in cleavage of the second, nontransferred strand) lie at amino acid residues that are immediately adjacent to mutations that render transposase unable to carry out any of the chemical steps, including cleavage of the first, transferred strand. It is particularly intriguing that a single active site might have the capacity to cleave strands of opposite polarities.

A protein monomer has been postulated to mediate cleavage of both strands of DNA in two other cases: pancreatic *DNase*I (SUCK et al. 1988) and restriction endonuclease *Fok*I (WAUGH and SAUER 1993). In neither case is it known whether the cleavages involve one or two active site(s).

The nonreplicative transposon Tn7 also makes double-strand cleavages at the two transposon ends (BAINTON et al. 1991). This element, which encodes five different genes involved in transposition, appears to use two different proteins to cleave the two strands at each end. Catalysis-minus mutants of TnsB and TnsA are defective in cleavage of the transferred and nontransferred strands, respectively, and accumulate the appropriate corresponding nicked species (N. Craig, personal communication).

3.9 Two Monomers of Transposase Appear to Carry Out All of the Chemical Steps in Tn10 Transposition

Reactions carried out by mixtures of a wild-type transposase and a catalytically inert transposase further suggest that the same two monomers which mediate double-strand cleavage also mediate subsequent strand transfer of cleaved ends to target DNA (S. Bolland and N. Kleckner, manuscript in preparation). In such mixed reactions, all of the synaptic complexes that have undergone cleavage at both ends subsequently go on to carry out strand transfer. If cleavage and strand transfer were promoted different monomers, only a fraction of the complexes containing doubly cleaved ends would also have contained wild-type monomers suitably positioned for the next step.

The notion that a single active site could accommodate both strand nicking and strand transfer has been proposed to account for integration of both Mu and HIV (MIZUUCHI and ADZUMA 1991; ENGELMAN et al. 1991). In both reactions the bond to be cleaved must be labilized; for nicking, the nucleophile is water, while for strand transfer, the nucleophile is the 3'OH group generated by nicking. Similarly, in DNA polymerases, the mechanisms of phosphoryl transfer during 3' to 5' exonuclease action and polymerization are thought to be closely analogous (JOYCE and STEITZ 1994).

The results obtained for Tn10 appear to imply that a single monomer must carry out three successive reactions: nicking of the transferred strand, nicking of the nontransferred strand, and finally, following target DNA capture by the synaptic complex, transfer of 3'OH terminus to the target. If these reactions are all mediated by a single active site, nicking of the first strand would be followed by a structural transition that moves the transferred strand out of the active site and moves the nontransferred strand into that site. Then, following the next cleavage, the nontransferred strand terminus would also move. Finally, after dissociation of flanking donor DNA, the active site would be occupied by target DNA. If uncleaved ends and target DNA successively occupy the same transposase active site, it would be easy to understand why stable noncovalent association of the synaptic complex with target DNA absolutely requires cleavage of both strands at both ends (above).

3.10 Two Roles for Divalent Metal Ions

Divalent cations have two distinguishable effects on the transposition process (S. Bolland, R. Chalmers, J. Sakai and N. Kleckner, manuscripts in preparation). First, Mg^{++} is specifically required for the catalytic steps of the reaction; Mn^{++} is tolerated but inhibits events subsequent to nicking of the first strand; Ca^{++} does not support any chemical events. Second, the presence of a divalent cation stabilizes the synaptic complex, particularly at the precleavage and target capture stages; any of the three metals will suffice. Also, several mutant phenotypes of IS10 transposase that show stability defects are affected by variations in the divalent metal ion, in a way that suggests a structural role for the metal ion (D. Haniford, personal communication; see below). A crucial role for the active site, and for metal binding at that site, in assembly of transposase synaptic complexes has been proposed for Mu (MIZUUCHI et al. 1992).

3.11 Mechanistic Basis for Nonreplicative Transposition

Many features of nonreplicative Tn10 transposition are closely analogous to those described for bacteriophage Mu, which transposes by a replicative process. In the case of Mu, replicative transposition is possible because the nontransferred strand maintains its connection to flanking donor DNA, and because the polarity of the strand transfer reaction is such as to leave a 3'OH target strand end suitably positioned to prime DNA replication across the transposon (Fig. 2). For Tn10, the chemistry of the strand transfer reaction appears to be the same as for Mu; thus it is the fact that the element undergoes efficient double-strand cleavage at the two ends of the element that precludes replicative transposition (BENJAMIN and KLECKNER 1989).

In this regard, it is interesting that for Tn10, nicking of the transferred strand precedes nicking of the nontransferred strand. Had the order of these steps been reversed, Tn10-promoted co-integrate formation would have been precluded as an intrinsic consequence of the reaction order: the 3'OH termini required for strand transfer would have been generated in the same step that cleaved the second strand. Instead, as the reaction actually occurs, the intermediate formed after the first nicking step is chemically identical to the nicked intermediate that undergoes strand transfer in the Mu reaction and yet does not undergo strand transfer at any reasonable level, either in vivo or in vitro.

We favor the view that strand transfer is precluded by the nature of the transitions that occur within the synaptic complex—specifically, that the synaptic complex becomes competent for strand transfer only by undergoing an ordered series of transitions that are tightly coupled to the normal progression of chemical events. The dependence of target capture on double-end cleavage suggests that strand transfer does require changes in the synaptic complex; it is not yet known, however, whether double-end nicking is sufficient to permit target capture. An alternative view is that the absence of strand transfer could be a simple kinetic effect: nicked complexes might be structurally competent for target capture and strand transfer but simply fail to do so because cleavage of the second strand is faster.

In the case of Tn7, catalysis-minus TnsA mutants that specifically block cleavage of the nontransferred strand are still capable of giving co-integrates, both in vivo and in vitro (N. Craig, personal communication). In this case, apparently, progression to the strand transfer stage is not dependent upon completion of the second chemical step (i.e., nicking).

3.12 Important Differences Between the Tn10 and Mu Reactions

Although Tn10 and Mu transposition exhibit many fundamental similarities, they also exhibit important differences:

Assembly of a precleavage synaptic complex and strand transfer are simpler processes in the case of Tn10/IS10 than in the case of Mu. For Tn10, these events require only transposase and sequences in the immediate vicinity of the transposon end as befits a small, modular transposable element. For Mu, instead, assembly of the synaptic complex requires multiple transposase-binding sites at the end plus a specific "enhancer" site located some distance from one Mu end. This enhancer requirement places the Mu transposition reaction under the control of Mu repressor, which competes with transposase for binding at the site (HARSHEY et al. 1985; MIZUUCHI 1992a).

Mu transposition is subject to "target immunity": Mu will not insert into itself or into a second molecule containing a Mu end (ADZUMA and MIZUUCHI 1989). This feature is essential, because Mu uses transposition for genomic replication during its lytic growth cycle. Target immunity is conferred via a second Mu protein, Mu B, which brings the target DNA into association with the synaptic complex. Target immunity is conferred by specific exclusion of Mu B from regions of DNA that are within striking distance of a transposase-bound Mu end.

Tn10 and IS10 do not exhibit target immunity in vivo or in vitro (BENNETT et al. 1977; R. Chalmers and N. Kleckner, manuscript in preparation; Fig. 3). The absence of these features reflects the fact that for any composite transposon such as Tn10, insertion of the element into itself is actually a crucially important biological activity: the Tn-promoted inversion events that generate new composite IS-based transposons are in effect attacks of the "inside" ends of the element on sequences within a transposon that comprises the entire bacterial chromosome (Fig. 1B). Indeed, a host-factor-mediated regulatory process actually promotes the formation of intratransposon events of this type for Tn10 (below).

The Mu reaction has two additional features that permit the effective functioning of target immunity, both of which are absent in the Tn10 reaction. First, for Mu, assembly of the precleavage synaptic complex is a very slow step; this kinetic feature provides adequate time for elimination of Mu B from regions near Mu ends. For Tn10, assembly of the precleavage complex is very rapid. Second, interaction of target DNA-bound Mu B with the precleavage complex dramatically increases the rate of cleavage and strand transfer, reducing the time required for cleavage from 15 min to less than 1 min; this effect ensures that Mu will not undergo even the first chemical steps of the transposition reaction until it is committed to interacting with an appropriate (nonimmune) target DNA. For Tn10, instead, there is no stable or effective interaction between the Tn10/IS10 synaptic complex and the target DNA until after double-strand cleavage is complete.

The above differences between Tn10 and Mu apply to transposon Tn7 as well. Although the Tn7 transposition process is nonreplicative in nature, it resembles that of bacteriophage Mu with respect to target interactions. Tn7 exhibits target immunity, mediated by a protein, TnsC, that is probably analogous to Mu B (ARCISZEWSKA et al. 1989). Furthermore, the dependence of chemical steps on interaction with target DNA is even stronger than in the case of Mu: in the case of Tn7, no cleavage is ever observed in the absence of target DNA and the proteins specifically devoted to target DNA interactions (BAINTON et al. 1993).

For Tn10, available evidence suggests that two monomers of transposase carry out all chemical steps. For Mu, instead, it appears that the nicking and strand-transfer reactions are carried out by different specific molecules within the synaptic complex, two monomers of transposase mediating nicking and two different monomers mediating strand transfer (BAKER et al. 1993, 1994). It seems possible that the two-monomer configuration of Tn10 is present in order to ensure that transposition is nonreplicative (above).

3.13 Minority Products Generated During Tn10 Transposition In Vitro

3.13.1 Unconstrained Synapses

During normal transposition, the two transposon ends that interact are present in inverted orientation on the same piece of DNA. For bacteriophage Mu, this orientation is highly preferred, and orientation dependence arises as a consequence of substrate molecule supercoiling (CRAIGIE and MIZUUCHI 1986). Unexpectedly, Tn10 transposition in vitro exhibits substantial levels of "unconstrained" synapses. Both bimolecular synapses (between ends on two different molecules) and synapses between directly repeated ends constitute important products, even in the absence of stabilizing agents such as glycerol (R. Chalmers and N. Kleckner, manuscript in preparation; Fig. 3).

3.13.2 Pseudo-End Events

Occasionally, transposition products arise by interaction between one wild-type end and a fortuitously positioned sequence that accidentally resembles a transposon end. Pseudo-end events have been described for Tn3 (ARTHUR et al. 1984) from in vivo analysis of transposition products generated by substrates containing only a single wild-type end.

3.13.3 Possible Non-replicative Mechanisms for the Formation of Co-integrates and Adjacent Deletions

For the replicative transposons identified thus far, co-integrates represent the primary intermolecular transposition product. Co-integration into a target site located on the same molecule as the transposon directly yields an adjacent deletion or the companion event, replicative inversion (SHAPIRO 1979; ARTHUR and SHERRATT 1979). IS10 also promotes the formation of adjacent deletions (ROBERTS et al. 1991) and, in a $RecA^+$ host, co-integrates (WEINERT et al. 1984; HARAYAMA et al. 1984). The same is true for IS10's closest relative, IS50, which in addition gives detectable levels of co-integrates in a $RecA^-$ host (reviewed in LICHENS-PARK and SYVANEN 1988; JILK et al. 1993).

Analysis of in vitro reactions suggests two mechanisms by which IS10 might promote formation of adjacent deletions and makes a third mechanism seem unlikely. (a) Bimolecular synapses between ends located on sister chromatids, followed by the usual events of nonreplicative transposition and appropriate degradation of extraneous segments, can yield a complementary pair of adjacent deletion circles, as originally proposed by ROBERTS et al. (1991) (Fig. 6). The occurrence of IS10-promoted intersister events should be favored by coupling of transposition to passage of the replication fork (ROBERTS et al. 1985). (b) An adjacent deletion can also arise via interaction between one normal transposon end and a pseudo-end located between the normal end and the target side. (c) A third possible pathway would involve synapsis of two normal ends followed by double-strand cleavage and strand transfer to an adjacent target site at only one of those ends. This does not seem to be a common mode for a wild-type element in vitro, as the requisite single-end cleavage/single-end strand transfer events do not occur at a detectable level.

Bimolecular synapses might also account for the ability of IS10 and IS50 elements to form co-integrates in certain circumstances. Interaction of ends on sister chromatids will yield a co-integrate in a reaction exactly analogous to that shown for adjacent deletions (Fig. 6). LICHENS-PARK and SYVANEN (1988) proposed such a pathway for co-integrate formation to account for the observation that, in a $RecA^-$ host, IS50-mediated co-integrates form more readily if the donor molecule is a replicating plasmid than if it is a nonreplicating λ phage. Such a pathway could also contribute to co-integrate formation in a $RecA^+$ host, although the occurrence of co-integrates in this case is usually attributed to the formation of "dimer donors" (BERG 1983).

A number of elements have been reported to promote a mixture of replicative co-integrate formation and nonreplicative transposition, with the occurrence of replicative events inferred from the appearance of co-integrates and/or adjacent deletions (MACHIDA et al. 1982; WEINERT et al. 1984). The finding that bimolecular synapses occur at significant levels raises the possibility that all IS elements are nonreplicative in nature, and that differences among elements with respect to the levels of the apparent replicative events reflect variations in the level of bimolecular synapsis and/or other aspects of the process.

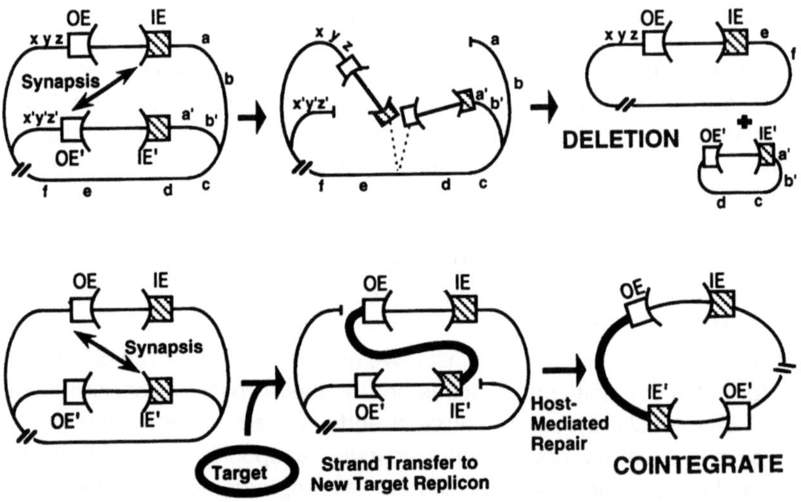

Fig. 6. Nonreplicative mechanisms for the formation of co-integrates and adjacent deletions. Synapsis of transposon ends from different chromatids is analogous to the bimolecular synapsis observed in vitro. The outside end (OE) from one chromatid would synapse with the inside end (IE) from the new replicated sister that has been activated by the hemimethylated state. Strand transfer to a new location on the same chromosome (*top half*) would produce a chromosomal deletion (loci *abcd*), while intermolecular transposition (*bottom half*) would produce a co-integrate by host-mediated repair

3.14 Host Factors IHF and HU Modulate Transposition In Vitro

A role for IHF and HU in Tn10/IS10 transposition was first revealed by analysis of semi-purified in vitro extracts (MORISATO and KLECKNER 1987). Subsequent analysis with highly purified transposase has revealed that these host factors can influence the transposition reaction in two separate ways.

First, IHF can serve as a positive factor for transposition. Under optimal reaction conditions, transposition proceeds very efficiently on both outside and inside end substrates in the absence of any host factor (CHALMERS and KLECKNER 1994 and unpublished results). Under suboptimal conditions, however, such as the absence of supercoiling, outside end activity is dramatically stimulated by and/or absolutely dependent upon addition of IHF. In the most extreme situations, at least, HU will not substitute for IHF in this regard (J. Sakai and N. Klecker, manuscript in preparation).

Second, IHF and HU both serve efficiently as negative modulators of transposition. At physiologically sensible concentrations of either protein, the outcome of the strand-transfer process is altered. Formation of intermolecular and intratransposon collision products is suppressed, and essentially, the only species observed is a topologically simple (unknotted) inversion circle (Fig. 3; H. Benjamin, R. Chalmers and N. Kleckner, unpublished results). We refer to this change in the outcome of the transposition process as "channeling".

The negative effects of IHF are clearly distinct from the positive effects: at low IHF levels, collision events are stimulated without any strong evidence of channeling, while at higher IHF concentrations, channeling occurs without any major change in the total level of products.

The existence of channeling implies that, in the presence of host factors, transposon ends no longer select a target DNA by random collision but instead interact with target DNA in a topologically constrained way (e.g., Fig. 3). The specific geometry of the DNA during this interaction is not established. It is attractive, however, to suppose that intratransposon events arise via synaptic complexes that are being held in the "folded" form by host factors, with target interactions constrained by wrapping of transposon sequences around the synaptic complex.

4 Protein/DNA Components

4.1 Insertion Specificity

Although Tn10 is capable of inserting at a great many different sites in the bacterial genome, it does exhibit a significant preference for particular sites over others. In the 10-kb histidine operon, for example, ~100 insertions occurred at a total of ~20 sites, and half of all insertions occurred at one specific site (KLECKNER

1979; Kleckner et al. 1979b). At any given "hot spot", insertions of Tn10 are recovered at roughly the same frequency in both orientations, as expected from the structural symmetry of the element.

Tn10 insertion specificity reflects the sum of two independent components: sequence-specific information is provided by a symmetrical 6-bp consensus sequence located within the 9-bp target site sequence duplicated during insertion, and structural information is provided by basepairs located ~10bp to either side of the 9-bp target site sequence (Halling and Kleckner 1982; Bender and Kleckner 1992a). Variations in either of these determinants singly can alter the probability of insertion into a particular site by several orders of magnitude. Thus, for example, non-consensus target sites are often preferentially utilized over perfect consensus sites, due to the effects of flanking DNA.

The consensus sequence for insertion is 5'n*GCT*n *AGC*n3'. At each "half site" of the consensus sequence, the first position is sometimes A instead of G and the second position is sometimes T instead of C. Overall, information at the third position is less important, as all four basepairs occur at significant hot spots. For major hot spots, however, T strongly predominates, and the 5-methyl group of that residue has been shown to be the important determinant in this effect (Lee et al. 1987). The nature of the structural information conferred by flanking sequences has not been determined, but a tendency to assume an appropriate bent configuration is a likely possibility. An important effect of a particular bent configuration of the target DNA has been reported for IS231A (Hallet et al. 1994).

The specificity of Tn10 insertion can be significantly reduced by mutations at two specific cysteine residues of transposase (Bender and Kleckner 1992b). These ATS ("altered target specificity") mutations have only a minor (less than two-fold) effect on the frequency of transposition but result in a much broader distribution of insertions at sites that exhibit a substantially reduced match to the normal consensus sequence. Apparently, some aspect of the transposition reaction has been altered in such a way that the interaction with target DNA is less dependent upon sequence-specific interactions with the target site consenus sequence (below).

Global factors affecting Tn10 insertion have not been investigated systematically. Tn10 is not known to exhibit any strong region- or replicon-specific differences, in contrast to certain other elements that exhibit strong regional specificity (e.g., Grinsted et al. 1978). On the other hand, Tn10's more prominent preference for particular specific sites may tend to obscure any global effects.

4.2 Structure of IS10 Transposase

IS10 transposase is a 402-amino acid polypeptide encoded by a single open reading frame from base pair 108 to base pair 1313 of IS10 (Fig. 1). Primary sequence comparisons suggest that IS10 transposase belongs to the IS4 family, which also includes IS50, IS231, and IS903 (Mahillon et al. 1985; Galas and Chandler 1989; Rezsöhazy et al. 1993). Members of this family share two major regions of sequence conservation: a carboxy terminal region (aa266–326 in IS10 transposase) that includes the signature Y-2-R-3-E-6-K and a central region (aa157–187 in IS10 transposase) that includes the motif D-1-G/A-Y/F; also, a third region of weak conservation is found at the amino terminus, positions 93–132 in IS10 transposase (Fig. 7; Rezsöhazy et al. 1993). Each of the two conserved motifs and the third region contains a single acidic residue that could potentially be part of a D D E motif, like the one that has been observed for the IS3 family members retroelements, Tn7, Mu, and transposon-like elements from protozoa (e.g., Fayet et al. 1990; Doak et al. 1994; Baker and Luo 1994). Indeed, mutations in the conserved residues D161 and E292 each completely abolish catalytic activity of IS10 transposase.

Analysis by limited proteolysis has shown that IS10 transposase (46kD) is organized into two principle structural domains (Fig. 7; Kwon et al. 1995) a 28-kD amino terminal domain (domain Nαβ) and a 17-kD carboxy-terminal domain (domain C). These two domains are connected by a 1-kD protease-sensitive loop region (aa247–255). Domain Nαβ can be subdivided by a weakly protease-sensitive site into a 6-kD amino terminal fragment (domain Nα) and a 22-kD region (domain Nβ). Nonspecific DNA binding has been localized to domain Nβ through southwestern analysis. None of the individual isolated domain fragments has thus far exhibited specific DNA binding.

Remarkably, full transposition activity can be reconstituted by mixing purified domains Nαβ and C without the intervening linker region; the same is true for Nαβ

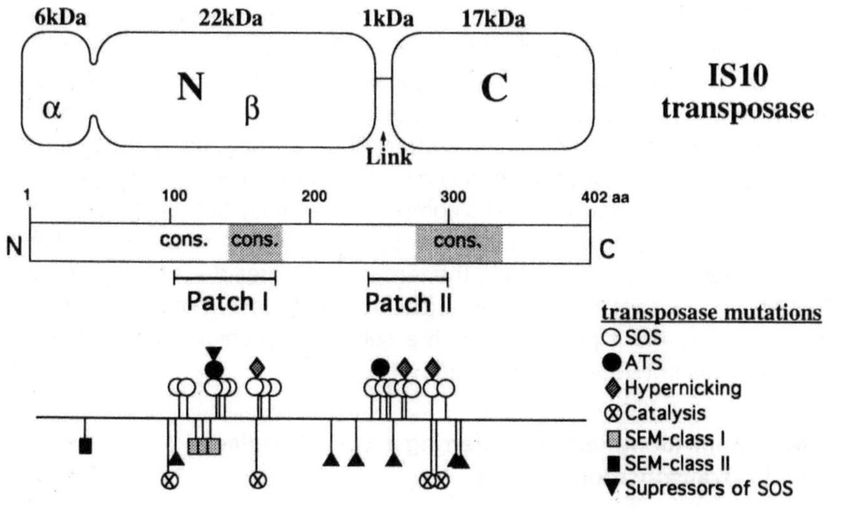

Fig. 7. Domain Structure of IS10 transposase and location of mutations. *Top*: domain map determined by limited proteolysis. Nαβ and C are the two main domains; the link is a nine-residue protease-sensitive region. *Middle*: scale representation of the primary sequence of transposase. Areas of sequence conservation with other bacterial transposases are shaded. *Patch I* and *Patch II* are two regions defined by the mapping of SOS mutations. *Bottom*: locations of mutations of various classes. Multiple symbols appearing one above another indicate different classes of mutations that result from changes in the same residue, although the specific change is not necessarily identical

and a fragment comprising C plus the linker. The amino-terminal Nα segment is essential for activity in this assay: mixtures of purified Nβ and C with or without the linker region form synaptic complexes that do not carry out cleavage.

The fact that in vitro steps of transposition occur normally in the absence of the 9-amino acid linker region that joins the two domains implies that structural integrity of transposase is maintained by direct contacts between the two regions. We infer either that the fundamental interactions within and between transposase molecules remain relatively constant during transposition and/or that important transitions involve establishment of new interactions prior to or in the absence of complete elimination of pre-existing interactions.

Interestingly, although the 9-residue linker region is not essential for transposition, it is the site of two mutations that block the reaction after excision but prior to strand transfer (the "SOS" mutations PL252 and SL253) and one mutation that affects insertion specificity (CY249). Perhaps an aberrant structure in this region can prevent necessary conformational changes from occurring.

4.3 Transposase Mutants

4.3.1 Catalysis-minus Mutants

Strand cleavage and strand transfer by Mu and retroviral elements, and by analogy by Tn10, are proposed to involve standard acid-base catalysis (Mizuuchi 1992b). Residues of IS10 transposase that might potentially be involved in catalysis were identified by oligonucleotide-directed mutagenesis of residues highly conserved among related transposases. Of 30 D and E residues in IS10 transposase, 15 have been mutated individually to alanine or lysine, and among these only three exhibit the phenotype expected for a catalysis-minus mutation. A catalysis-minus phenotype is also conferred by a mutation at a conserved arginine (RA288). The four corresponding mutant transposases, DA97, DA161. RA288, and EA292, all form synaptic complexes normally but are defective for nicking, for double-strand cleavage, and for strand transfer when presented with ends precleaved to make accessible the terminal 3'OH residues (S. Bolland and N. Kleckner, manuscript in preparation). The ability of these mutants to nick the nontransferred strand if presented with a substrate "prenicked" on the transferred strand has not yet been analyzed.

4.3.2 Mutations Identified as Conferring a Specific Defect in the Transposition Process

4.3.2.1 "SOS" Mutants

Mutants proficient for excision but defective for strand transfer were obtained by a genetic screen for mutations that blocked transposition but did not block transposase-dependent induction of host-encoded SOS functions (Haniford et al. 1989). These "SOS" mutants represent a relatively abundant subset of all transposition-defective mutants, about 5% of the total, suggesting that the transition from excision to strand transfer is an important one for the transposition process and that a number of different residues are involved in that transition. Also, the phenotypes of SOS mutants are somewhat variable, suggesting that multiple factors are involved. Recent analysis demonstrates that many SOS mutants fail to show noncovalent target capture (D. Haniford, personal communication). Such a defect could be sufficient to account for the SOS phenotype.

4.3.2.2 Mutants with Altered Target Specificity ("ATS")

ATS mutants were identified in a genetic screen for transposases that increase the frequency of Tn10 insertion into a small target region lacking a wild-type consensus insertion sequence (Bender and Kleckner 1992). These mutants exhibit a general relaxation in target specificity, inserting efficiently at a much wider variety of sites than is normally the case, and only a mild defect in transposition. ATS mutations appear to represent rather specific changes in transposase: these are quite rare as compared with SOS mutations; moreover, all the mutations isolated mapped to two specific codons. Recent analysis demonstrates that ATS mutant CY134 is capable of noncovalent target binding under conditions where wild-type transpose is defective, i.e., in the absence of divalent cation (D. Haniford, personal communication). Such a defect is consistent with the previous proposal that the ATS mutations stabilize some important interaction in such a way that target interactions are less dependent upon interactions with specific target-site consensus-sequence base pairs (Bender and Kleckner 1992).

4.3.2.3 Suppressors of SOS Mutations

Intragenic second-site suppressors have been isolated that restore transposition activity to two different SOS mutants in Patch I (Junop et al. 1994). Many of the suppressor mutations map in either Patch I or Patch II areas. Interestingly, seven different SOS mutations, in both Patch I and Patch II, are suppressed by the previously isolated ATS mutation in Patch I. This supports the idea that the defects in SOS and ATS mutations are in some way opposite in sign.

4.3.2.4 SOS and ATS Phenotypes Appear to Be Opposites

It has been proposed that SOS and ATS mutants might represent opposite phenotypes, with the more abundant SOS mutations conferring a loss of some important interaction and the rarer ATS mutations conferring a gain of function which restores or compensates for that same interaction (Bender and Kleckner 1992b). This proposal is supported by three observations: (a) Two ATS codons map immediately adjacent to codons that have yielded SOS mutations (Fig. 7); moreover, at codon 134, CW134 is an SOS mutation while CY134 is an ATS mutation (D. Haniford, personal communication). (b) The reversion of the SOS mutations by an ATS mutation (above). (c) SOS and ATS mutations analyzed thus far have opposite effects on noncovalent target binding (see above).

4.3.3 Hypernicking Mutants

Mutations have been sought that result in elevated levels of nicking at transposon ends in vitro. Approximately 100 mutants that exhibit strong transposition defects in vivo were screened in in vitro mini protein preparations for production of nicks. Among these, three exhibited a hypernicking phenotype (BOLLAND and KLECKNER 1995). In vitro, these mutant transposases form synaptic complexes normally but accumulate roughly equal proportions of two types of complexes, one containing nicks at both transposon ends and one containing one fully cleaved end and one nicked end.

Remarkably, all three hypernicking mutants had previously been identified as SOS mutants whose transposition defects were strong but not absolute (AT162, EK263, and MI289; HANIFORD et al. 1989). The occurrence of double-strand cleavage at one transposon end should account for the induction of SOS functions in vivo. The hypernicking mutants may represent a different type of SOS mutant from those defective in target capture per se (above). The latter mutants tend to have much stronger transposition defects; also, at least one of the latter mutants, PS167, does not have a hypernicking phenotype in vitro. The failure of these mutants to undergo double-strand cleavage at both ends could well be sufficient to confer a defect in target capture and strand transfer (i.e., transposition).

4.3.4 "Interaction" Mutants

4.3.4.1 Mutations that Suppress Transposon End Mutations: "SEM"

Mutations in transposase were sought which increased transposition of Tn10 elements bearing symmetrical mutations at the transposon termini (J. Sakai and N. Kleckner, manuscript in preparation). Two classes of mutations have been identified.

Class-I mutations were isolated based on suppression of a mutation at bp 8, which lies in the region of strongest contact between transposase and the terminus (below). The mutations recovered improved transposition of elements carrying several different mutations at bp 8 or mutations at a subset of other base pairs in the bp 6–13 region. Thus, suppression appears to be a general and relatively indirect effect rather than a specific alteration in transposase/terminus specificity. These mutations appear to define an alpha helical region that is involved in some type of protein/protein interaction: all class-I SEM mutations map to three hydrophobic residues that are spaced at roughly seven codon intervals from one another, and all of the mutations substitute a new hydrophobic residue for the original one. Apparently, the SEM phenotype results from a rather subtle change in the nature of the proposed interaction. The region affected by these mutations, and thus the proposed interaction per se, appears to be important for synaptic complex stability: the putative alpha helical region defined by these mutations, aa114–127 within Patch I, includes the *trans* dominant mutations RH119 and SF120 described above, which exhibit defects in synaptic complex stability.

Class-II mutations were isolated as suppressors of mutations at both bp 9 and bp 13 of the outside end terminal inverted repeat and are found to improve transposition of elements mutant at essentially any base pairs examined, as well as transposition of wild type. All three independently isolated class-II mutations occur at codon 35, normally a D. Analysis of multiple amino acid substitutions at this position suggests that some suppression is achieved by elimination removal of the negative charge by alanine substitution, but additional suppression is conferred by residues that result in a gain of positive charge. Codon 35 lies within the amino-terminal Nα segment shown by reconstitution experiments to be required after synaptic complex formation and before or during the first chemical steps.

4.3.4.2 Negative Transdominant Mutants

Eight mutants have been found which inhibit wild-type transposase for transposition in vivo: seven missense mutations (WR98, IS101, RH119, SF120, AV162, EA292, RQ296) and one nonsense mutation (Qstop329) (D. Haniford, personal communication). Two of these mutants make normal precleavage complexes but are defective in double-strand cleavage and all subsequent steps. One of these, EA292, was identified previously as a catalysis-minus mutation (above) and is expected to have this phenotype. RQ296 appears to be another such mutation. All six of the remaining *trans* dominant mutants are defective in the formation and/or stability of precleavage synaptic complexes and also in later steps, in vitro and/or in vivo with considerable variation among their detailed phenotypes.

4.4 Inferences About Transposase Structure and Function Based on Mutant Phenotypes

Many of the transposase mutations identified thus far lie in two broadly delimited regions of the gene, Patch I and Patch II, which are identified as the regions in which SOS mutations occur. Moreover, Patch I and Patch II substantially overlap the regions of sequence conservation described above, which should correspond primarily to segments of the protein involved in forming and modulating the active site(s). The localization of mutations to Patch I and Patch II probably reflects the fact that all were isolated by mutant screens that required the mutant transposase, either alone or in combination with wild-type transposase, to form synaptic complexes. Such screens will preferentially recover mutations that affect residues of functional importance to the reaction per se, including residues involved in communication between chemical and conformational changes, rather than residues responsible for overall folding of protein monomers.

Another remarkable feature of the transposase mutations identified thus far is that many of the in vitro mutant phenotypes are dramatically affected by variations in the divalent metal ion. For example, several strong SOS mutations and several negative *trans*-dominant mutations exhibit strong defects if Mg^{++} is present and little or no defect if Mn^{++} is present instead. These effects provide further evidence that divalent metal ions play a crucial role in the progression of the reaction, with respect not only to the chemical steps, but also to the structural integrity of the synaptic complex and its ability to undergo the necessary transitions as the reaction progresses.

In IS10 transposase, as in most other proteins, the catalytic active site appears to be comprised of residues from widely separated regions. Three regions conserved among elements of the IS4 family map at three separate locations; the family can be further subdivided into two classes distinguished by the spacing between the two most highly conserved regions (REZSÖHAZY et al. 1993). Furthermore, the three acidic residues of IS10 transposase at which mutations completely abolish catalysis occur in these three conserved regions (Fig. 7).

Taken at face value, existing evidence supports the more general possibility that Patch I and Patch II interact with one another. Such a feature follows directly from the convergence of diverse regions into the active site because catalysis-minus mutations map in Patch I and Patch II. More generally, both regions have yielded not only SOS and catalysis-minus mutations, but also ATS and hypernicking mutations. Finally, a mutation in Patch I can suppress a mutation in Patch II (JUNOP et al. 1994).

Interestingly, the strong SOS mutation PS167, which is defective in target capture, and the RH119 mutation, which is defective in synaptic complex stability during transposon excision, efficiently suppress one another's defects in mixed in vitro reactions (D. Haniford, personal communication). Since the phenotypes of both individual mutations are also suppressed by Mn^{++}, it seems likely that co-suppression reflects mutually beneficial changes in the conformation of the synaptic complex. The alternative possibility, that PS167 monomer(s) carry out the first step of the reaction while RH119 monomer(s) carry out the second step of the reaction, is difficult to reconcile with the view that two monomers suffice for all of the chemical steps in the reaction.

4.5 IS10 Termini

The "inside" and "outside" ends of IS10 are functionally equivalent: the same types of products arise from interaction between two outside ends, two inside ends, or one inside and one outside end. Inside and outside ends share a 23-bp, nearly perfect terminal inverted repeat; outside ends also encode a specific binding site for Integration Host Factor, whose consensus binding site is located immediately internal to the terminal inverted repeat.

The detailed organization of the terminal inverted repeat has been elucidated by analysis of outside ends (HUISMAN et al. 1989; HANIFORD and KLECKNER 1994; SAKAI et al. 1995). Base pairs 6–13 are most critical: mutations at these positions confer strong transposition defects prior to transposon excision both in vivo and in vitro; moreover, major groove methylation of purines at bp 6, 10, 11, 12, and 13 interferes strongly with transposon excision in vitro (Fig. 8). Base pairs 1–3 constitute a separate functional domain which is less essential overall and which contributes sequence-specific information primarily at steps subsequent to interaction between the transposon ends (Haniford and Kleckner, in preparation).

The nature of bp 1–3 appears to be particularly critical during the chemical steps of the reaction, cleavage and strand transfer. Consistent with this view, transposon cousins Tn10 and Tn5 have the same sequences at these positions and substantially different sequences in bp 6–13 (Huisman et al. 1989). Furthermore, mutations at these positions can have important effects on the level and/or rate at which these chemical steps occur (Haniford and Kleckner 1994; Sakai et al. 1995). In addition, however, such mutations can affect the amount and/or stability of precleavage complexes and the ability of those complexes to remain intact through the chemical steps. These observations are consistent with the view that the transposase active site is directly involved in holding the synaptic complex together.

Transposase appears to make intimate contacts with the terminal inverted repeat along its entire length. Precleavage synaptic complexes (the unfolded form; above) exhibit strong DNase protection throughout this region (Fig. 8). Methylation interference analysis further suggests that transposase lies along one face of the DNA helix on outside ends: interference by methylation in the major groove occurs at bp 6, 10–13, and 18 (Fig. 8). Phosphate ethylation appears not to interfere, at least with transposon excision. Ethylation at two positions (bp 12 and bp 24) enhances reactivity. Similar enhancement has been observed for HIV excision; such effects could reflect creation of an additional contact or elimination of an inhibitory negative charge (Bushman and Craigie 1992). Interaction of transposase with the inside ends of IS10 appears to be generally similar.

All of the basic features of IS10 termini described above are likely to be shared by the entire IS4 family of transposases (Rezsöhazy et al. 1993), as judged by analogous features observed for the ends of two other family members, IS50

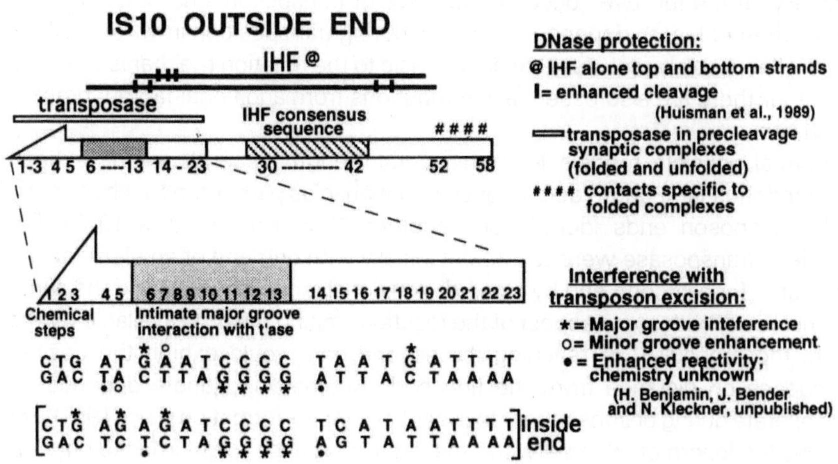

Fig. 8. Structure of IS10 outside end

and IS903 (see Huisman et al. 1989 and Derbyshire and Grindley 1992 for detailed discussion).

These general organizational features are also shared by another well-characterized family of transposons that includes Mu and HIV. In all of these cases, terminal base pairs are highly conserved and appear to be important for the chemical steps but can influence earlier steps as well (e.g., Surette et al. 1991), and the tightest contacts with transposase are several base pairs removed from bp 1 (Craigie et al. 1984; Bushman and Craigie 1992).

5 Additional Interesting Features

5.1 Transposition Length Dependence

IS10-based composite transposons exhibit transposition length dependence: in vivo, the frequency of transposition decreases exponentially with increasing transposon length; an increase of 1kb results in a 40% decrease in transposition (Morisato et al. 1983). This phenomenon was originally reported for IS1-based composite transposons, for which the effect is quantitatively similar to that seen with IS10 (Chandler et al. 1982).

The mechanistic basis for length dependence is not understood. The constraint appears to arise prior to strand transfer: in the presence of a mutant transposase that is proficient for transposon excision but defective in strand transfer, the level of excised transposon fragment generated by chromosomal elements in vivo exhibits length dependence (Flick 1991). Whatever the mechanism, it must involve some kinetic barrier to completion of the transposition reaction; if the reaction were able to continue indefinitely, transposition would eventually, at the full level, occur irrespective of transposon length. It is not yet known whether length dependence occurs during transposition in vitro, in which case it reflects solely features that are intrinsic to the reaction mechanism per se, or whether there are also essential contributions from additional factors present only in vivo.

Several possible models for transposition length dependence have been suggested thus far. Length dependence might arise as part of the mechanism by which transposon ends identify one another (Way and Kleckner 1985). For example, if transposase were to interact initially with one end of an element and then search for a second end by a one-dimensional random walk along the same DNA molecule, with some aspect of the reaction limiting the time available for the search, the probability of reaching the second end would exhibit the desired dependence on distance from the first end. Alternatively, length dependence might operate during or subsequent to synaptic complex formation (Flick 1991). For example, the length of DNA between the interacting ends might impose physical stress on the (nascent) complex, inhibiting its formation and/or causing it to decay.

5.2 Preferential *cis* Action of Transposase

IS10 transposase, like the transposases of other IS elements, acts preferentially in *cis* (MORISATO et al. 1983). *cis* action has been demonstrated rigorously by analysis of constructs in which the transposase gene was placed at increasing distances from a pair of transposon ends. When the intervening distance is 50 kb, the frequency of transposition drops to about 1% of the level observed when the two determinants are located adjacent to one another. A second manifestation of *cis* action is inefficient complementation of a pair of transposon ends by a transposase gene located on another replicon.

The existence of *cis* action implies the existence of two important constraints. First, the initial physical contact between transposase and the DNA occurs in the vicinity of the transposase gene itself, and second, movement of transposase after that initial contact is limited in some way. *cis* action does not reflect the preferential titration of a limited amount of transposase by the transposon ends; the same degree of *cis* action is observed over a 10^4-fold range in transposase concentration.

There is some additional information regarding each of the two component features of *cis* action: in prokaryotic systems, restricted initial contact is possible because translation and transcription are temporally and spatially coupled. In fact, since the amino-terminal portion of transposase binds DNA nonspecifically, it is not improbable that this region of the protein may often contact the DNA before the carboxy-terminus of the protein has even been synthesized, while the message is still tethered to the template DNA.

In addition, the degree of *cis* action is reduced or enhanced by mutations that increase or decrease the efficiency of transposase gene translation, not because of changes in the level of transposase but as a consequence of some change in the gene expression process per se (JAIN and KLECKNER 1993); a modest decrease in *cis* preference is also conferred by host mutations that should increase the stability of transposase mRNA. One proposed model suggests that there is a competition between contact of a single transposase molecule with DNA, which increases *cis* preference, and dimerization of transposase prior to DNA binding, which reduces *cis* preference (JAIN and KLECKNER 1993). By this model, any change in the rate of translation, or in the lifetime of the mRNA as a consequence of a change in the rate of translation, would change *cis* preference by changing the probability that two transposases will be made in rapid succession from the same message. Moreover, since DNA-binding activity resides in the amino-terminal portion of transposase, the balance could be tipped in favor of *cis* preference by contact of nascent protein molecules with DNA prior to synthesis of an oligomerization domain in the carboxy-terminus.

Additional observations suggest that the degree of *cis* preference can be affected by qualitative changes in transposase protein (JAIN and KLECKNER 1993). *cis* action of IS50 transposase can be substantially reduced by point mutations within the transposase gene that cause pleiotropic changes in the properties of transposase protein (DELONG and SYVANEN 1991; WEINRICH et al. 1994; WEIGAND and

REZNIKOFF 1994). In this case, a model has been proposed in which multimerization favors *cis* action (WEINRICH et al. 1994).

The failure of transposase to move very far once it has contacted DNA can be achieved in any of several ways. Most generally, however, it should reflect the combined effects of strong nonspecific DNA binding and a balance between the rate of one-dimensional diffusion and the rate at which the protein becomes inactive via proteolysis or other factors. IS10 transposase in vitro is, in fact, always bound to nonspecific DNA, although the "off rate" of this binding is not known. Different transposases may achieve preferential *cis* action through different balances between DNA binding parameters and protein inactivation rates. IS903 transposase is very unstable, and elimination of *lon*-mediated proteolysis concomitantly eliminates preferential *cis* action (DERBYSHIRE et al. 1990). The stability of IS10 transposase is low when the protein is present at low concentration and high when the protein is overexpressed but exhibits normal *cis* action under both conditions (J. Sakai and N. Kleckner, unpublished observations).

6 Regulation of Tn10/IS10 Transposition

Tn10 and IS10 transposition are regulated and modulated at several different levels by many different mechanisms. All of these regulatory features can be rationalized by supposing that the long-term evolutionary success of an IS element depends upon the balance between two factors: maintaining a quantity and quality of transposase-promoted events sufficient to ensure propagation of the element itself and modulation of transposition activity so as to minimize deleterious effects to the host. Many of the known regulatory features of Tn10/IS10 have been reviewed in detail previously (KLECKNER 1990 a, b) and are discussed only briefly here. Recent insights into the role of host proteins IHF and HU are presented here in more detail.

6.1 Inhibition of Transposon Accumulation

One threat faced by an IS10-containing cell is death by accumulation of IS elements. Since each IS10 transposition event results in an increase in the number of transposable elements in the genome (above), the number of IS elements present in the genome will increase with time. Moreover, in the absence of any control mechanism, this increase will be exponential: as the number of elements increases, the rate of accumulation of additional elements will increase as a consequence both of the number of available copies and of the increased level of transposase available to those copies. IS10 has addressed this problem primarily by the combined effects of two features. Preferential *cis* action of transposase eliminates the second threat by ensuring that the effective level of transposase available to any individual element does not increase as the number of element copies increases. Furthermore, IS10 encodes an antisense RNA molecule that binds to the transposase mRNA and blocks transposase gene translation (SIMONS et al. 1983; SIMONS and KLECKNER 1983; CASE et al. 1989). The effectiveness of this antisense RNA depends upon its concentration, which is arranged in such a way that there is virtually no effect when the genome contains a single copy of IS10 but a steadily increasing effect as the copy number increases above that point. The combined effects of preferential *cis* action and IS10 antisense RNA are such that the total frequency of transposition events per cell per generation remains essentially constant as the element copy number increases.

6.2 Coupling of Transposition to Passage of a Replication Fork

A second important aspect of IS10 regulation is modulation of IS10 transposition by *dam*-mediated methylation of GATC sites (ROBERTS et al. 1985; KLECKNER 1989). IS10 contains two GATC sites, one located within the transposase gene promoter and a second located within the transposase-binding site at the inside end of the element. When IS10 is fully methylated, both transposase expression and inside end activity are strongly inhibited. Immediately after passage of the chromosomal replication fork, however, two chemically different hemimethylated forms of the element are generated. The pattern of methylation effects on these two forms is such that one version is strongly activated for transposition while the other is activated only weakly; the difference may be as great as several hundredfold. These effects result from the fact that both hemimethylated versions of the promoter are moderately activated, but one version of the hemimethylated terminus is strongly activated while the other is still inert. This regulation should have the effect of coupling IS10 transposition to passage of the chromosomal replication fork. Such coupling should be advantageous because it ensures that a gapped donor molecule will have a sister chromatid substrate available for repair in case of donor molecule degradation (above). Such coupling might also increase the opportunity for an individual IS10 element to capture another replicon by increasing the probability of bimolecular synapses (above). Finally, activation of hemimethylated species should also increase the probability of transposition for elements that have just been transferred to a new host by plasmid-mediated conjugation, which involves transfer of a single strand from the donor cell with concomitant replication in the recipient.

6.3 Blocks to Fortuitous Activation of Transposase Gene Expression

A transposon is at risk from fortuitous activation as a consequence of insertion into an actively expressed gene or operon. Previous work has also shown that IS10 encodes specific sequence features which prevent such fortuitous activation: secondary structures specific to readthrough transcripts block transposase gene translation (DAVIS et al. 1985). In addition, the possibility of fortuitous activation is minimized by the tendency of Tn10 to avoid inserting into actively transcribed genes (CASADESUS and ROTH 1989).

6.4 IHF and HU Regulate Tn10/IS10 Activity In Vivo

Genetic analysis reveals that IHF and HU modulate Tn10/IS10 activity in vivo in several ways (SIGNON and KLECKNER 1995). First, host factors specifically suppress intermolecular transposition off of multicopy plasmid replicons but not off of the chromosome. This effect results in part from the fact that IHF specifically inhibits transposase expression from multicopy plasmids but not from chromosomal elements. Most importantly, transposition events from multicopy plasmid donors are dominated by very strong negative effects which are not observed for transposition from the chromosome: in the plasmid case, elimination of both IHF and the IHF-binding site at the outside end of IS10 results in an increase of ~20-fold, while in the chromosomal case the increase is approximately twofold.

Many different observations suggest that the negative effect observed in the plasmid case reflects the occurrence of channeling. One such observation is the fact that plasmid substrates in vivo give rise only to intratransposon inversions and not to intratransposon deletions (BENJAMIN and KLECKNER 1989). Genetic effects suggest that channeling can be mediated at outside ends by either IHF or HU acting at the IHF-binding site or by IHF (but not HU) even in the absence of a specific binding site at the outside end or on inside ends.

Second, specific interaction of IHF at the outside-end IHF-binding site positively activates Tn10-promoted chromosome rearrangements. In constructs where "outside" IS10 ends are placed in the relative orientation corresponding to the "inside" ends of IS10 in Tn10, elimination of IHF and/or the IHF-binding site causes a five- to tenfold decrease in deletions and inversions. Positive activation of chromosome rearrangements is particularly interesting in light of the fact that elimination of IHF and its binding site does not inhibit chromosomal transpositions (above). This difference implies that the positive effect of IHF acting at its binding site is designed to specifically favor or protect chromosome rearrangements. This feature provides a strong evolutionary rationale for positive activation: inversion events promoted by composite transposons are especially important evolutionarily because they give rise to new transposable elements (Fig. 1B).

Further consideration suggests that positive activation may promote formation of new transposable elements in yet another way. Transposition and chromosome rearrangements (deletions and inversions that arise by a random collision pathway) are mechanistically equivalent; they differ only in the length of the segment between the two interacting transposon ends, which is short in the case of transposition and is the size of the bacterial genome in the case of chromosome rearrangements (Fig. 1B). Yet, the positive effects of IHF are apparent only for chromosome rearrangements. Thus, it has been proposed that IHF should positively activate transposition of very long transposable elements. This feature would again favor evolution of new transposable elements, which should first exist in primordial form with fortuitously arising IS elements flanking some gene of selective value at rather great distances.

Finally, there are hints that at least half of all chromosomal events are also channeled specifically into the inversion pathway. If so, the channeling reaction directly and specifically promotes the evolution of new transposons bearing chromosomal genes at the expense of two deleterious events, intermolecular transposition and deletions.

Acknowledgments. R.C., D.K., J.S., and S.B. and all research on Tn10/IS10 transposition in vivo and in vitro in the laboratory of N.K. are funded by NIH Grant R37GM25326 to N.K. The work of S.B. was supported by a postdoctoral fellowship from the Spanish Ministry of Education. We thank D. Haniford and N. Craig for communicating results prior to publication.

References

Adzuma K, Mizuuchi K (1989) Interaction of proteins located at a distance along DNA: mechanism of target immunity in the Mu DNA strand-transfer reaction. Cell 57: 41–47

Arciszewska LK, Drake D, Craig NL (1989) Transposon Tn7 *cis*-acting sequences in transposition and transposition immunity. J Mol Biol 207: 35–52

Arthur A, Sherratt D (1979) Dissection of the transposition process: a transposon-encoded site-specific recombination system. Mol Gen Genet 175: 267–274

Arthur A, Nimmo E, Hettle S, Sherratt D (1984) Transposition and transposition immunity of transposon Tn3 derivatives having different ends. EMBO J 3: 1723–1730

Bainton R, Gamas P, Craig NL (1991) Tn7 transposition in vitro proceeds through an excised transposon intermediate generated by staggered breaks in DNA. Cell 65: 850–816

Bainton RJ, Kubo KM, Feng JN, Craig NL (1993) Tn7 transposition: target DNA recognition is mediated by multiple Tn7-encoded proteins in a purified in vitro system. Cell 72: 931–943

Baker TA, Luo L (1994) Identification of residues in the Mu transposase essential for catalysis. Proc Natl Acad Sci USA 91: 6654–6658

Baker TA, Mizuuchi M, Savilahti H, Mizuuchi K (1993) Division of labor among monomers within the Mu transposase tetramer. Cell 74: 723–733

Baker TA, Kremenstova E, Luo L (1994) Complete transposition requires four active monomers in the Mu transposase tetramer. Genes Dev 8: 2416–2428

Bender J, Kleckner N (1986) Genetic evidence that Tn10 transposes by a nonreplicative mechanism. Cell 45: 801–815

Bender J, Kleckner N (1992a) Tn10 insertion specificity is strongly dependent upon sequences immediately adjacent to the target site consensus sequence. Proc Natl Acad Sci USA 89: 7996–8000

Bender J, Kleckner N (1992b) IS10 transposase mutations that specifically alter target site recognition. EMBO J 11: 741–750

Bender J, Kuo J, Kleckner N (1991) Genetic evidence against intramolecular rejoining of the donor DNA molecular following IS10 transposition. Genetics 128: 687–694

Benjamin H, Kleckner N (1989) Intramolecular transposition by Tn10. Cell 59: 373–383

Benjamin H, Kleckner N (1992) Tn10 transposase excises Tn10 from flanking donor DNA by flush double-strand cleavages at the transposon termini. Proc Natl Acad Sci USA 89: 4648–4652

Bennett PM, Robinson MK, Richmond MH (1977) R-factors: their properties and possible control. In: Drews J, Högenauer G (eds) Topics in infectious diseases, vol 2. Springer, Berlin Heidelberg, New York, p81

Berg DE (1983) Structural requirements for IS50-mediated gene transposition. Proc Natl Acad Sci USA 80: 792–796

Bolland S, Kleckner N (1995) The two single strand cleavages at each end of Tn10 occur in a specific order during transposition. Proc Natl Acad Sci USA (in press)

Bushman FD, Craigie R (1992) Integration of human immunodeficiency virus DNA: adduct interference analysis of required DNA sites. Proc Natl Acad Sci USA 89: 3458–3462

Casadesus J, Roth JR (1989) Transcriptional occlusion of transposon targets. Mol Gen Genet 216: 204–209

Case CC, Roels SM, Jensen PD, Lee J, Kleckner N, Simon RW (1989) The unusual stability of the IS10 anti-sense RNA is critical for its function and is determined by the structure of its stem domain. EMBO J 13: 4297–4305

Chalmers RM, Kleckner N (1994) Tn10/IS10 transposase purification activation and in vitro reaction. J Biol Chem 269: 8029–8035

Chandler M, Clerger M, Galas DJ (1982) The transposition frequency of IS1-flanked transposons is a function of their size. J Mol Biol 154: 229–243

Chao L, Vargas C, Spear BB, Cox EC (1983) Transposable elements as mutator genes in evolution. Nature 303: 633–635

Craigie R, Mizuuchi K (1986) Role of DNA topology in Mu transposition: mechanism of sensing the relative orientation of two DNA segments. Cell 45: 793–800

Craigie R, Mizuuchi M, Mizuuchi K (1984) Site-specific recognition of the bacteriophage Mu ends by the Mu A protein. Cell 39: 387–394

Davis MA, Simons RW, Kleckner N (1985) Tn10 protects itself at two levels against fortuitous activation by external promoters. Cell 43: 379–387

DeLong A, Syvanen M (1991) Trans-acting transposase mutant from Tn5. Proc Natl Acad Sci USA 88: 6072–2076

Derbyshire KM, Grindley NDF (1992) Binding of the IS903 transposase to its inverted repeat in vitro. EMBO J 11: 3449–3455

Derbyshire KM, Kramer M, Grindley NDF (1990) Role of instability in the *cis* action of the insertion sequence IS903 transposase. Proc Natl Acad Sci USA 87: 4048–4052

Doak TG, Doerder FP, Jahn CL, Herrick G (1994) A proposed superfamily of transposase genes: transposon-like elements in ciliated protozoa and a common "D35E" motif. Proc Natl Acad Sci USA 91: 942–946

Engelman A, Mizuuchi K, Craigie R (1991) HIV–1 DNA integration: mechanism of viral DNA cleavage and DNA strand transfer. Cell 67: 1211–1221

Fayet O, Ramond P, Polard P, Prére MF, Chandler M (1990) Functional similarities between retroviruses and the IS3 family of bacterial insertion sequences? Mol Microbiol 4: 1771–1777

Flick K (1991) The relationship between element length and transposition frequency of Tn10 in vivo and in vitro. Honor thesis. Biochemical Sciences, Harvard University

Foster T, Davis MA, Takeshita K, Roberts DE, Kleckner N (1981a) Genetic organization of transposon Tn10. Cell 23: 201–213

Foster T, Lundblad V, Hanley–Way S, Halling S, Kleckner N (1981b) Three Tn10-associated excision events: relationship to transposition and role of direct and inverted repeats. Cell 23: 215–227

Galas DJ, Chandler M (1989) Bacterial insertion sequences. In: Berg DE, Howe MM (eds) Mobile DNA. American Society for Microbiology, Washington DC, pp 109–162

Grinsted J, Bennett PM, Higginson S, Richmond MH (1978) Regional preference of insertion Tn501 and Tn802 into RP1 and its derivatives. Mol Gen Genet 166: 313

Hagemann AT, Craig NL (1993) Tn7 transposition creates a hotspot for homologous recombination at the transposon donor site. Genetics 133: 9–16

Hallet B, Rezsöhazy R, Mahillon J, Delcour J (1994) IS213A insertion specificity: consensus sequence and DNA bending at the target site. Mol Microbiol 14: 131–139

Halling SM, Kleckner N (1982) A symmetrical six-basepair target site sequence determines Tn10 insertion specificity. Cell 28: 155–163

Haniford D, Kleckner N (1994) Tn10 transposition in vivo: temporal separation of cleavages at the two transposon ends and roles of terminal basepairs subsequent to interaction of ends. EMBO J 13: 3401–3411

Haniford DB, Chelouche AR, Kleckner N (1989) A specific class of IS10 transposase mutants are blocked for target site interactions and promote formation of an excised transposon fragment. Cell 59: 385–394

Haniford DB, Benjamin HW, Kleckner N (1991) Kinetic and structural analysis of a cleaved donor intermediate and strand transfer product in Tn10 transposition. Cell 64: 171–179

Harayama S, Oguchi T, Iino T (1984) Does Tn10 transpose via the cointegrate molecule? Mol Gen Genet 194: 444–450
Harshey RM, Getzoff ED, Baldwin DL, Miller JL, Chaconas G (1985) Primary structure of phage Mu transposase: homology to Mu repressor. Proc Natl Acad Sci USA 82: 7676–7680
Huisman O, Errada PR, Signon L, Kleckner N (1989) Mutational analysis of IS10's outside end. EMBO J 8: 2101–2109
Jain C, Kleckner N (1993) Preferential *cis* action of IS10 transposase depends upon its mode of synthesis. Mol Microbiol 9: 249–260
Jilk RA, Makris JC, Borchardt L, Reznikoff WS (1993) Implications of Tn5-associated adjacent deletions. J Bacteriol 175: 1264–1271
Joyce CM, Steitz TA (1994) Function and structure relationships in DNA polymerases. Annu Rev Biochem 63: 777–822
Junop MS, Hockman D, Haniford D (1994) Intragenic suppression of integration-defective IS10 transposase mutants. Genetics 137: 343–352
Kleckner N (1979) DNA sequence analysis of Tn10 insertions: origin and role of 9-bp flanking repetitions during Tn10 translocation. Cell 16: 711–720
Kleckner N (1989) Transposon Tn10. In: Berg DE, Howe MM (eds) Mobile DNA. American Society for Microbiology, Washington DC, pp 227–268
Kleckner N (1990a) Regulating Tn10 and IS10 transposition. Genetics 124: 449–454
Kleckner N (1990b) Regulation of transposition in bacteria. Annu Rev Cell Biol 6: 297–327
Kleckner N, Reichardt K, Botstein D (1979a) Inversions and deletions of the *Salmonella* chromosome generated by the translocatable tetracycline-resistance element Tn10. J Mol Biol 127: 89–115
Kleckner N, Steele D, Reichardt K, Botstein D (1979b) Specificity of insertion by the translocatable tetracycline-resistance element Tn10. Genetics 92: 1023–1040
Kwon D, Chalmers RM, Kleckner N (1995) Structural domains of IS10 transposase and reconstitution of transposition activity from proteolytic fragments lacking an inter-domainal linker. Proc Natl Acad Sci USA (in press)
Lee S, Butler D, Kleckner N (1987) Efficient Tn10 transposition into a DNA insertion hot spot in vivo requires the 5-methyl groups of symmetrically disposed thymines within the hot-spot consensus sequence. Proc Natl Acad Sci USA 84: 7876–7880
Leung PC, Teplow DB, Harshey RM (1989) Interaction of distinct domains in Mu transposase with Mu DNA ends and an internal transpositional enhancer. Nature 338: 656–658
Lichens-Park A, Syvanen M (1988) Cointegrate formation by IS50 requires multiple donor molecules. Mol Gen Genet 211: 244–251
Machida Y, Machida H, Ohtsubo H, Ohtsubo E (1982) Factors determining frequency of plasmid cointegration mediated by insertion sequence IS1. Proc Natl Acad Sci USA 79: 277–281
Mahillion J, Seurinck J, Van Rompuy L, Delcour J, Zabeau M (1985) Nucleotide sequence and structural organization of an insertion sequence element (IS231) from *Bacillus thuringiensis* strain Berlin 1715. EMBO J 4: 3895–3899
Mizzuchi K (1992a) Transpositional recombination: mechanistic insights from studies of Mu and other elements. Annu Rev Biochem 61: 1011–1051
Mizuuchi K (1992b) Polynucleotidyl transfer reactions in transpositional DNA recombination. J Biol Chem 276: 21273–21276
Mizuuchi K, Adzuma K (1991) Inversion of the phosphate chirality at the target site of Mu DNA strand transfer: evidence for a one-step transesterification mechanism. Cell 66: 129–140
Mizuuchi M, Mizuuchi K (1989) Efficient Mu transposition requires interaction of transposase with a DNA sequence at the Mu operator: implications for regulation. Cell 58: 399–408
Mizuuchi M, Baker TA, Mizuuchi K (1992) Assembly of the active form of the transposase-Mu DNA complex: a critical control point in Mu transposition. Cell 70: 303–311
Morisato D, Kleckner N (1984) Transposase promotes double strand breaks and single strand joints at Tn10 termini in vivo. Cell 39: 181–190
Morisato D, Kleckner N (1987) Tn10 transposition and circle formation in vitro. Cell 51: 101–111
Morisato D, Way JC, Kim H-J, Kleckner N (1983) Tn10 transposase acts preferentially on nearby transposon ends in vivo. Cell 32: 799–807
Navas J, Garcia-Lobo JM, Leon J, Ortiz JM (1985) Structural and functional analyses of the fosfomycin resistance transposon Tn 2921. J Bacteriol 162: 1061–1067
Raleigh EA, Kleckner N (1984) Multiple IS10 rearrangements in *Escherichia coli*. J Mol Biol 173:437–461
Reif HJ, Saedler H (1977) Chromosomal rearrangements in the *gal* region of *E. coli* K-12 after integration of IS1. In: Bukhari AI, Shapiro JA, Adhya SL (eds) DNA insertion elements, plasmids, and episomes. Cold Spring Harbor Laboratory Press, Cold Spring Harbor, New York, pp 81–91

Rezsöhazy R, Hallet B, Delcour J, Mahillon J (1993) The IS4 family of insertion sequences: evidence for a conserved transposase motif. Mol Microbiol 9: 1283–1295

Roberts D, Kleckner N (1988) Tn10 transposition promotes RecA-dependent induction of a lambda prophage. Proc Natl Acad Sci USA 85: 6037–6041

Roberts DE, Hoopes BC, McClure WR, Kleckner N (1985) IS10 transposition is regulated by DNA adenine methylation. Cell 43: 117–130

Roberts D, Ascherman D, Kleckner N (1991) IS10 promotes formation of adjacent deletions at low frequency. Genetics 128: 37–43

Ross D, Swan J, Kleckner N (1979) Physical structures of Tn10-promoted deletions and inversions: role of 1400 basepair inverted repetitions. Cell 16: 721–731

Sakai J, Chalmers RM, Kleckner N (1995) Identification and characterization of a precleavage synaptic complex that is an early intermediate in Tn10 transposition. EMBO J (in press)

Shapiro JA (1979) Molecular model for the transposition and replication of bacteriophage Mu and other transposable elements. Proc Natl Acad Sci USA 76: 1933–1937

Shen M, Raleigh EA, Kleckner N (1987) Physical analysis of IS10-promoted transpositions and rearrangements. Genetics 116: 359–369

Signon L, Kleckner N (1995) Negative and positive regulation of Tn10/IS10-promoted recombination by IHF: two distinguishable processes inhibit transposition off of multicopy plasmid replicons and activate chromosomal events that favor evolution of new transposons. Genes Dev 9: 1123–1136

Simons RW, Kleckner N (1983) Translational control of IS10 transposition. Cell 34: 683–691

Simons RW, Hoopes B, McClure W, Kleckner N (1983) Three promoters near the ends of IS10: p-IN p-OUT and p-III. Cell 34: 673–682

Starlinger P, Saedler H (1976) IS-elements in microorganisms. In: Compans RW, Cooper M, Koprowski H et al. (eds) Current Topics in Microbiology and Immunology, Vol. 75. Springer, Berlin Heidelberg, New York, p111

Suck D, Lahm A, Oefner C (1988) Structure refined to 2A of a nicked DNA octanucleotide complex with DNase I. Nature 332: 464–468

Surette MG, Chaconas G (1992) The Mu transpositional enhancer can function in *trans*: requirement of the enhancer for synapsis but not strand cleavage. Cell 68: 1101–1108

Surette MG, Buch SJ, Chaconas G (1987) Transpososomes: stable protein-DNA complexes involved in the in vitro transposition of bacteriophage Mu DNA. Cell 49: 253–262

Surette MG, Harkness T, Chaconas G (1991) Stimulation of the Mu A protein-mediated strand cleavage reaction by the Mu B protein, and the requirement of DNA nicking for stable type 1 transpososome formation. J Biol Chem 266: 3118–3124

Thaler DS, Stahl MM, Stahl FW (1987) Tests of the double-strand-break repair model for Red-mediated recombination of phage λ and plasmid λdv. Genetics 116: 501–511

Waugh DS, Sauer RT (1993) Single amino acid substitutions uncouple the DNA binding and strand scission activities of FokI endonuclease. Proc Natl Acad Sci USA 90: 9596–9600

Way J, Kleckner N (1985) Transposition of plasmid-borne Tn10 elements does not exhibit simple length dependence. Genetics 111: 705–713

Weinert TA, Derbyshire K, Hughson FM, Grindley NDF (1984) Replicative and conservative transpositional recombination of insertion sequences. Cold Spring Harbor Symp Quant Biol 49: 251–260

Weigand TW, Reznikoff WS (1992) Characterization of two hypertransposing Tn5 mutants. J Bacteriol 174: 1229–1239

Weinreich MD, Gasch A, Reznikoff WS (1994) Evidence that the *cis* preference of the Tn5 transposase is caused by nonproductive multimerization. Genes Dev 8: 2363–2374

Transposition of Phage Mu DNA

B.D. Lavoie and G. Chaconas

1 Overview of the Mu In Vitro Strand-Transfer Reaction

The temperate bacteriophage Mu has been an extremely useful model system for studies on the mechanistic aspects of DNA transposition. Mu was the first element for which a soluble in vitro transposition system was established (Mizuuchi 1983). Furthermore, some of the features of the Mu system, such as the polarity of strand transfer and the chemical steps of the reaction, appear to be conserved in a wide variety of elements from bacterial transposons through

---

Department of Biochemistry, University of Western Ontario, London, Ontario N6A 5C1, Canada

mammalian viruses. The purpose of this review is to summarize the biochemical details of Mu DNA transposition which have been gleaned over recent years. Due to space limitations, an exhaustive review of the literature cannot be performed here and the reader is also referred to several other recent reviews (PATO 1989; MIZUUCHI 1992a, b; HANIFORD and CHACONAS 1992).

The primary focus of in vitro studies on Mu DNA transposition has been the strand-transfer reaction, an early step in the process. This reaction can be performed in vitro (see Fig. 1) with purified DNA substrates and proteins (CRAIGIE et al. 1985). Nicks are introduced at the 3'ends of the Mu DNA, and the 3'-OH Mu termini are covalently linked to 5'-phosphates, 5 bp apart in the target DNA (MIZUUCHI 1984). Strand transfer into the target DNA occurs at near random sites. The product of the strand-transfer reaction is a θ-shaped molecule (CRAIGIE and MIZUUCHI 1985; MILLER and CHACONAS 1986) which can be processed in a crude cell extract by nuclease cleavage and repair to give a simple insertion in the target molecule, or by replication to give a co-integrate molecule carrying two copies of Mu in a directly repeated orientation and sandwiched between the vector and target sequences. The replicative pathway for processing strand-transfer intermediates is the predominant, if not exclusive, pathway which is operative following prophage induction (CHACONAS et al. 1981); however, the replication reaction has not yet been characterized in vitro.

1.1 DNA Sites

One of the distinguishing features of transposable elements is the inverted-repeat DNA sequences which are normally present at their termini and which are

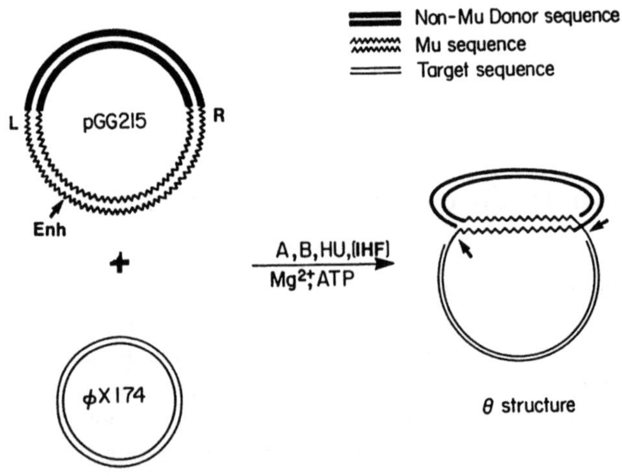

Fig. 1. Strand-transfer step of mu DNA transposition in vitro. *L* and *R*, left and right ends of the Mu genome; *Enh*, enhancer. *Arrows* in the θ structure point to the 5 bp gaps in the target DNA. (Modified from SURETTE et al. 1987)

required for transposition. In phage Mu the simple inverted-repeat end scenario is replaced by a more complex arrangement of DNA sites, as shown in Fig. 2. At both the left and right ends are three consensus binding sites for the Mu A protein (discussed below) (CRAIGIE et al. 1984). The left and right end sites are inverted relative to each other, with the exception of the R3 site. The spacing between the repeats also differs at the two ends; the reasons for this are not clear but may be related to the fact that Mu is a phage, and the *pac* site involved in DNA packaging is located in the spacer between L1 and L2 (GROENEN and VAN DE PUTTE 1985; HAREL et al. 1990). Proper spacing and phasing, at least between the R1 and R2 binding sites, is critical (NAMGOONG et al. 1994). The requirement for multiple Mu A binding sites at the ends is related to the assembly of a highly cooperative higher-order nucleoprotein complex or transpososome, which mediates the reaction. Furthermore, the binding affinity for Mu A protein is not the same for the six sites, nor is their relative importance in the transposition reaction (CRAIGIE et al. 1984; LAVOIE et al. 1991; ALLISON and CHACONAS 1992).

In addition to the unusual arrangement of terminal repeats, Mu is distinguished as an eccentric transposon by the presence of a transpositional enhancer (MIZUUCHI and MIZUUCHI 1989; LEUNG et al. 1989; SURETTE et al. 1989). The enhancer region stimulates the in vivo and in vitro reaction 100-fold or more. The enhancer is located about 1 kb from the left end (although it functions in a distance-independent manner) and is a very busy piece of DNA since it also functions in regulation of the early Mu promoter. The early operator sites 01 and 02 and the integration host factor (IHF) binding site between them comprise the functional enhancer element. The Mu A protein binds to sequences within 01 and 02 via a site-specific DNA-binding domain distinct from that which recognizes the Mu ends (LEUNG et al. 1989; MIZUUCHI and MIZUUCHI 1989). The enhancer can be coaxed to function in *trans* on an unlinked DNA molecule when provided in high

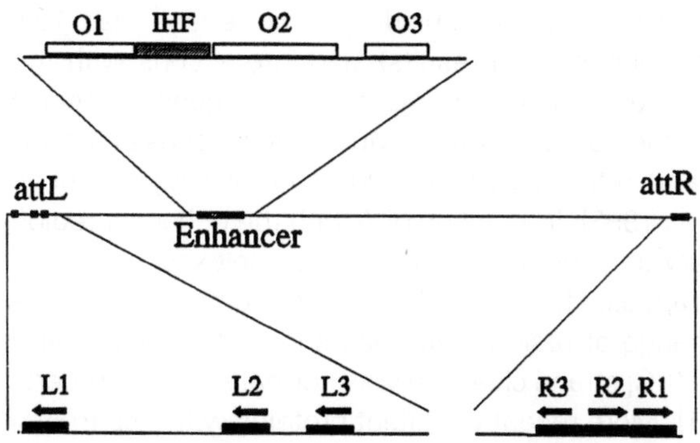

Fig. 2. Required regions for the Mu DNA strand-transfer reaction. Our standard mini-Mu contains approximately 1.7 kbp of both the left and right ends of bacteriophage Mu cloned into pUC19 in their natural orientations. The required regions *attL*, *attR*, and the transpositional *enhancer* are enlarged. The enhancer sites 01, 02, and 03 are indicated by the operators *01* and *02* separated by the *IHF*-binding site. The Mu A protein-binding sites *L1*, *L2*, and *L3* at the left end and *R1*, *R2*, and *R3* at the right end are indicated. *Arrows* above the sites indicate the orientation of the site. (Reprinted from ALLISON and CHACONAS 1992)

enough concentration (SURETTE and CHACONAS 1992). The role of the enhancer is to facilitate synapsis of the two Mu ends; it is not required for the actual strand cleavage event (SURETTE and CHACONAS 1992; MIZUUCHI et al. 1992). The enhancer region appears to be involved in a complex circuit of protein-protein and protein-DNA interactions which includes the two Mu ends (ALLISON and CHACONAS 1992). Although several lines of evidence implicate interactions between the Mu ends and the enhancer, a three-site synaptic complex involving these regions has been difficult to observe biochemically, probably because of the transient nature of such a complex.

1.2 Phage-encoded Proteins Involved in Mu Transposition

The in vitro Mu DNA strand transfer reaction as depicted in Fig. 1 (an intermolecular reaction) requires two Mu-encoded proteins: A and B. The Mu A protein can promote both donor cleavage and ligation to an intramolecular target in the absence of Mu B protein. The B protein is responsible for target capture, for stimulating A to initiate strand transfer, and for conferring target immunity.

1.2.1 The Mu A Protein

The Mu A protein is a 663-amino acid polypeptide which binds as a monomer (KUO et al. 1991) to the six consensus binding sites at the Mu ends (CRAIGIE et al. 1984). Mu A can be divided into three domains by protease-hypersensitive sites at residues 247 and 574 (NAKAYAMA et al. 1987). The 30-kD N-terminal domain is responsible for site-specific binding to the Mu ends and to the transpositional enhancer through different subdomains; residues 1–76 bind the enhancer (LEUNG et al. 1989; MIZUUCHI and MIZUUCHI 1989), and the region from 99–235 binds the ends (K.Kim and R.M. Harshey, personal communication). The half-life of Mu A complexes with L1, one of the strongest end-binding sites, is less than 10 s (T. Harkness and G. Chaconas, unpublished results), and interactions with the enhancer are substantially weaker. Mu A monomers make sequence-specific contacts in the major groove of the nonsymmetrical Mu A-binding sites at the left and right ends and impart a bend in the DNA of 60°–90° upon binding (KUO et al. 1991; ZOU et al. 1991; DING et al. 1993). Interaction of A with the enhancer region is still ill-defined due to the very transient nature of these complexes.

The 35-kD central domain appears to contain the catalytic components of the Mu A protein. A Thr to Ile change at residue 548 was shown to block in vitro strand transfer at 42° but not 30°C; strand cleavage was normal at both temperatures (LEUNG and HARSHEY 1991). More recently, mutant proteins with Asp to Asn at position 269 or Glu to Gln at residue 392 were shown to block both strand cleavage and strand transfer but can still form the uncut, stable synaptic (type-0) complex (BAKER and LUO 1994; refer to Sect. 3.2). In addition, strand transfer of precleaved DNA does not occur, suggesting that both cleavage and strand transfer use at least some of the same active-site residues. These mutants may be defective in the binding of divalent metal ion in the active-site pocket, and they contain mutations in a region which may be analogous to the conserved D-D_{35}-E motif found in a variety of retroviral and bacterial transposases (KULKOSKY et al. 1992). The central domain of Mu A also possesses sequence-independent DNA-binding activity (NAKAYAMA et al. 1987).

The C-terminal 10-kD domain is involved in protein-protein interactions with the Mu B protein (LEUNG and HARSHEY 1991; BAKER et al. 1991). Mu A lacking residues 616–663 can promote the strand cleavage reaction but cannot perform intermolecular strand transfer or other functions dependent upon interactions with Mu B. Removal of the entire C-terminal domain, however, results in an inactive protein. Hence, other structural and or functional components must be present on this domain. The purified domain can promote ATP-dependent release of bound DNA by the Mu B protein and inhibits intermolecular strand transfer. (WU and CHACONAS 1994).

1.2.2 The Mu B Protein

Although Mu A alone can promote strand cleavage and intramolecular strand transfer in vitro, in vivo experiments suggest that even nicking of the Mu ends does not occur at biochemically detectable levels in the absence of Mu B in *Escherichia coli* (B.S. Chan and G. Chaconas, unpublished results). The reasons for this discrepancy are not yet known. Mu B is a 312-residue polypeptide which displays nonspecific DNA-binding activity (CHACONAS et al. 1985) and a weak ATPase activity, which is stimulated by DNA and Mu A protein (MAXWELL et al. 1987; ADZUMA and MIZUUCHI 1991). The most notable duty of the B protein in vitro is capture of the target DNA for Mu integration. Target binding is ATP dependent, and stimulation of Mu B to hydrolyse ATP by Mu A results in release of the target molecule (ADZUMA and MIZUUCHI 1988). This complex set of interactions plays a pivotal role in the process of transposition immunity, whereby self-integration by Mu rarely occurs (see MIZUUCHI 1992).

Another important role of Mu B is to stimulate the strand-transfer reaction by Mu A. This normally occurs when B is bound to target DNA. However, chemical modification of Mu B to block target DNA binding still results in stimulation of strand transfer by Mu A; in this case intramolecular strand transfer results, with integration occurring somewhere in the donor plasmid (BAKER et al. 1991; SURETTE et al. 1991). The B protein has also been shown to greatly stimulate the strand cleavage reaction in vitro under certain conditions. When a terminal base pair mutation is present at the left or right end of Mu, significant nicking does not occur unless Mu B is present (SURETTE and CHACONAS 1991). Similarly, the presence of inhibitory flanking sequences in the vector can be overcome by Mu B stimulation (WU and CHACONAS 1992).

The physical structure of the B protein has been difficult to study due to its propensity to form insoluble aggregates. The protein also oligomerizes in the presence of ATP and shows a positive cooperativity with respect to ATP concentration (ADZUMA and MIZUUCHI 1991). Mu B has been shown by partial proteolysis to be made up of two domains: an N-terminal 25-kD polypeptide and a C-terminal

8-kD region (TEPLOW et al. 1988). The N-terminal domain contains a putative helix-turn-helix at residues 19–40 (MILLER et al. 1984) and homology with the nucleotide binding folds of a number of ATPases at residues 86–115 and 156–185 (CHACONAS 1987).

1.3 Host Protein Requirements

1.3.1 The HU Protein

In addition to Mu-encoded proteins, the *E. coli* HU protein, a heterodimer of 18 kD, is required for an early step in Mu DNA transposition in vitro (CRAIGIE et al. 1985; SURETTE et al. 1987; CRAIGIE and MIZUUCHI 1987). HU typifies a family of small, basic, and heat-stable DNA-binding proteins which stimulate a variety of DNA metabolic reactions (for reviews see DRLICA and ROUVIÈRE-YANIV 1987; PETTIJOHN 1988; SCHMID 1990) possibly by engineering DNA deformations such as bending, wrapping, or melting (for reviews see SERRANO et al. 1993; BOOCOCK et al. 1992; NASH 1990). HU, a sequence-independent DNA-binding protein, is highly conserved among prokaryotes (DRLICA and ROUVIÈRE-YANIV 1987) and was originally identified as a stimulatory factor in lambda transcription (ROUVIÈRE-YANIV and GROS 1975). Although the crystal structure of the related *Bacillus thermophilus* HU has been solved and refined to 2.1 Å (TANAKA et al. 1984; WHITE et al. 1989, respectively), the molecular details of the HU-DNA interaction have not been elucidated. HU is believed to bind DNA via the minor groove (WHITE et al. 1989) by analogy to the other members of the HU family (see DRLICA and ROUVIÈRE-YANIV 1987), namely IHF (CRAIG and NASH 1984; YANG and NASH 1989; PANIGRAHI and WALKER 1991) and TF-1 (SCHNEIDER et al. 1991), and to modulate DNA flexibility (HODGES-GARCIA et al. 1989; HAYKINSON and JOHNSON 1993). Although its precise role in stimulating the Mu in vitro transposition reaction has remained elusive, HU appears to function in at least three capacities: at the transpositional enhancer, at the Mu left end, and possibly at the Mu-host junction.

1.3.2 Integration Host Factor

Under standard in vitro reaction conditions, HU is the only host factor required for the strand-transfer reaction (CRAIGIE et al. 1985); however, at lower "physiological" levels of supercoiling an additional protein factor is required, *E. coli* integration host factor (IHF) (SURETTE and CHACONAS 1989), which was initially identified as a factor stimulating phage lambda integration (NASH and ROBERTSON 1981). IHF exists as a heterodimer of ~20 kD, binds DNA sequence specifically (CRAIG and NASH 1984) and dramatically bends DNA (ROBERTSON and NASH 1988; THOMPSON and LANDY 1988). IHF stimulates the Mu in vitro strand-transfer reaction (SURETTE and CHACONAS 1989) by action at the transpositional enhancer (SURETTE et al. 1989), located some 950 bp from the Mu left end. IHF acts stoichiometrically at this site and is believed to induce a sharp bend (HIGGINS et al. 1989) required for enhancer function (SURETTE et al. 1989; ALLISON and CHACONAS 1992; SURETTE and CHACONAS 1992).

2 Transpososomes: Structural Aspects

The Mu in vitro transposition reaction can be divided into two distinct steps, the first being the strand transfer of the 3' ends of Mu to a target DNA to form a θ structure and the second involving the replication of this product to yield a cointegrate molecule. Higher-order nucleoprotein structures are likely to mediate both of these reactions.

Noncovalent protein-DNA intermediates of the strand-transfer reaction have been identified: the type 0, 1, and 2 transpososomes (SURETTE et al. 1987; CRAIGIE and MIZUUCHI 1987; MIZUUCHI et al. 1992; Figs. 3 and 4). The type-1 and -2 complexes are the products of the strand cleavage and transfer reactions, respectively. The type-0 complex is an uncut synaptic complex which accumulates in Ca^{+2} (see Sect. 3.2).

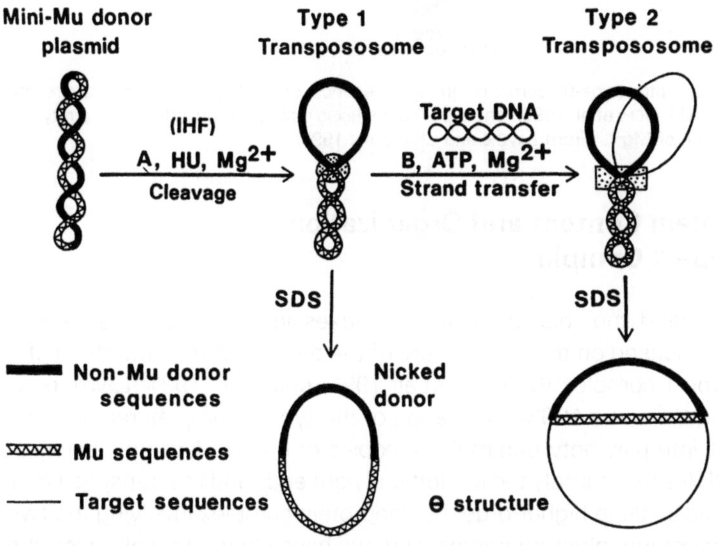

Fig. 3. Transpososomes in the strand-transfer reaction. A type-1 complex, which is an intermediate in the strand-transfer reaction, is formed when a supercoiled mini-Mu donor plasmid is incubated with the Mu A protein and *E. coli* HU protein. The Mu ends are held together in a higher-order protein-DNA complex (type-1 transpososome) defining two topological domains, a relaxed non-Mu domain and a supercoiled Mu domain. Disruption of the complex with SDS results in the liberation of a nicked donor plasmid. The type-2 complex is the product of the strand-transfer reaction which remains complexed with protein (type-2 transpososome). In addition to type-1 reaction requirements, Mu B protein, ATP, and target DNA are required for type-2 complex formation. The type-2 complex can be generated in a single reaction mixture or by conversion of a preformed type-1 into a type-2 complex. Once a type-1 complex has been formed, supercoiling is not required for its conversion into the strand-transferred product. Disruption of the type-2 complex with SDS liberates the protein-free strand-transferred product or θ structure (CRAIGIE and MIZUUCHI 1985; MILLER and CHACONAS 1986). Although the θ structure above has been drawn as a relaxed molecule for simplicity of presentation, it is important to note that the Mu DNA sequences, but not the vector or target DNA, are topologically constrained (CRAIGIE and MIZUUCHI 1985). Hence, the Mu DNA stem retains the Mu supercoils originally present in the type-2 complex. (From SURETTE and CHACONAS 1989)

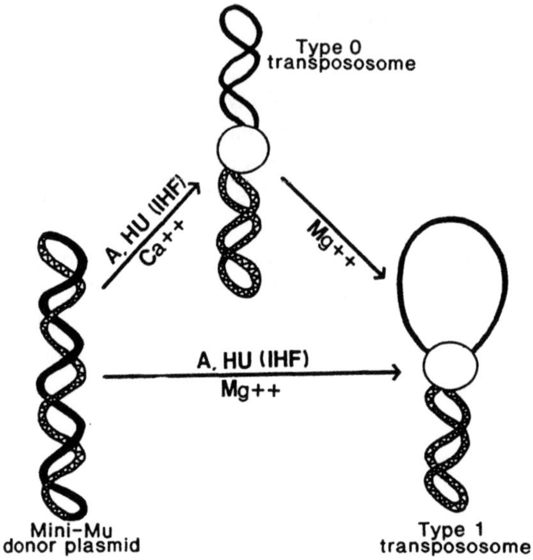

Fig. 4. The stable synaptic (type-0) complex. In the presence of Ca^{2+}, the Mu ends are stably synapsed but uncut (MIZUUCHI et al. 1992). The type-0 complex is rapidly converted into the cut type-1 complex upon addition of Mg^{2+}. (From LAVOIE and CHACONAS 1993)

2.1 Mu A Protein Content and Organization in the Type-1 Complex

To better understand the role of these complexes in effecting transposition, recent work has focused on the architecture of the product of the strand-transfer reaction, the type-1 complex (MIZUUCHI et al. 1991; KUO et al. 1991; LAVOIE et al. 1991; LAVOIE and CHACONAS 1993). Formation of the type-1 complex requires the highly regulated interplay between multiple copies of the Mu A protein acting at three distinct DNA sites, namely the Mu left and right ends and the transpositional enhancer, to yield a stable higher-order nucleoprotein complex involving the two Mu ends. Using immunoelectron microscopy, we have shown that although the *E. Coli* HU protein is also required for this reaction and localizes to greater than 90% of the complexes, it can be efficiently removed from the transpososome core without compromising the structural and functional integrity of the complex (LAVOIE and CHACONAS 1990; MIZUUCHI et al. 1992). Thus the essential protein component of the transpososome is Mu A (LAVOIE and CHACONAS 1990), which oligomerizes into a stable tetramer during complex formation (LAVOIE et al. 1991; MIZUUCHI et al. 1992).

Nuclease and chemical protection assays have shown that only three Mu A-binding sites (L1, R1, and R2) are occupied by the A tetramer in all three transpososome complexes (LAVOIE et al. 1991; KUO et al. 1991; MIZUUCHI et al. 1991, 1992); furthermore, the protection extends from the two end-most sites, L1 and R1, to encompass the 3' ends of Mu, where strand cleavage occurs, and an additional 10–13 bp (LAVOIE et al. 1991; MIZUUCHI et al. 1991, 1992). The source of the fourth A monomer in the complex remains uncertain at this time. It is tempting to speculate that it originates from the weak L2-binding site, but other possibilities cannot be excluded. Interestingly, no HU-specific protection could be identified using standard footprinting methods (LAVOIE et al. 1991; LAVOIE and CHACONAS 1993), despite the presence of stably bound protein. Recent results using an HU "chemical nuclease" have revealed that HU binds stably to the L1–L2 spacer region (LAVOIE and CHACONAS 1993) and will be discussed below.

2.2 Role of HU

2.2.1 Site-specific Binding of HU at the Mu Left End

The first direct evidence for site-specific HU binding in any system has recently been obtained at the Mu left end, between the L1 and L2 Mu A-binding sites, using HU chemically coupled to an iron-EDTA-based DNA cleavage reagent (LAVOIE and CHACONAS 1993). Specific DNA cleavages deriving from a single dimer of the HU nuclease (Lavoie and Chaconas, unpublished results) cluster to the 83-bp spacer region separating the L1 and L2 Mu A-binding sites. Interestingly, site-specific HU binding was detected only in the transpososome, suggesting an affinity for DNA structure rather than sequence; our naked supercoiled mini-Mu donor substrate was ineffective in stably recruiting HU to this or any site.

The HU-nuclease cleavage was restricted to one side of the DNA helix and is consistent with the presence of a tight, HU-assisted loop at the Mu left end. Such HU-dependent loops have been postulated in the *Hin* system, where HU is required to mediate communication between the enhancer and the end sites when the enhancer is moved to less than 104 bp from the *hixL* site (HAYKINSON and JOHNSON 1993). The presence of a DNA loop would allow communication of Mu A monomers on both sides of the spacer. The A monomers brought into contact need not be confined to the Mu left end, and may be involved in pairing of the ends with the transpositional enhancer to generate the stable A tetramer. Alternatively, the HU protein may play a more architectural role by introducing a DNA deformation such as a sharp bend or kink. Such conformational changes in DNA are energetically costly and therefore likely to be transient in the absence of proteins.

2.2.2 A Second Site for HU Binding in the Transpososome

Although the HU nuclease identified one specific HU-binding site, neither removal of the L1–L2 spacer nor increase of its size to reduce any DNA strain abolished the HU requirement for efficient reaction (LAVOIE and CHACONAS 1994). Furthermore, HU-depleted complexes were competent for efficient HU reassembly at a site other than the L1–L2 spacer, suggesting that at least one more high-affinity HU-binding site resides in the transpososome. HU binding to this second site is unlikely to rely on protein-protein contacts with Mu A, DNA supercoiling, or the presence of cleaved Mu ends in the complex (LAVOIE and CHACONAS 1994). We have postulated that an altered DNA structure, possibly at the Mu-host junction, confers high-affinity binding of HU to the transpososome (LAVOIE et al. 1991). Considering that HU efficiently binds and reassembles to uncut type-0 as well as cut type-1 complexes, it seems unlikely that the 5' or 3' termini resulting from cleavage play an important role in HU binding. Rather, we prefer a model invoking HU binding to a sharp kink of the DNA helix, possibly as a prelude to the Mu A-mediated strand cleavage step. Alternatively, HU could contribute to a localized unpairing of the strands at or near the junction, similar to HU stimulation of dnaA strand opening at the *oriC* origin of replication (HWANG and KORNBERG 1992). Evidence supporting a role for HU at or near the Mu-host junction must await further investigation.

2.3 An Altered DNA Structure at the Mu-Host Junction

Further probing of the Mu-host junction region has revealed enhanced hydroxyl radical cleavage sites on the continuous strand and has led us to propose the existence of an altered DNA structure, such as a sharp bend or kink, which could facilitate the additional DNA protection beyond the junction (LAVOIE et al. 1991). Moreover, this localized enhancement in hydroxyl radical reactivity is not specific to the (cleaved) type-1 complex and occurs in the uncleaved type-0 complex, although the most susceptible sugar moiety is shifted by one position towards the Mu sequences on the uncut strand (Lavoie and Chaconas, unpublished results). In addition, a region of DNA hypersensitivity to spontaneous cleavage has been noted on the "cut" strand near the Mu-host junction and may also reflect DNA strain (LAVOIE and CHACONAS 1993; refer to Sect. 3.3.1).

Interestingly, structural changes in the Mu-host junction have also been observed in model transpososome complexes using short oligonucleotides. H. Savilahti and K. Mizuuchi (personal communication) have noted destabilization of the duplex in the non-Mu DNA just beyond the Mu-host junction. Thus DNA structural alterations in the neighborhood of the Mu-host junction, where cleavage occurs, are likely to be important for complex assembly or function. Consistent with this idea, "bad" flanking DNA sequences possibly related to conformational flexibility have previously been shown to inhibit transpososome formation (WU and CHACONAS 1992).

3 Transpososomes: Mechanistic Aspects

The Mu A tetramer is intrinsic to all three transpososome complexes, types 0, 1, and 2, and assembles at the Mu ends of a supercoiled molecule in the presence of the transpositional enhancer, *E. coli* HU and IHF, and divalent metal ions;

however, not all of the reaction components are required at all stages. The in vitro strand-transfer reaction can be divided into multiple steps: (a) presynaptic events involving multiple interactions of Mu A with the Mu ends and the enhancer in the presence of host factors; (b) stable synapsis and cleavage of the Mu ends to generate 3'OH termini; and (c) strand transfer.

3.1 Requirements for Stable Synapsis

Despite intensive study, the multiple protein-protein and protein-DNA interactions which govern tetramer assembly and complex formation are ill-defined. The conversion from A monomer (Kuo et al. 1991) to tetramer (Lavoie et al. 1991; Mizuuchi et al. 1992; Baker and Mizuuchi 1992) is likely to result from a complex set of interactions between the two Mu ends and the transpositional enhancer. A transient three-site synaptic complex between the two Mu ends and the transpositional enhancer is believed to be a precursor to stable complex formation, as for the *hin* and *gin* site-specific recombination systems (Heichman and Johnson 1990; Kanaar and Cozzarelli 1992). In the case of Mu transposition, such an intermediate has recently been observed (Watson and Chaconas, unpublished results).

3.1.1 Enhancer-End Interactions in Transpososome Formation: Three-Site Synapsis

A complex circuit of end-enhancer interactions has been described (Allison and Chaconas 1992) involving at least *five* Mu A-binding sites at the Mu ends (Lavoie et al. 1991) and an unknown number at the transpositional enhancer. Deletion analysis has recently identified five potential A-binding sites within the functional enhancer (R.G. Allison and G. Chaconas, unpublished data; Fig. 5) which have been named with regard to the IHF-binding site (proximal, middle, or distal) in the operators 01 and 02. In the simplest scenario, single Mu A monomers would specifically bridge end and enhancer sites via their two distinct DNA-binding domains (refer to Sect. 1.2) in a three-site synaptic complex. This complex was recently observed in our laboratory (M.A. Watson and G. Chaconas, unpublished results) and is likely to involve a myriad of protein-protein and protein-DNA interactions. Recent data indicate that all four A monomers present in the stable tetramer must be competent for enhancer interaction (M. Mizuuchi and K. Mizuuchi, personal communication), although the molecular details of these interactions have been difficult to elucidate biochemically. Specific end-enhancer interactions have been mapped genetically (Fig. 5) by coupling individual end and enhancer mutations, essentially as described in Allison and Chaconas (1992). Using this approach, four of the six Mu A end-type binding sites have been assigned tentative enhancer partners (Fig. 5); these interactions are likely to be important for transpososome assembly. The indirect nature of these experiments warrants caution in their interpretation, however, and the noted assignments await biochemical confirmation.

A

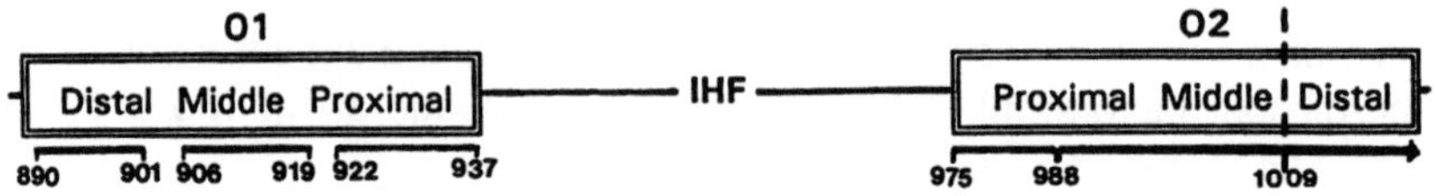

B

Assignments of end-enhancer interactions	
Enhancer region	Mu end site
01 distal	L3
01 middle	none detected
01 proximal	R1 or R2
02 proximal	L1
02 middle	R3

Fig. 5A, B. A complex circuit of end-enhancer interactions. **A** A schematic of the enhancer region showing the IHF site and the 01 and 02 regions. *Numbers* below them are the nucleotide positions used for substitutions in the proximal, middle, and distal operator regions. **B** End-enhancer interactions mapped genetically using double mutants (R.G. Allison and G. Chaconas, unpublished results). For simplicity, all interactions are listed as primary (direct), but we cannot exclude the possibility that some are indirect through cooperative interactions (refer to ALLISON and CHACONAS 1992). The Mu end-binding site L2 was not found to participate in any enhancer interactions based upon these experiments

3.1.2 Accessory Factors at the Mu Enhancer

The Mu transpositional enhancer is believed to adopt a bent structure important for function (SURETTE et al. 1989; SURETTE and CHACONAS 1992). Bending is likely to result from IHF binding (SURETTE et al. 1989; SURETTE and CHACONAS 1992), or it may be facilitated by HU interactions when supercoiling levels are high (σ=–0.06) and IHF is not required (SURETTE and CHACONAS 1989; R.G. Allison and G. Chaconas, unpublished results). The recent demonstration that the Mu enhancer promotes end synapsis either in *cis* or in *trans* and is not required for the strand cleavage event (SURETTE and CHACONAS 1992) emphasizes its essential role in assembly and may provide a useful experimental system for the study of geometrical parameters in Mu transposition.

3.1.3 DNA Supercoiling

The requirement for DNA supercoiling is a feature which is common to many DNA metabolic events. Under standard reaction conditions a supercoiled mini-Mu donor substrate is required for the in vitro strand-transfer reaction. DNA supercoiling is not required for the actual chemical steps of the reaction (Surette et al. 1987; Craigie and Mizuuchi 1987; Mizuuchi et al. 1992; Wang and Harshey 1994) but appears to be required primarily in the assembly of a stable tetrameric complex. DNA supercoiling does, however, also greatly stabilize the type-0 complex generated in Ca^{++} (R.G. Allison and G. Chaconas, unpublished results). How, then, does DNA supercoiling promote the formation of and stabilize the uncut synaptic complex? This probably occurs through multiple effects exerted by negative supercoiling (see Nash 1990; Kanaar and Cozzarelli 1992) as noted below:

Protein binding – DNA supercoiling is known to facilitate protein-DNA interactions with proteins which alter DNA twist or writhe in the appropriate direction or with protein which bend DNA. Both simple protein-DNA interactions and the formation of wrapped higher-order complexes may be driven by the free energy of negatively supercoiled DNA. In the Mu system the affinities of Mu A (Kuo et al. 1991), IHF (M.G. Surette and G. Chaconas, unpublished results), and HU (unpublished results) are all substantially increased for supercoiled DNA. Whether or not negative supercoils are constrained in the type-0 and type-1 complexes is not yet known; if so, this would also help drive the reaction.

Increase in local concentration of DNA sites – The local concentration of two DNA sites on a supercoiled versus a relaxed circular molecule is estimated to increase by a factor of roughly 100 (Vologodskii et al. 1992). This effect would be even greater for a three-site reaction. Such a dramatic increase in the local concentration of sites would be expected to have a large stimulatory effect on the reaction kinetics. This is illustrated by recent experiments of the Mu reaction with the enhancer in *trans*. A 40-fold molar excess of enhancer was used and resulted in a reaction rate 25 times slower than with the enhancer in *cis* on a supercoiled plasmid (Surette and Chaconas 1992).

Site alignment and interwinding – Negative supercoiling has been shown to kinetically and/or thermodynamically favour alignment of the two Mu ends in the proper orientation required for stable complex formation. Elegant experiments by Craigie and Mizuuchi (1986) have shown that proper interwrapping of the two Mu ends to give a fixed geometry is an essential feature for complex formation and can be provided by a negatively supercoiled substrate or by a multiply catenated or knotted donor which favours appropriate interwrapping of the ends.

DNA conformational changes – The free energy of supercoiling is used to facilitate the induction of a bend between 01 and 02 in the transpositional enhancer. At low levels ($\sigma = -0.025$) of supercoiling this bend is not effectively generated unless IHF protein is present (Surette et al. 1989; Surette and Chaconas 1989; Higgins et al. 1989). DNA supercoiling may also be required for the formation of a tight bend or loop between L1 and L2 where stable HU binding occurs opon/during establishment of a synaptic complex. The DNA unwinding, bending, and looping mentioned above, as well as other possible conformational changes in the Mu complex (refer to Sect. 2.3), may be driven by the free energy of supercoiling in the donor substrate molecule.

Protein conformational changes– The tremendous stability exhibited by the Mu complex upon tetramerization of Mu A likely results from large conformational changes in the protein. Although it is entirely speculation at this point, the free energy of supercoiling could be used to drive such conformational changes in Mu A.

Finally, a recent study by WANG and HARSHEY (1994) on the role of DNA supercoiling in the Mu reaction has suggested that the free energy of supercoiling associated with DNA outside the Mu ends is required to drive complex formation. They propose that the external supercoiling energy is used to overcome the activation barrier of the reaction, which they estimate to be about 67 kcal/mol. The use of torsional strain specifically from outside the Mu ends has not been previously considered and is an interesting and provocative proposal. Based upon the high energy of activation for the reaction, Wang and Harshey argue for substantial DNA and/or protein conformational changes to be effected by the superhelical strain outside the ends. The recent observation of duplex destabilization in the flanking host sequences in a type-1 complex made with synthetic oligonucleotides (H. Savilahti and K. Mizuuchi, personal communication), as well as the existence of inhibitory flanking sequences outside the Mu ends (WU and CHACONAS 1992), underscores the possible importance for DNA conformational changes outside the Mu ends.

3.2 The Stable Synaptic (Type-0) Complex

The stable synaptic (type-0) complex is an uncut complex which accumulates when Ca^{+2} is substituted for Mg^{+2} in the strand cleavage reaction (MIZUUCHI et al. 1992). The ability to generate this uncut complex has been invaluable in distinguishing between factors required for the actual cleavage step versus those needed only for synapsis. A key question which remains to be answered is whether this complex is actually an obligatory intermediate on the pathway to type-1 formation under normal reaction conditions (Mg^{+2}), or whether the stable type-0 represents an alternative pathway. Isolation and characterization of a type-0 complex in Mg^{+2} has not been reported to date. Moreover, two different types of experiments are consistent with the type-0 complex being part of an alternate pathway, and proof for either hypothesis has not yet been obtained.

The first type of experiment involves the use of terminal base pair mutations at the Mu ends. A single base pair change at the cleavage position blocks the formation of the type-1 complex, although cleavage can be stimulated by the Mu B protein (SURETTE et al. 1991). If the same reactions are run in $Ca^{+2,}$ however, significant amounts of type-0 complex are not detectable, either in the presence or absence of Mu B (E. Passi and G. Chaconas, unpublished data). This suggests that the type-0 complex is not an obligatory intermediate in the formation of a type-1 complex. Another possible explanation for the data is that the type-0 complex is unstable with the terminal base pair mutants.

The second type of experiment involves the substitution of HU with HMG-1 (LAVOIE and CHACONAS 1994). In the presence of Mg^{+2} the rate of type-1 formation is some 65 times slower with HMG-1 than with HU. In contrast, the rates of type-0 formation by both proteins in Ca^{+2} are within a factor of two of each other. Moreover, once formed, the type-0 generated with HMG-1 in calcium is rapidly cleaved in Mg^{+2}. These data suggest that the presence of Ca^{+2} perturbs the normal reaction and may help funnel the reactants through an alternative pathway.

In summary, although the role of the stable synaptic complex in the Mu transposition pathway has not been unambiguously demonstrated, it has clearly been a very useful tool in dissecting reaction mechanisms and will play an important role in future studies of Mu transposition.

3.3 Chemical Steps

3.3.1 DNA Cleavage

Unlike the lambda integrase family of recombinases, which are type-1 topoisomerases and utilize a covalent intermediate to conserve the energy of the sugar phosphate bond, Mu A is thought to mediate strand transfer through a one-step transesterification mechanism (MIZUUCHI and ADZUMA 1991). This reaction occurs in the context of the stable Mu A tetramer in the type-1 complex and has led to suggestions that the preceding strand cleavage reaction, which exposes the 3' hydroxyls required for direct attack, may proceed in an analogous manner but utilizing a water molecule as the primary nucleophile (MIZUUCHI 1992a, b). This hypothesis has not been directly tested in Mu but has been addressed during the endonuclease step promoted by the HIV integrase (IN) protein. This reaction involves the removal of a dinucleotide off each blunt long terminal repeat end of a linear viral double-stranded DNA (ENGLEMAN et al. 1991; VINK et al. 1991). A key feature of the IN cleavage model involves the activation of the phosphodiester bonds prior to hydrolysis and strand transfer (ENGLEMAN et al. 1991). In Mu, strand-specific DNA hyper-reactivity at/near the left end cleavage site (Mu-host junction) has been noted under conditions where Mu A-dependent cleavage is blocked (LAVOIE and CHACONAS 1993) and is consistent with DNA strain near the cleavage site (refer to Sect. 2.3). Along these lines, DNA hydrolysis may be mediated by activation of the nucleophile, as in the catalytic triad of DNAse I (WESTON et al. 1992; SUCK et al. 1988; SUCK and OEFNER 1986 and references therein) or the 3'–5' exonuclease of klenow fragment (BEESE and STEITZ 1991; FREEMONT et al. 1988 and references therein), which both appear to distort DNA structure and activate water as a prelude to bond hydrolysis via in-line displacement mechanisms.

Recent studies have indicated that donor cleavage requires only two active Mu A monomers in the catalytic tetramer (BAKER et al. 1993; 1994).

3.3.2 Strand Transfer

Following strand cleavage, the Mu ends become coupled to 5'p of the target DNA. The preferred target site for integration is the consensus sequence N-Py-G/

C-Pu-N (Mizuuchi and Mizuuchi 1993). Strand transfer into intramolecular or intermolecular target sites is determined largely by the absence or presence of Mu B, respectively (Maxwell et al. 1987), which interacts directly with Mu A (Baker et al. 1991; Surette and Chaconas 1991; Surette et al. 1991), presumably within the C-terminal 48 amino acids (Baker et al. 1991; Leung and Harshey 1991; Harshey and Cuneo 1986). Although ATP is required for the reaction, the energy of ATP hydrolysis is not used for strand transfer (Adzuma and Mizuuchi 1988) but instead plays a role in transposition immunity (Adzuma and Mizuuchi 1988, 1989). Rather, the reactive 3' hydroxyl groups uncovered at the ends of Mu attack the target DNA molecule in a direct (one-step) transesterification reaction (Mizuuchi and Adzuma 1991; for recent reviews see Mizuuchi 1992a, b). Briefly, DNA containing chiral phosphorothioates was used to probe the stereochemical outcome of the Mu strand-transfer reaction; the inversion of chirality of the strand-transfer target site is consistent with a single-step in-line displacement mechanism which does not require a covalent protein-DNA intermediate. More complex models invoking an odd number of chiral inversions cannot yet be ruled out but seem unlikely (Mizuuchi and Adzuma 1991).

The chemical steps of the Mu strand-transfer reaction may be carried out by only two of the four A monomers in the catalytic tetramer, although cooperative interactions between all four monomers are likely to be critical (M. Mizuuchi and K. Mizuuchi, personal communication; Baker et al. 1993, 1994).

4 Perspectives

The past several years have been an exciting period in the study of the Mu DNA strand-transfer reaction in vitro. In the years to come we still have a number of important questions to answer, but, in general, future work will be much more concerned with structure and function studies. The three-dimensional structures of A and B are still not known and are a prerequisite for the determination of transpososome structure. In addition, the active site(s) in both A (are there one or two?) and B remain uncharacterized, as do the protein-protein contacts and conformational changes involved in tetramer formation and A-B interactions. Furthermore, the role of metal ions in the reaction and the details of the chemical catalysis with respect to the A active site(s) need to be sorted out. This isolation and characterization of three-site synaptic complexes is also a piece of unfinished business, as are the origin of the fourth A monomer in the Mu transpososome and the issue of whether cleavage occurs in *cis* or in *trans*.

The role of HU in complex formation is also not fully understood, and the importance of DNA distortions in complex formation and catalysis awaits further experimentation, as does the possible role of supercoiling outside the Mu ends. The DNA replication step and, in particular, initiation and disassembly of the extremely stable type-2 complex have also yet to be investigated. Moreover, the integration of infecting Mu phage DNA occurs by a non-replicative pathway to give only simple inserts. The similarities and differences between this pathway and the replicative pathway which has been reviewed in this article remain to be established; this will require the development of a faithful in vitro integration system for infecting Mu DNA. Finally, the relationship of the in vitro findings discussed here to Mu development in vivo will remain the final test of our attempts to dismantle and reassemble this complex and fascinating process in vitro.

Acknowledgments. We are indebted to K. Mizuuchi, M. Mizuuchi, H. Savilahti, Tania Baker, and Rasika Harshey for the communication of unpublished results. We are also grateful to K.M., M.M., T.B. and Z. Wu for their helpful comments on the manuscript.

Note Added in Proof. Since submission of the manuscript in March 1994, a number of papers on Mu DNA transposition have appeared:
Kim K, Namgoong S-Y, Jayaram M, Harshey RM (1995) J Biol Chem 270: 1472–1479
Wu Z, Chaconas G (1995) EMBO J 14 (in press)
Yang J-Y, Kim K, Jayaram M, Harshey RM (1995) EMBO J 14: 2374–2384

References

Adzuma K, Mizuuchi K (1988) Target immunity of Mu transposition reflects a differential distribution of Mu B protein. Cell 53: 257–266

Adzuma K, Mizuuchi K (1989) Interaction of proteins located at a distance along DNA: mechanism of target immunity in the Mu DNA strand-transfer reaction. Cell 57: 41–47

Adzuma K, Mizuuchi K (1991) Steady-state kinetic analysis of ATP hydrolysis by the B protein of bacteriophage Mu. J Biol Chem 266: 6159–6167

Allison RG, Chaconas G (1992) Role of the A protein-binding sites in the in vitro transposition of Mu DNA: a complex circuit of interactions involving the Mu ends and the transpositional enhancer. J Biol Chem 267: 19963–19970

Allison RG, Chaconas G (1995) Assignment of Mu end-enhancer partners in an early stage of Mu DNA transposition (submitted)

Baker TA, Luo L (1994) Identification of residues in the Mu transposase essential for catalysis. Proc Natl Acad Sci USA 91: 6654–6658

Baker TA, Mizuuchi K (1992) DNA-promoted assembly of the active tetramer of the Mu transposase. Genes Dev 6: 2221–2232

Baker TA, Mizuuchi M, Mizuuchi K (1991) MuB protein allosterically activates strand transfer by the transposase of phage Mu. Cell 65: 1003–1013

Baker TA, Mizuuchi M, Savilahti H, Mizuuchi K (1993) Division of labor among monomers within the Mu transposase tetramer. Cell 74: 723–733

Baker TA, Kremenstova E, Luo L (1994) Complete transposition requires four active monomers in the Mu transposase tetramer. Genes Dev 8: 2416–2428

Beese L, Steitz TA (1991) Structural basis for the 3'–5' exonuclease activity of *Escherichia coli* DNA Polymerase I=a two metal ion mechanism. EMBO J 10: 25–33

Boocock MR, Rowland S-J, Stark WM, Sherratt DJ (1992) Insistent and intransigent: a phage Mu enhancer functions in *trans.* TIG 8: 151–153

Chaconas G (1987) Transposition of Mu DNA in vivo. In: Symonds N, Toussaint A, van de Putte P, Howe MM (eds) Phage Mu. Cold Spring Harbor Laboratory, Cold Spring Harbor, New York, pp 137–157

Chaconas G, Harshey RM, Sarvetnick N, Bukhari AI (1981) Predominant end-products of prophage Mu DNA transposition during the lytic cycle are replicon fusions. J Mol Biol 150: 341–359

Chaconas G, Gloor G, Miller JL (1985) Amplification and purification of the bacteriophage Mu-encoded B transposition protein. J Biol Chem 260: 2662–2669

Craig NL, Nash HA (1984) *E. coli* integration host factor binds to specific site in DNA. Cell 39: 707–716
Craigie R, Mizuuchi K (1985) Mechanism of transposition of bacteriophage Mu: structure of a transposition intermediate. Cell 41: 867–876
Criaigie R, Mizuuchi K (1986) Role of DNA topology in Mu transposition: mechanism of sensing the relative orientation of two DNA segments. Cell 45: 793–800
Craigie R, Mizuuchi K (1987) Transposition of Mu DNA: joining of Mu to target DNA can be uncoupled from cleavage at the ends of Mu. Cell 51: 493–501
Craigie R, Mizuuchi M, Mizuuchi K (1984) Site-specific recognition of the bacteriophage Mu ends by the Mu A protein. Cell 39: 387–394
Craigie R, Arndt-Jovin DJ, Mizuuchi K (1985) A defined system for the DNA strand-transfer reaction at the initiation of bacteriophage Mu transposition: protein and DNA substrate requirements. Proc Natl Acad Sci USA 82: 7570–7574
Ding Z-M, Harshey RM, Hurley LH (1993) (+)-CC-1065 as a structural probe of Mu transposase-induced bending of DNA: overcoming limitations of hydroxyl-radical footprinting. Nucleic Acids Res 21: 4281–4287
Drlica K, Rouvière-Yaniv J (1987) Histonelike proteins of bacteria. Microbiol Rev 51: 310–319
Engleman A, Mizuuchi K, Craigie R (1991) HIV–1 DNA integration: mechanism of viral DNA cleavage and DNA strand transfer. Cell 67: 1211–1221
Freemont PS, Friedman JM, Beese LS, Sanderson MR, Steitz TA (1988) Cocrystal structure of an editing complex of klenow fragment with DNA. Proc Natl Acad Sci USA 85: 8924–8928
Groenen MAM, van de Putte P (1985) Mapping of a site for packaging of bacteriophage Mu DNA. Virology 144: 520–522
Groenen MAM, van de Putte P (1986) Analysis of the attachment sites of bacteriophage Mu using site-directed mutagenesis. J Mol Biol 189: 597–602
Haniford DB, Chaconas G (1992) Mechanistic aspects of DNA transposition. Curr Opin Genes Dev 2: 698–704
Harel J, Duplessis L, Kahn JS, DuBow MS (1990) The *cis*-acting DNA sequences required in vivo for bacteriophage Mu helper-mediated transposition and packaging. Arch Microbiol 154: 67–72
Harshey RM, Cunèo SD (1986) Carboxyl-terminal mutants of phage Mu transposase. J Genet 65: 159–164
Haykinson MJ, Johnson RC (1993) DNA looping and the helical repeat in vitro and in vivo: effect of HU protein and enhancer location on Hin invertasome assembly. EMBO J 12: 2503–2512
Heichman KA, Johnson RC (1990) The Hin invertasome: protein-mediated joining of distant recombination sites at the enhancer. Science 249: 511–517
Higgins NP, Collier DA, Kilpatrick MW, Krause HM (1989) Supercoiling and integration host factor change the DNA conformation and alter the flow of convergent transcription in phage Mu. J Biol Chem 264: 3035–3042
Hodges-Garcia Y, Hagerman PJ, Pettijohn DE (1989) DNA ring closure mediated by protein HU. J Biol Chem 264: 14621–14623
Hwang DS, Kornberg A (1992) Opening of the replication origin of *Escherichia coli* by DnaA protein with protein HU or IHF. J Biol Chem 15: 23083–23086
Kanaar R, Cozzarelli NR (1992) Roles of Supercoiled DNA structure in DNA transactions. Curr Opin Struct Biol 2: 369–379
Kulkosky J, Jones KS, Katz RA, Mack JPG, Skalka AM (1992) Residues critical for retroviral integrative recombination in a region that is highly conserved among retroviral/retrotransposon integrases and bacterial insertion sequence transposases. Mol Cell Biol 12: 2331–2338
Kuo C-F, Zou A, Jayaram M, Getzoff E, Harshey R (1991) DNA-protein complexes during attachment-site synapsis in mu DNA transposition. EMBO J 10: 1585–1591
Lavoie BD, Chaconas G (1990) Immunoelectron microscopic analysis of the A, B, and HU protein content of bacteriophage Mu transpososomes. J Biol Chem 265: 1623–1627
Lavoie BD, Chaconas G (1993) Site-specific HU binding in the Mu transpososome: conversion of a sequence-independent DNA-binding protein into a chemical nuclease. Genes Dev 7: 2510–2519
Lavoie BD, Chaconas G (1994) A second high-affinity HU-binding site in the phage Mu transpososome. J Biol Chem 269: 15571–15576
Lavoie BD, Chan BS, Allison RG, Chaconas G (1991) Structural aspects of a higher order nucleoprotein complex: induction of an altered DNA structure at the Mu-host junction of the Mu type 1 transpososome. EMBO J 10: 3051–3059
Leung PC, Harshey RM (1991) Two mutations of phage Mu transposase that affect strand transfer or interactions with B protein lie in distinct polypeptide domains. J Mol Biol 219: 189–199

Leung PC, Teplow DB, Harshey RM (1989) Interaction of distinct domains in Mu transposase with Mu DNA ends and an internal transpositional enhancer. Nature 338: 656–658
Maxwell A, Craigie R, Mizuuchi K (1987) B protein of bacteriophage Mu is an ATPase that preferentially stimulates intermolecular DNA strand transfer. Proc Natl Acad Sci USA 84: 699–703
Miller JL, Chaconas G (1986) Electron microscopic analysis of in vitro transposition intermediates of bacteriophage Mu DNA. Gene 48: 101–108
Miller JL, Anderson SK, Fujita DJ, Chaconas G, Baldwin DL, Harshey RM (1984) The nucleotide sequence of the B gene of bacteriophage Mu. Nucleic Acids Res 12: 8627–8638
Mizuuchi K (1983) In vitro transposition of bacteriophage Mu: a biochemical approach to a novel replication reaction. Cell 35: 785–794
Mizuuchi K (1984) Mechanism of transposition of bacteriophage Mu: polarity of the strand–transfer reaction at the initiation of transposition. Cell 39: 395–404
Mizuuchi K (1992a) Transpositional recombination: mechanistic insights from studies of Mu and other elements. Annu Rev Biochem 61: 1011–1051
Mizuuchi K (1992b) Polynucleotidyl transfer reactions in transpositional DNA recombination. J Biol Chem 267: 21273–21276
Mizuuchi K, Adzuma K (1991) Inversion of the phosphate chirality at the target site of Mu DNA strand transfer: evidence for a one-step transesterification mechanism. Cell 66: 129–140
Mizuuchi M, Mizuuchi K (1989) Efficient Mu transposition requires interaction of transposase with a DNA sequence at the Mu operator: implications for regulation. Cell 58: 399–408
Mizuuchi M, Baker TA, Mizuuchi K (1991) DNase protection analysis of the stable synaptic complexes involved in Mu transposition. Proc Natl Acad Sci USA 88: 9031–9035
Mizuuchi M, Baker TA, Mizuuchi K (1992) Assembly of the active form to the transposase-Mu DNA complex: a critical control point in Mu transposition. Cell 70: 303–311
Nakayama C, Teplow DB, Harshey RM (1987) Structural domains in phage Mu transposase: identification of the site-specific DNA-binding domain. Proc Natl Acad Sci USA 84: 1809–1813
Namgoong S-Y, Jayaram M, Kim K, Harshey RM (1994) DNA-protein cooperativity in the assembly and stabilization of Mu strand transfer complex: relevance of DNA phasing and *att* site cleavage. J Mol Biol 238: 514–527
Nash HA (1990) Bending and supercoiling of DNA at the attachment site of bacteriophage lambda. TIBS 15: 222–227
Nash HA, Robertson CA (1981) Purification and properties of the *E. coli* protein factor required for λ integrative recombination. J Biol Chem 256: 9246–9253
Panigrahi GB, Walker IG (1991) Use of monoacetyl-4-hydroxyaminoquinoline 1-oxide to probe contacts between guanines and protein in the minor and major grooves of DNA: interaction of *Escherichia coli* integration host factor with its recognition site in the early promoter and transposition enhancer of bacteriophage Mu. Biochemistry 30: 9761–9767
Pato ML (1989) Bacteriophage Mu. In: Berg DE, Howe MM (eds) Mobile DNA. American Society for Microbiology, Washington DC, pp 23–52
Pettijohn DE (1988) Histone-like proteins and bacterial chromosome structure. J Biol Chem 263: 12793–12796
Robertson CA, Nash HA (1988) Bending of the bacteriophage λ attachment site by *Escherichia coli* integration host factor. J Biol Chem 263: 3554–3557
Rouvière-Yaniv J, Gros F (1975) Characterization of a novel, low-molecular-weight DNA-binding protein from *Escherichia coli*. Proc Natl Acad Sci USA 72: 3428–3432
Schmid MB (1990) More than just "histone-like" proteins. Cell 63: 451–453
Schneider GJ, Sayre MH, Geiduschek EP (1991) DNA-binding properties of TF1. J Biol Chem 221: 777–794
Serrano M, Salas M, Hermoso JM (1993) Multimeric complexes formed by DNA-binding proteins of low sequence specificity. TIBS 18: 202–206
Shore D, Langowski J, Baldwin R (1981) DNA flexibility studied by covalent closure of short fragments into circles. Proc Natl Acad Sci USA 78: 4833–4837
Suck D (1992) Nuclease structure and catalytic function. Curr Opin Struct Biol 2: 84–92
Suck D, Oefner C (1986) Structure of DNase I at 2.0 Å resolution suggests a mechanism for binding to and cutting DNA. Nature 321: 620–625
Suck D, Lahm A, Oefner C (1988) Structure refined to 2 Å of a nicked DNA octanucleotide complex with DNase I. Nature 332: 464–468
Surette MG, Chaconas G (1989) A protein factor that reduces the negative supercoiling requirement in the Mu strand-transfer reaction is *Escherichia coli* integration host factor. J Biol Chem 264: 3028–3034

Surette MG, Chaconas (1991) Stimulation of the Mu DNA strand cleavage and intramolecular strand-transfer reactions by the Mu B protein is independent of stable binding of the Mu B protein to DNA. J Biol Chem 266: 17306–17313

Surette MG, Chaconas G (1992) The Mu transpositional enhancer can function in *trans:* requirement of the enhancer for synapsis but not strand cleavage. Cell 68: 1101–1108

Surette MG, Buch SJ, Chaconas G (1987) Transpososomes: stable protein-DNA complexes involved in the in vitro transposition of bacteriophage Mu DNA. Cell 49: 253–262

Surette MG, Lavoie BD, Chaconas G (1989) Action at a distance in Mu DNA transposition: an enhancer-like element is the site of action of supercoiling relief activity by integration host factor [IHF]. EMBO J 8: 3483–3489

Surette MG, Harkness T, Chaconas G (1991) Stimulation of the Mu A protein-mediated strand cleavage reaction by Mu B protein, and the requirement of DNA nicking for stable type-1 transpososome formation. J Biol Chem 266: 3118–3124

Tanaka I, Appelt D, Dijk J, White SW, Wilson KS (1984) 3-Å resolution structure of a protein with histone-like properties in prokaryotes. Nature 310: 376–381

Teplow DB, Nakayama C, Leung PC, Harshey RM (1988) Structure-function relationships in the transposition protein B of bacteriophage Mu. J Biol Chem 263: 10851–10857

Thompson JF, Landy A (1988) Empirical estimation of protein-induced DNA binding angles: applications to λ site-specific recombination complexes. Nucleic Acids Res 16: 9687–9705

Vink C, Yeheskiely E, van der Marel GA, van Boom JH, Plasterk RHA (1991) Site-specific hydrolysis and alcoholysis of human immunodeficiency virus DNA termini mediated by the viral integrase protein. Nucleic Acids Res 19: 6691–6698

Vologodskii AV, Levene SD, Klenin KV, Frank-Kamenetskii M, Cozzarelli NR (1992) Conformational and thermodynamic properties of supercoiled DNA. J Mol Biol 227: 1224–1243

Wang Z, Harshey RM (1994) Crucial role for DNA supercoiling in Mu transposition: kinetic study. Proc Natl Acad Sci USA 91: 699–703

Weston SA, Lahm A, Suck D (1992) X-ray structure of the DNAse I-d(GGTATACC)2 complex at 2.3 angstrom resolution. J Mol Biol 226: 1237–1256

White SW, Appelt K, Wilson KS, Tanaka I (1989) A protein structural motif that bends DNA. Proteins 5: 281–288

Wu Z, Chaconas G (1992) Flanking host sequences can exert an inhibitory effect on the cleavage step of the in vitro Mu DNA strand transfer reaction. J Biol Chem 267: 9552–9558

Wu Z, Chaconas G (1994) Characterization of a region in phage Mu transposase that is involved in interaction with the Mu B protein. J Biol Chem 269: 28829–28833

Yang C-C, Nash HA (1989) The interaction of *E. coli* IHF protein with its specific binding sites. Cell 57: 869–880

Zou A, Leung PC, Harshey RM (1991) Transposase contacts with Mu DNA ends. J Biol Chem 266: 20476–20482

P Elements in *Drosophila*

W.R. ENGELS

The *Drosophila* genome has many families of transposable elements; some of them have been studied in detail, and others are known only superficially (BERG and HOWE 1989). Particular attention has been given to the P family (reviewed by ENGELS 1989), which has been the subject of intensive research for nearly two decades. There are two reasons for this special interest. First, the population biology and recent evolutionary history of P elements suggest a remarkable scenario of horizontal transfer from another species into *D. melanogaster*, followed by rapid spread through the global population. Second, a wide array of technical applications have made P elements an indispensable tool for manipulating the *Drosophila* genome.

Genetics Department, University of Wisconsin, 445 Henry Mall, Madison, WI 53706, USA

1 P-Element Structure

The structure of an autonomous P element is shown in Fig. 1. The 2907-bp sequence features a perfect 31-bp terminal inverted repeat and an 11-bp subterminal inverted repeat (O'HARE and RUBIN 1983). These repeats are needed in *cis* for efficient transposition, but they are not sufficient for it (MULLINS et al. 1989). Internally, there are other repeat units of unknown function plus a transposase gene composed of four exons. This gene is required in *trans* for transposition, and part of the gene is also involved in regulation of P mobility (RIO and RUBIN 1988; RIO 1990).

Nonautonomous P elements also exist. Some occur naturally through internal deletions of the autonomous elements, as shown in Fig. 1. Such elements lack the transposase gene but retain the parts of the sequence required in *cis* for transposition. Mobilization of nonautonomous P elements occurs only if there is at least one autonomous P element present to supply transposase. Many artificial nonautonomous P elements have also been created in which the transposase

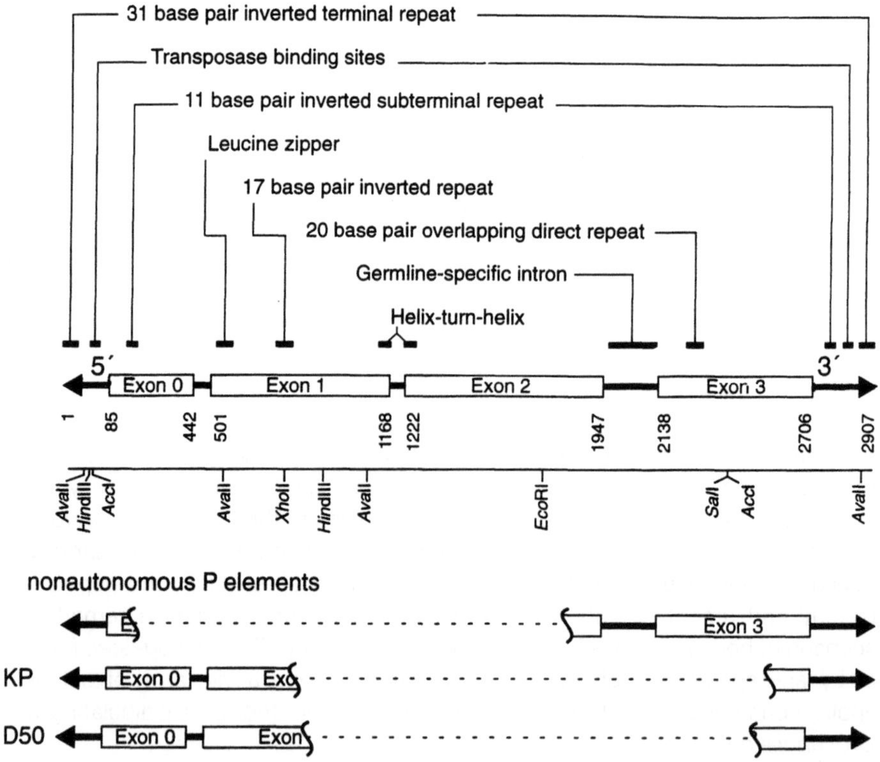

Fig. 1. P-element anatomy. An autonomous P element is shown with some of its sequence features and a restriction map. Three examples of nonautonomous P elements are given, including the type-II repressor-making elements, *KP* and *D50*. (Modified from LINDSLEY and ZIMM 1992)

gene has been replaced by another gene of interest, often functioning as a marker or reporter. Such elements are discussed further below.

2 Hybrid Dysgenesis

When P elements are mobilized they produce a syndrome of traits known collectively as hybrid dysgenesis (KIDWELL et al. 1977). These traits include temperature-dependent sterility, elevated rates of mutation, chromosome rearrangement, and recombination. The syndrome is usually seen only in the progeny of males with autonomous P elements and females that lack P elements. These two kinds of strains are called "P" and "M" because they contribute paternally and maternally, respectively, to hybrid dysgenesis. The reciprocal cross, P♀ × M♂, yields hybrids in which the dysgenic traits are much reduced, due to the maternal component of P-element regulation by cytotype, as will be discussed below.

The dysgenic traits can be explained largely by genomic changes due to P-element transposition and excision in developing germ cells. The sterility is due to loss of germ cells early in development (ENGELS and PRESTON 1979; KIDWELL and NOVY 1979; NIKI 1986; NIKI and CHIGUSA 1986; WEI et al. 1991). It is more pronounced in females, where there are fewer germ cells to spare than in males, and at temperatures above 25°C. The mutations come about through several mechanisms but are primarily P insertions into genes and imprecise excision of P elements near genes (RUBIN et al. 1982; TSUBOTA et al. 1985; SALZ et al. 1987). Chromosome rearrangements usually result from breakage at the sites of two or more P-element insertions, followed by rejoining of the chromosome segments in a different order (ENGELS and PRESTON 1981, 1984; ROIHA et al. 1988). P-induced recombination occurs preferentially in the genetic intervals containing mobile P elements (SVED et al. 1990, 1991) and usually within 2 kb of the insertion site C. Preston, J. Sved and W. Engels, unpublished).

P-element mobilization happens throughout development of the germline. Most mutations, rearrangements, and recombination events occur prior meiosis (ENGELS 1979c; HIRAIZUMI 1979), but some meiotic events have also been detected (DANIELS and CHOVNICK 1993). Premeiotic events tend to be recovered in clusters of two or more aberrant individuals among the progeny of a single dysgenic parent. The premeiotic timing of these events places a limitation on the genomic changes that can be recovered in the next generation, since the product must be cell viable in the germline. Thus, mutations or rearrangements that do not yield viable germ cells in the parent will not result in functional gametes, and therefore will not be recovered in the next generation, regardless of whether the hypothetical progeny bearing these changes would have been viable. In addition, premeiotic events result in frequency data that cannot be analyzed reliably by standard statistical methods based on Poisson or binomial distributions, and more robust alternatives must be employed (ENGELS 1979a).

3 Population Biology

It is now widely believed that P elements have existed in the *D. melanogaster* genome for less than 100 years. According to this view, the elements were introduced through a rare horizontal transmission event in which one or more autonomous P-element copies were acquired by *D. melanogaster* from another *Drosophila* species. The elements then spread strictly by heredity and transposition to become ubiquitous in natural populations within a few decades. This startling scenario was proposed by KIDWELL (1979, 1983) and recently reviewed (see ENGELS 1992; KIDWELL 1993) to explain the observation that the only true M strains were old laboratory stocks dating back to the early days of *Drosophila* genetics. Such strains would be reproductively isolated from natural populations, and thus escaped the P-element invasion.

An alternative way to explain M strains was to postulate that the elements have existed in *D. melanogaster* over evolutionary time, but that some aspect of laboratory culture conditions, such as small population size, acted to remove the P elements from the genome over several thousand generations (ENGELS 1981). The question was resolved when P elements from other *Drosophila* species were examined (Fig. 2). Some species closely related to *melanogaster* lacked P elements, but several much more distant relatives had them. In particular, the DNA sequence of a P element from *D. willistoni* was nearly identical to the *melanogaster* sequence, differing by only 1 bp among 2907. Such conservation would be impossible over the 60 million years that the two species have diverged. It implies that P elements in the two species had a common ancestor in recent historical times.

D. melanogaster is now a cosmopolitan species, but it is thought to have evolved in Western Africa (LACHAISE et al. 1988). The species became established elsewhere only when human commercial shipping provided a means for long-distance migration (JOHNSON 1913; STURTEVANT 1921). Meanwhile, *D. willistoni* and related species evolved primarily in Central and South America, and are still endemic to these regions (ASHBURNER 1989a). Therefore, they had no contact with *melanogaster* until the latter species arrived in the Americas. By examining antique insect collections, JOHNSON (1913) estimated that the first appearance of *D. melanogaster* in the New World occurred in the early 1800s, and that they became widespread by the end of the century.

The horizontal transfer event could have occurred at any time since *melanogaster* and *willistoni* became sympatric, but the spread of P elements through *melanogaster* was presumably not yet complete by the 1930s when the last laboratory M populations were established (KIDWELL 1983).

Currently, natural populations of *D. melanogaster* all appear to have P elements, including populations in such remote sites as the mountainous regions of central Asia (S. Nuzhdin and W. Engels, unpublished). However, the type and number of P elements show geographical differences (KIDWELL et al. 1983; ANXOLABÉHÈRE et al. 1984, 1985b; 1988; BOUSSY et al. 1988). For example, a North

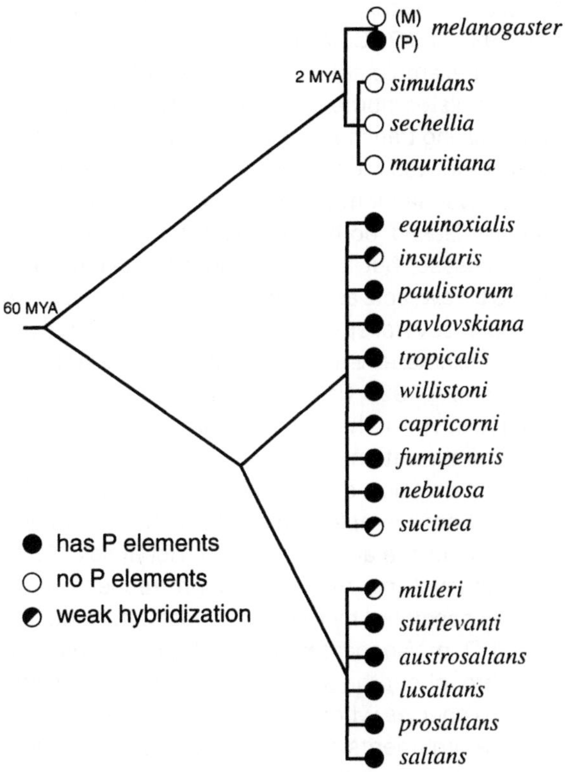

Fig. 2. Species distribution of P elements. Southern blotting was used to determine the presence or absence of P-element sequences in various *Drosophila* species to illustrate the evidence for recent invasion (DANIELS et al. 1990 and references cited therein). Estimates of divergence time, given in millions of years ago, are averages derived from LACHAISE et al. (1988) and SPICER (1988). (Modified from ENGELS 1992)

American population had 30–50 P elements in scattered chromosomal locations, with approximately two thirds of them nonautonomous (O'HARE and RUBIN 1983; O'HARE et al.1992). Samples from other parts of the world, however, especially near the Mediterranean, show fewer P elements in the genome and a higher proportion of nonautonomous ones (ANXOLABÉHÈRE et al. 1985b, 1988; BLACK et al. 1987). It is not known how such population differences are related to the invasion history of P elements.

The mechanism of the horizontal transfer between *D. willistoni* and *D. melanogaster* is also unknown. Speculation has focused on vector organisms, such as viruses and mites (HOUCK et al. 1991; ENGELS 1992). One parasitic mite species, *Proctolaelaps regalis*, has been studied as a possible DNA vector, and shows various features that make it compatible with this role (HOUCK et al. 1991). Once an autonomous P element had been introduced into *D. melanogaster* by whatever means, its ability to spread through the species was undoubtedly facilitated by a transposition mechanism in which DNA gap repair acts to increase the P-element copy number, as discussed below.

P elements have a long evolutionary history in Diptera prior to the invasion of *D. melanogaster* (LANSMAN et al. 1985, 1987; CLARK et al. 1994; HAGEMANN et al. 1994). In the *D. willistoni* genome, for example, there are many "dead" P elements whose DNA sequences have accumulated numerous frameshifts and substitutions preventing them from either making transposase or serving as its substrate (DANIELS et al. 1990). One species group was found to have a single genomic site where a portion of the P sequence was tandemly repeated (PARICIO et al. 1991; MILLER et al. 1992). This sequence lacked both P-element termini and was incapable of encoding the P transposase, but it did encode a truncated transposase protein which, as discussed below, acts as a repressor of P mobility. P-like elements have also been identified in several species of other genera and even outside the *Drosophilidae* family (ANXOLABÉHÈRE et al. 1985a; ANXOLABÉHÈRE and PÉRIQUET 1987; SIMONELIG and ANXOLABÉHÈRE 1991; PERKINS and HOWELLS 1992). There is preliminary evidence that distant relatives of the P element are common among Diptera (H. Robertson, personal communication).

Horizontal transfer and genomic invasion are probably not unusual in the world of transposable elements. The best example is that of the *mariner* element, which seems to have spread throughout the animal kingdom (ROBERTSON 1993). However, horizontal transfer events involving *mariner*, though frequent on an evolutionary time scale, are typically separated by millions of years. It is notable, therefore, that P elements invaded *D. melanogaster* within a few decades after the opportunity arose and spread throughout the species in less than 200 years.

What are the consequences to a species when a new transposable element invades its genome? Despite some arguments to the contrary (SYVANEN 1984; MCDONALD 1993), most evidence suggests that the harmful mutations and chromosome rearrangements produced by transposition far outweigh any beneficial mutations that might also arise (CHARLESWORTH and LANGLEY 1989). In one series of experiments, P-element invasion and rapid expansion in inbred laboratory M strains led to extinction of the lines within 20 generations (PRESTON and ENGELS 1989). The only exception was a case in which the population was expanded sufficiently to allow natural selection to eliminate deleterious insertions more efficiently. The ability of P elements to produce a negative regulator of their own mobility (discussed below) undoubtedly gave *D. melanogaster* a better chance of surviving its recent P-element invasion by reducing the equilibrium copy number (CHARLESWORTH and LANGLEY 1989; BROOKFIELD 1991).

4 Transposition

Several lines of evidence show that P elements transpose nonreplicatively and without an RNA intermidiate (ENGELS et al. 1990; KAUFMAN and RIO 1992). The donor element is excised and reinserted into a recipient site, creating a direct duplication of 8 bp at the site of insertion (O'HARE and RUBIN 1983). The transposition reaction can be carried out in a cell-free system with partially purified transposase (Kaufman and Rio 1992), but host-encoded factors might also facilitate the reaction in vivo (Kaufman et al. 1989).

4.1 Insertion Site Preference

P element insertions have been found at thousands of genomic positions, but not all sites are equally likely to be hit. The mechanism of insertion site selection is not known, but several generalizations can be made. (a) Euchromatic sites are hit more often than the heterochromatin (Engels 1989; Berg and Spradling 1991). (b) Some euchromatic loci are much more susceptible to P mutagenesis than others. For example, the *signed* gene is hit at frequencies approaching 10^{-2} (Green 1977; Simmons et al. 1984; Robertson et al. 1988), whereas the *vestigial* gene has a rate of less than 10^{-6} (Williams and Bell 1988). Despite this variability, there is no evidence that any loci are immune from P-element mutagenesis, given a sufficiently large sample size. (c) Within genes there is a preference for insertion in the noncoding upstream sequences (Kelley et al. 1987). (d) Target sites with close matches to the consensus octamer GGCCAGAC are more likely to receive P-element insertions (O'Hare and Rubin 1983; O'Hare et al. 1992). (e) P elements tend to insert into or near other P-elements, with a particular preference for base pairs 19–26 of the target P element (Eggleston 1990). (f) Some P elements have been observed to jump preferentially to sites closely linked to the donor site (Tower et al. 1993; Golic 1994).

4.2 Transposase

The P-element transposase is an 87-kD protein encoded by autonomous P elements (Fig. 1). It binds to subterminal regions at both ends of the element and represses transcription (Kaufman et al. 1989; Kaufman and Rio 1991). GTP is also bound by the transposase, and is required for transposition in vitro (Kaufman and Rio 1992).

4.3 Fate of the Donor Site

There is now considerable evidence that P-element transposition leaves behind a double-strand DNA break. Sequences homologous to the flanking DNA are then copied in to repair the break (Engels et al. 1990; Gloor et al. 1991). The relative frequencies of transposition and excision suggest that approximately 85% of these repair events utilize a sister chromatid for the template (Engels et al. 1990). Thus the donor element is replaced by an identical P element copied in from the sister chromatid (Fig. 3). In such cases, the end result of transposition is a net gain of one P-element copy. This net gain is probably responsible for the ability of P elements to increase their copy number in nature and in experimental

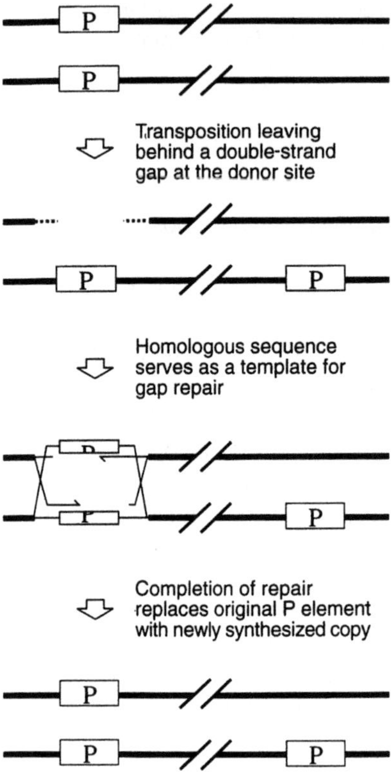

Fig. 3. P-element transposition and gap repair: The P element on one sister strand jumps to a new site, leaving behind a double-strand gap. Repair utilizes the other sister strand, resulting in a net gain of one P-element copy (ENGELS et al. 1990). (Modified from ENGELS 1992)

populations (KIYASU and KIDWELL 1984; GOOD et al. 1989; PRESTON and ENGELS 1989; ENGELS 1992; MEISTER and GRIGLIATTI 1993).

In the remaining 15% of repair events in which the sister chromatid is not used, the template can be either the homologous chromosome or an ectopic sequence, such as a transgene (ENGELS et al. 1990; GLOOR et al. 1991). The homolog is used preferentially in such cases, especially if it contains a P element or a fragment of a P element at the site corresponding to the break (JOHNSON-SCHLITZ and ENGELS 1993). The tendency to copy from a P-bearing template might account for some of the preferential use of the sister chromatid, which necessarily contains a P element at the site. There is high sensitivity to mismatches between the sequence flanking the break and the template. Even 0.5% mismatches is sufficient to decrease the rate of repair threefold (NASSIF and ENGELS 1993). Templates located on the same chromosome as the break are also used preferentially, even if the template lies on the opposite end of the chromosome (ENGELS et al. 1994).

Various types of imprecise excision can be attributed to aberrant repair. The most frequent such events are internal deletions resulting in structures similar to the nonautonomous P elements shown in Fig. 1. Breakpoints commonly occur at direct repeats of three or more base pairs, resulting in loss of one copy and the intervening sequence (O'HARE and RUBIN 1983; ENGELS 1989; EGGLESTON 1990). When longer direct repeats are present, the frequency is greatly increased (PAQUES and WEGNEZ 1993). The most frequent kind of internal deletion leaves only 10–20 bp from each terminus, resulting in a nonmobile "footprint" (SEARLES et al. 1982; TAKASU-ISHIKAWA et al. 1992; JOHNSON-SCHLITZ and ENGELS 1993). The term "internal deletion" might be a misnomer, since the events are probably due to a complete deletion of the P element, followed by incomplete gap filling from the sister chromatid (GLOOR et al. 1991). This interpretation is strengthened by the observation that when the P element resides on an extrachromosomal plasmid, and therefore lacks a sister chromatid for a template, the resulting footprints rarely contain more than 4 bp from each terminus (O'BROCHTA et al. 1991). The four or fewer bases that remain from each end could be explained if excision occurs by a staggered cut. Finally, imprecise P excisions that remove flanking DNA (SALZ et al. 1987) can also be interpreted as aberrant repair events by assuming the gap left by a P-element excision can be subsequently widened to varying degrees.

5 Regulation of P Element Mobility

P elements are not normally mobile in somatic cells, and their germline mobility does not occur within P strains. These two restrictions come about by different mechanisms.

5.1 Tissue Specificity

Repression of P-element transposition in somatic cells occurs on the level of RNA processing (LASKI et al. 1986). The 2–3 intron (Fig.1) is spliced only in the germ cells, resulting in the absence of transposase in somatic cells. Splicing of this intron is prevented in the somatic cells by a 97-kD protein that binds to a site in exon 2 located 12–31 bases from the 5'splice site (SIEBEL and RIO 1990; CHAIN et al. 1991; TSENG et al. 1991). When the 2–3 intron is removed artificially, the resulting transposase gene, designated Δ2–3, produces functional transposase in both somatic and germline cells, and P elements are mobile in all tissues (LASKI et al. 1986). This mobility results in pupal lethality if several mobile P elements are also present (ENGELS et al. 1987), especially in a DNA repair-deficient background (BANGA et al. 1991).

5.2 Cytotype and Repressors

The quiescent state of P elements within established P strains is best attributed to element-encoded repressor products. These repressors fall into at least two discrete categories, designated type I and type II.

5.2.1 Type-I Repressor

As noted earlier, P elements are repressed not only within P strains, but also in the hybrids of P♀ × M♂ crosses. They are not repressed in hybrids from the reciprocal cross (Kidwell et al. 1977), suggesting that repressor products in the P-strain germline can be inherited maternally. Interestingly, this effect goes beyond simple, maternal inheritance. Progeny from the cross MP♀ × P♂ show more P mobility than those from the cross PM♀ × P♂ (Engels 1979d), where "MP" and "PM" represent the two reciprocal F_1 hybrids with the female component shown first. Thus, the repressed state, called the P cytotype, is jointly determined by chromosomal and maternal components. One explanation proposed for this unusual inheritance was that the repressor-making P elements in the MP hybrids are more likely to be excised than those of the PM hybrids (Misra and Rio 1990; Misra et al. 1993). However, that explanation was ruled out by the finding that a single repressor-producing P element was sufficient for this mode of inheritance (Ronsseray et al. 1993). An alternative model (Lemaitre et al. 1993), involving differential splicing of the same intron that is involved in tissue-specific regulation, provides an adequate explanation for available data and will be discussed further below.

Various lines of evidence indicate that the 66-kD truncated protein that is made when the 2–3 intron is unspliced functions as a repressor of P mobility (Rio et al. 1986; Misra and Rio 1990; Gloor et al. 1993; Handler et al. 1993; Misra et al. 1993). Several other truncated transposase molecules with breakpoints in slightly different places can also function in this way (Robertson and Engels 1989; Gloor et al. 1993). Fine-structure deletion mapping revealed that the minimal 3' boundary for this kind of repressor was between nucleotides 1950 and 1956 of the P-element sequence (Gloor et al. 1993). These repressors are designated type I to distinguish them from a class of much smaller truncated transposase proteins called type II, discussed below.

The use of a reporter gene fused to the P element's transposase promoter showed that this repression acted on the level of transcription (Lemaitre and Coen 1991; Lemaitre et al. 1993). This effect also provided an explanation for several observations that P repressors affected expression of genes neighboring P-element insertion sites (Engels 1979; Williams et al. 1988; Robertson and Engels 1989; Coen 1990; Gloor et al. 1993). Finally, the transcriptional regulation revealed by the reporter gene showed a maternal effect in the germline but not in somatic cells (Lemaitre et al. 1993).

These observations led Lemaitre et al. (1993) to propose a model in which the 97-kD protein previously shown to prevent splicing of the 2–3 intron in somatic cells was also present germinally, but in much reduced quantities. Thus, in the P cytotype, transcription of the transposase message is relatively low, and the available germline splice blocker is sufficient to ensure that only a 66-kD product is made. This 66-kD product serves as a transcriptional repressor in the germline to perpetuate the P cytotype through the female lineage as long as there are sufficient repressor-producing P elements on the chromosomes. In the M cytotype, the level of expression is increased sufficiently to overwhelm the splice-blocking agent in the germline, and functional 87-kD transposase is made. This model requires a nonlinear relationship between transcript level and the ratio of 66-kD to 87-kD products in order to explain the maintenance of M cytotype.

5.2.2 Type-II Repressor Elements

The number of distinct nonautonomous P elements in nature is so large that few have been observed in more than one population (O'Hare and Rubin 1983; O'Hare et al. 1992). The first and most conspicuous exception to this rule is the KP element, which is very common worldwide (Black et al. 1987). It was therefore suggested that KP elements might function as P-element repressors and thus be favored by natural selection (Black et al. 1987; Jackson et al. 1988). This possibility was verified when KP elements were isolated genetically and tested for repressor (Rasmusson et al. 1993). KP elements did not fit the paradigm of type-I elements for two reasons: First, they had a deletion for nucleotides 808–2560 (Fig. 1.), and thus lacked the minimal sequence previously shown to be required for type-I repressors (Gloor et al. 1993). Second, the repression in KP lines showed none of the maternal inheritance associated with cytotype (Raymond et al. 1991; Rasmusson et al. 1993). This lack of maternal inheritance is expected under the splice-blocking model of cytotype described above, because the 2–3 intron is not present in KP elements.

It soon became clear that KP was not the only other element of this kind. The D50 element had similar repressor properties but slightly different deletion endpoints (Rasmusson et al. 1993). To date, five such elements, designated type-II repressor-makers, have been identified, at least four of which are geographically widespread (C. Preston, G. Gloor and W. Engels, unpublished). All have large deletions whose endpoints are similar (within 300 bp) to the KP deletion endpoints. A base substitution at either nucleotide position 32 or 33 is also present in most type-II repressors.

It is not known whether type-II regulation works by the same mechanism as type I, but they do share several features in common. Type-II regulation probably also acts on the level of transcription, since the elements can reduce the expression of a reporter gene fused to the P promoter (Lemaitre et al. 1993), and they display the secondary effects on expression of genes closely linked to P insertions (C. Preston and W. Engels, unpublished). For both types there is a pronounced position effect, such that the ability of the element to function as a repressor is highly sensitive to its genomic insertion site (Robertson and Engels 1989; Ronsseray et al. 1991; Higuet et al. 1992; Gloor et al. 1993; Misra et al. 1993).

6 P Elements as Molecular Biological Tools

Drosophila has long been a favorite organism for genetic and developmental research, but it was largely through the use of P elements that the powerful tools of molecular biology were fully employed. P elements are used for identifying genes of interest, for cloning them, and for placing them back into the genome. There are several key features of P-element biology that make them especially well suited for those roles. The existence of M strains allows experimenters to create stocks containing only selected P elements. Transposase can easily be added or removed genetically. The high mobility of P elements and their retention of this mobility despite drastic modifications to their internal sequences are also essential features. Most recently, the double-strand DNA breaks created by P-element excision have been used to effect gene replacement and to study the repair process.

6.1 Mutagenesis

A basic problem in genetics has long been how to obtain molecular information for a gene known only by the phenotype of its mutations. The reverse problem, obtaining mutations in a gene known only by its DNA sequence, is also becoming increasingly common. Oddly enough, both problems often have the same solution for *Drosophila* geneticists: obtain a P-element insertion into the gene. In the former case, the P insertion allows cloning by transposon tagging (Bingham et al. 1982; Searles et al. 1982), and in the latter, P-element insertion mutations can be selected by PCR-based methods without knowledge of the phenotype (see below). For these reasons, the search for P element, insertion mutations engages much of the effort of *Drosophila* workers, regardless of their particular biological focus (Kidwell 1986).

The most efficient approach to P-element mutagenesis utilizes an immobile copy of the transposase gene combined with one or more, mobile, but nonautonomous, P elements to serve as "ammunition". The P[*ry*$^+$ Δ2–3](99B) element is an example of an immobile transposase source that has been widely used for P mutagenesis (Robertson et al. 1988). Its transposase gene lacks the 2–3 intron, thus precluding the production of any 66-kD protein, which, as discussed above, functions as a repressor. This element cannot transpose due to a deletion of one of its termini (H. Robertson, personal communication). A cross with the transposase source coming from one parent and the ammunition elements from the other yields progeny in whose germ cells mutagenesis occurs. The transposase source is then eliminated by segregation in the next generation to stabilize any mutations obtained.

There are two general strategies in the selection of ammunition elements. One is to use a chromosome carrying as many highly mobile elements as possible, such as the Birm2 chromosome (Engels et al. 1987; Robertson et al. 1988), which has

17 small nonautonomous P elements from nature. This approach maximizes the likelihood of obtaining the desired mutation, but it is often laborious to isolate the mutation from rest of the P elements in the genome. Alternatively, one can use a smaller number of artificially constructed P elements (COOLEY et al. 1988). This method usually requires a larger screen, but any mutation obtained is easier to isolate, especially if the ammunition element(s) carries a bacterial origin of replication and selectable marker to permit cloning by plasmid rescue.

PCR can be employed to screen for P insertions if no phenotypic screen is available but the target gene has been cloned. DNA is extracted from a pool of potential mutants and amplified with one primer in the P-element sequence and an other in the targeted DNA. A P-element insertion close to the targeted site is required to bring sites for these two primers together and yield amplification BALLINGER and BENZER 1989; KAISER and GOODWIN 1990). The target of this approach is necessarily small, because PCR cannot amplify more than a few kilobases. An alternative approach that permits a much larger target is provided by inverse PCR (SENTRY and KAISER 1994). Here, the two primers are both within the P-element sequence, but directed away from other. Amplification can occur when the newly inserted P element, along with some of the flanking DNA, is circularized following digestion with a restriction enzyme and ligation (OCHMAN et al. 1988). DNA amplified in this way from a pool of potential mutants can then be probed with target DNA of arbitrary length, and insertion mutations can be identified.

Once a P insertion has been obtained in or near the gene of interest, additional genetic variability can be generated readily by the reintroduction of transposase. Internal deletions or flanking deletions can be selected (TSUBOTA and SCHEDL 1986; SALZ et al. 1987). In some cases, transposase can catalyze an event in which one P element in the genome is substituted for another P element elsewhere in the genome by an unknown mechanism (STAVELEY et al. 1994). This process can be useful for putting a reporter gene into a specific site.

6.2 P-Element-mediated Transformation

The most important use of P elements is undoubtedly that of making transgenic flies (RUBIN and SPRADLING 1982; SPRADLING and RUBIN 1982). The gene of interest is placed between P-element ends, usually within a plasmid, and injected into preblastoderm embryos in the presence of transposase. This P element, with the gene as cargo, then transposes from the plasmid to a random chromosomal site. Technical aspects of the method have been described elsewhere (SPRADLING 1986; ASHBURNER 1989b). In a typical experiment, 10–20% of the fertile injected flies produce transformant progeny.

The P element may also carry a second gene used to identify transformants. The frequency of transformation is usually sufficiently great that a visible marker, such as an eye-color gene (RUBIN and SPRADLING 1982; PIRROTTA 1988), is more efficient than a selectable marker, such as neomycin resistance. The size of the inserted sequence can exceed 40 kb (HAENLIN et al. 1985), but such large vectors come at a cost of decreased transformation frequency. In some cases, the sequence carried by the P element can influence the transformation rate (SPRADLING 1986) or the insertion site specificity (KASSIS et al. 1992).

There are several options for providing transposase to the injected DNA. One way is to bind purified transposase protein to the element prior to injection (KAUFMAN and RIO 1992). However, the difficulty in obtaining transposase in sufficient quantities usually makes this method impractical. Alternatively, one can co-inject a transposase-making "helper" plasmid, preferably one that is unable to integrate into the chromosomes itself (KARESS and RUBIN 1984). A third approach is to inject directly into embryos that have an endogenous transposase source, such as the P[ry^+ Δ2–3](99B) element mentioned previously (ROBERTSON et al. 1988). The transposase-bearing chromosome can be marked with a dominant mutation, and stable transformants lacking the transposase gene are then selected among the progeny. This procedure is probably more efficient than co-injection, since it does not require the embryonic nuclei to take up two independent plasmids.

6.3 Enhancer Trapping

P-element mobility also provides a way to sample the genome for loci whose expression matches a particular pattern (O'KANE and GEHRING 1987). A *lacZ* reporter gene is fused to a weak promoter and mobilized within a P element to produce a collection of *Drosophila* lines, each with a single insertion of the "enhancer trap" element at a random site. The expression pattern of *lacZ* in each line tends to reflect the expression of nearby genes. Thus, one can identify genes that are active in specific tissues and developmental periods. The power of this technique increases rapidly with time, as large collections of enhancer-trap lines become available (eg., HARTENSTEIN and JAN 1992), thus eliminating the need for each worker to produce a new collection.

6.4 P Vectors for Transgene Expression

Several P-element vectors are available to facilitate expression of a given gene in a particular tissue through fusion of the gene to a specific promoter. For example, one set of vectors includes a promoter for strong expression in the developing egg and early embryo (SERANO et al. 1994). A particularly versatile system employs a two-element combination to allow a given gene to be expressed in any of a wide variety of patterns (BRAND and PERRIMON 1993). One element in this combination is similar to the enhancer-trap construct discussed above, except that the reporter gene is the yeast transcriptional activator, GAL4. The second element carries the gene of interest driven by a promoter containing GAL4-binding sites. The gene is then activated only in the cells where GAL4 is expressed. Thus, the expression of the gene of interest depends on the insertion site of the P[GAL4] element.

6.5 Vectors for Site-specific Recombination

Another yeast system that has proven useful in *Drosophila* is the FLP site-specific recombinase and its target site, FRT (GOLIC and LINDQUIST 1989). This system is particularly useful for generating mosaics. One P-element carries the FLP recombinase gene driven by a heat-shock promoter, and a second element has a gene with two FRT sites embedded. When heat shock is applied to such flies, FLP-mediated recombination causes somatic loss of the gene carrying FRT sites. More recently, FLP has been used to generate somatic mosaics with sectors homozygous for an entire chromosome arm (XU and RUBIN 1993). A homozygous P element near the base of a chromosome arm and bearing an FRT site undergoes mitotic recombination when FLP is expressed. The result is a somatic sector that is homozygous for all genes distal to the FRT-bearing P element. Such sectors can be identified by absence of a cell-autonomous marker on one of the homologs. This method allows identification and analysis of genes that are lethal when homozygous in the whole organism.

6.6 Gene Replacement

With P-element-mediated transformation, as described above, the researcher has no control over where in the genome the construct goes. In many instances, however, what is needed is to replace genes in situ. For example, some genes are too large to manipulate in vitro and return to the genome by transformation. Others are too sensitive to position effects. In addition, some genes have no null alleles to provide a suitable background to test transgenes.

To achieve gene replacement, *Drosophila* geneticists can make use of P-induced double-strand breaks (GLOOR et al. 1991). The method requires construction of an altered version of the gene which will be used as the template for gap repair. This construct must contain the sequences flanking the P insertion site. As discussed earlier, the gaps produced by P-element excision are usually repaired by copying in sequences from the sister strand (Fig. 3). However, in approximately 15% of the cases, the homolog or an ectopic sequence can provide the template (ENGELS et al. 1990). This method has been tested most extensively in the *white* gene, where hundreds of gene replacement events have been analyzed, but it has also been used successfully at *forked* (D. Lankenau, V. Corces, unpublished) and at least two other loci (PAPOULAS et al. 1994). The frequencies of gene replacement measured with *white* were dependent on the genomic position of the template, averaging 1% for autosomal sites (GLOOR et al. 1991) and 6% for X-linked sites (ENGELS et al. 1994). Extrachromosomal templates have also been used (G. Gloor, personal communication; BANGA and BOYD 1992; PAPOULAS et al. 1994). Insertions and deletions could be copied into the gap just as efficiently as single base-pair changes (JOHNSON-SCHLITZ and ENGELS 1993; NASSIF et al. 1994).

The primary limitation of this technique is its requirement for a P insertion close to the site being modified. A site within 8 bp of the P insertion in *white* was replaced in close to 100% of the gene replacement events, but one 2 kb away was replaced less than 10% of the time. To a good approximation, the replacement frequency of a site *n* bp away from the P insertion was 0.99855^n, expressed as a proportion of the gap-repair events (GLOOR et al. 1991). Therefore, the P-mutagenesis techniques discussed above, especially the PCR-based screens (BALLINGER and BENZER 1989; KAISER and GOODWIN 1990; SENTRY and KAISER 1994), are particularly valuable as preliminarily steps toward gene replacement.

7 Conclusions

P elements are relative newcomers in the *Drosophila melanogaster* genome, probably arriving through a horizontal transfer event less than 200 years ago. Their invasion of the genome was almost certainly harmful to the species, lowering the average fitness throughout natural populations. However, P elements have undoubtedly enhanced the fitness of *Drosophila* geneticists, forming the basis for a variety of techniques that have become essential to most current research with this species.

Acknowledgment. Christine R. Preston provided many comments and suggestions in preparation of this review.

References

Anxolabéhère D, Périquet G (1987) P-homologous sequences in Diptera are not restricted to the Drosophilidae family. Genet Iber 39: 211–222

Anxolabéhére D, Kai H, Nouaud D, Périquet G, Ronsseray S (1984) The geographical distribution of P-M hybrid dysgenesis in *Drosophila melanogaster*. Genet Sel Evol 16: 15–26

Anxolabéhère D, Nouaud D, Périquet G (1985a) Sequences homologues à l' élément P chez des espèces de *Drosophila* du groupe *obscura* et chez *Scaptomyza pallida* (Drosophilidae). Genet Sel Evol 17: 579–584

Anxolabéhère D, Nouaud D, Périquet G, Tchen P (1985b) P-element distribution in Eurasian populations of *Drosophila melanogaster*: a genetic and molecular analysis. Proc Natl Acad Sci USA 82: 5418–5422

Anxolabéhére D, Kidwell MG, Périquet G (1988) Molecular characteristics of diverse populations are consistent with a recent invasion of *Drosophila melanogaster* by mobile P-elements. Mol Biol Evol 5(3): 252–269

Ashburner M (1989a) *Drosophila*, a laboratory handbook. Cold Spring Harbor Press, Cold Spring Harbor, New York

Ashburner M (1989b) *Drosophila*, a laboratory manual. Cold Spring Harbor Press, Cold Spring Harbor, New York

Ballinger DG, Benzer S (1989) Targeted gene mutations in *Drosophila*. Proc Nat Acad Sci USA 86: 9402–9406

Banga SS, Boyd JB (1992) Oligonucleotide-directed site-specific mutagenesis of *Drosophila melanogaster*. Proc Natl Acad Sci USA 89: 1735–1739

Banga SS, Velazquez A, Boyd JB (1991) P transposition in *Drosophila* provides a new tool for analyzing postreplication repair and double-strand break repair. Mutat Res 255: 79–88

Berg CA, Spradling AC (1991) Studies on the rate and site-specificity of P-element transposition. Genetics 127: 515–524

Berg DE, Howe MM (1989) Mobile DNA. American Society of Microbiology, Washington DC

Bingham PM, Kidwell MG, Rubin GM (1982) The molecular basis of P-M hybrid dysgenesis: the role of the P-element, a P strain-specific transposon family. Cell 29: 995–1004

Black DM, Jackson MS, Kidwell MG, Dover GA (1987) KP elements repress P-induced hybrid dysgenesis in *D. melanogaster.* EMBO J 6: 4125–4135

Boussy IA, Healy MJ, Oakeshott JG, Kidwell MG (1988) Molecular analysis of the P-M gonadal dysgenesis cline in eastern Australian *Drosophila melanogaster.* Genetics 119: 889–902

Brand AH, Perrimon N (1993) Targeted gene expression as a means of altering cell fates and generating dominant phenotypes. Development 118: 401–415

Brookfield J (1991) Models of repression of transposition in P-M hybrid dysgenesis by P cytotype and by zygotically encoded repressor proteins. Genetics 128: 471–486

Chain AC, Zollman S, Tseng JC, Laski FA (1991) Identification of a *cis*-acting sequence required for germ line specific splicing of theP-element ORF2-ORF3 intron. Mol. Cell. Biol.11: 1538–1546

Charlesworth B, Langley CH (1989) The population genetics of *Drosophila* transposable elements. Annu Rev Genet 23: 251–287

Clark JB, Maddison WP, Kidwell MG (1994) Phylogenetic analysis supports horizontal transfer of P transposable elements. Mol Biol Evol 11: 40–50

Coen D (1990) P-element regulatory products enhance *zeste*^{1} repression of a P[*white*^{duplicated}] transgene in *Drosophila melanogaster.* Genetics 126: 949–960

Cooley L, Kelley R, Spradling A (1988) Insertional mutagenesis of the *Drosophila* genome with single P elements. Science 239: 1121–1128

Daniels S, Peterson K, Strausbaugh, L Kidwell M, Chovnick A (1990) Evidence for horizontal transmission of the P transposable element between *Drosophila* species. Genetics 124: 339–355

Daniels SB, Chovnick A (1993) P-element transposition in *Drosophila melanogaster*: an analysis of sister-chromatid pairs and the formation of intragenic secondary insertions during meiosis. Genetics 133: 623–636

Eggleston WB (1990) P-element transposition and excision in *Drosophila:* interactions between elements. PhD thesis, University of Wisconsin

Engels WR (1979a) The estimation of mutation rates when premeiotic events are involved. Environ Mutag 1: 37–43

Engels WR (1979b) Extrachromosomal control of mutability in *Drosophila melanogaster.* Proc Natl Acad Sci USA 76: 4011–4015

Engels WR (1979c) Germline aberrations associated with a case of hybrid dysgenesis in *Drosophila melanogaster* males. Genet Res Camb 33: 137–146

Engels WR (1979d) Hybrid dysgenesis in *Drosophila melanogaster*: rules of inheritance of female sterility. Genet Res Camb 33: 219–236

Engels WR (1981) Hybrid dysgenesis in *Drosophila* and the stochastic loss hypothesis. Cold Spring Harb Symp Quant Biol 45: 561–565

Engels WR (1989) P elements in *Drosophila.* In: Berg D, Howe M (eds) P elements in *Drosophila,* American Society of Microbiology, Washington DC, pp 437–484

Engels WR (1992) The origin of P elements in *Drosophila melanogaster.* Bio essays 14: 681–686

Engels WR, Preston CR (1979) hybrid dysgenesis in *Drosophila melanogaster*: the biology of male and female sterility. Genetics 92: 161–175

Engels WR, Preston CR (1981) Identifying P factors in *Drosophila* by means of chromosome breakage hotspots. Cell 26: 421–428

Engels WR, Preston CR (1984) Formation of chromosome rearrangements by P factors in *Drosophila.* Genetics 107: 657–678

Engels WR, Benz WK, Preston CR, Graham PL, Phillis RW, Robertson HM (1987) Somatic effects of P-element activity in *Drosophila melanogaster*: Pupal lethality. Genetics 117: 745–757

Engels WR, Johnson-Schlitz DM, Eggleston WB, Sved J (1990) High-frequency P-element loss in *Drosophila* is homolog-dependent. Cell 62: 515–525

Engels WR, Preston CR, Johnson-Schlitz DM (1994) Long-range *cis* preference in DNA homology search extending over the length of a *Drosophila* chromosome. Science 263: 1623–1625

Gloor GB, Nassif NA, Johnson-Schlitz DM, Preston CR, Engels WR (1991) Targeted gene replacement in *Drosophila* via P-element-induced gap repair. Science 253: 1110–1117

Gloor GB, Preston CR, Johnson-Schlitz DM, Nassif NA, Phillis RW, Benz WK, Robertson HM, Engels WR (1993) Type I repressors of P-element mobility. Genetics 135: 81–95
Golic K, Lindquist S (1989) The FLP recombinase of yeast catalyzes site-specific recombination in the *Drosophila* genome. Cell 59: 499–509
Golic KG (1994) Local transposition of P elements in *Drosophila melanogaster* and recombination between duplicated elements using a site-specific recombinase. Genetics 137: 551–563
Good AG, Meister GA, Brock HW, Grigliatti TA, Hickey DA (1989) Rapid spread of transposable P elements in experimental populations of *Drosophila melanogaster*. Genetics 122: 387–396
Green MM (1977) Genetic instability in *Drosophila melanogaster*: de novo induction of putative insertion mutations. Proc Natl Acad Sci USA 74: 3490–3493
Haenlin M, Steller H, Pirrotta V, Mohier E (1985) A 43 kilobase cosmid P transposon rescues the $fs(1)^{K10}$ morphogenetic locus and three adjacent *Drosophila* developmental mutants. Cell 40: 827–837
Hagemann S, Miller WJ, Pinsker W (1994) Two distinct P-element subfamilies in the genome of *Drosophila bifasciata*. Mol Gen Genet 244: 168–175
Handler AM, Gomez SP, O'Brochta DA (1993) Negative regulation of P-element excision by the somatic product and terminal sequences of P in *Drosophila melanogaster*. Mol Gen Genet 237: 145–151
Hartenstein V, Jan Y-N (1992) Studying *Drosophila* embryogenesis with P-*lacZ* enhancer trap lines. Rouxxs Arch Dev Biol 201: 194–220
Higuet D, Anxolabéhère D, Nouaud D (1992) A particular P-element insertion is correlated to the P-induced hybrid dysgenesis repression in *Drosophila melanogaster*. Genet Res Camb 60: 15–24
Hiraizumi Y (1979) A new method to distinguish between meiotic and premeiotic recombinational events in *Drosophila melanogaster*. Genetics 92: 543–554
Houck MA, Clark JB, Peterson KR, Kidwell MG (1991) Possible horizontal transfer of *Drosophila* genes by the mite Proctolaelaps regalis. Science 253: 1125–1128
Jackson MS, Black DM, Dover GA (1988) Amplification of KP elements associated with the repression of hybrid dysgenesis in *Drosophila melanogaster*. Genetics 120: 1003–1013
Johnson CW (1913) The distribution of some species of *Drosophila*. Psyche 20: 202–204
Johnson-Schlitz DM, Engels WR (1993) P-element-induced interallelic gene conversion of insertions and deletions in *Drosophila*. Mol Cell Biol 13: 7006–7018
Kaiser K, Goodwin S (1990) "Site-selected" transposon mutagenesis of *Drosophila*. Proc Natl Acad Sci USA 87: 1686–1690
Karess RE, Rubin GM (1984) Analysis of P transposable element functions in *Drosophila*. Cell 38: 135–146
Kaufman PD, Rio DC (1991) *Drosophila* P-element transposase is a transcriptional repressor in vitro. Proc Natl Acad Sci USA 88: 2613–2617
Kaufman PD, Rio DC (1992) P-element transposition in vitro proceeds by a cut-and-paste mechanism and uses GTP as a cofactor. Cell 69: 27–39
Kaufman PD, Doll RF, Rio DC (1989) *Drosophila* P-element transposase recognizes internal P-element DNA sequences. Cell 59: 359–371
Kassis JA, Noll E, vanSickle EP, Odenwald WF, Perrimon N (1992) Altering the insertional specificity of a *Drosophila* P-element transposable element. Proc Natl Acad Sci USA 89: 1919–1923
Kelley MR, Kidd S, Berg RL, Young MW (1987) Restriction of P-element insertions at the *Notch* locus of *Drosophila melanogaster*. Mol Cell Biol 7:1545–1548
Kidwell MG (1979) Hybrid dysgenesis in *Drosophila melanogaster*: The relationship between the P–M and I–R interaction systems. Genet Res Camb 33: 105–117
Kidwell MG (1983) Evolution of hybrid dysgenesis determinants in *Drosophila melanogaster*. Proc Natl Acad Sci USA 80: 1655–1659
Kidwell MG (1986) P–M mutagenesis, In: Roberts DB (ed) P–M mutagenesis, IRL Press, Oxford, pp 59–82
Kidwell MG (1993) Horizontal transfer of P elements and other short inverted repeat transposons. In: McDonald JF (ed) Horizontal transfer of P elements and other short inverted repeat transposons, Kluwer Academic, London
Kidwell MG, Novy JB (1979) Hybrid dysgenesis in *Drosophila melanogaster*: sterility resulting from gonadal dysgenesis in the P–M system. Genetics 92: 1127–1140
Kidwell MG, Kidwell JF, Sved JA (1977) Hybrid dysgenesis in *Drosophila melanogaster*: syndrome of aberrant traits including mutation, sterility, and male recombination. Genetics 86: 813–833
Kidwell MG, Frydryk I, Novy JB (1983) The hybrid dysgenesis potential of *Drosophila melanogaster* strains of diverse temporal and geographical natural origins. Dros Inform Serv 59: 63–69

Kiyasu PK, Kidwell MG (1984) Hybrid dysgenesis in *Drosophila melanogaster*: the evolution of mixed P and M population maintainted at high temperature. Genet Res Camb 44: 251–259

Lachaise D, Cariou ML, David JR, Lemeunier F,Tsacas L, Ashburner M (1988) Historical biogeography of the *Drosophila melanogaster* species subgroup. Evol Biol 22: 159–225

Lansman RA, Stacey SN, Grigliatti TA, Brock HW (1985) Sequences homologous to the P mobile element of *Drosophila melanogaster* are widely distributed in the subgenus *Sophophora*. Nature 318: 561–563

Lansman RA, Shade RO, Grigliatti TA, Brock HW (1987) Evolution of P transposable elements: Sequences of *Drosophila nebulosa* P-elements, Proc Natl Acad Sci USA 84: 6491–6495

Laski FA, Rio DC, Rubin GM (1986) Tissue specificity of *Drosophila* P-element transposition is regulated at the level of mRNA splicing. Cell 44: 7–19

Lemaitre B, Coen D (1991) P regulatory products repress in vivo the P promoter activity in P–*lacZ* fusion genes. Proc Natl Acad Sci USA 88: 4419–4423

Lemaitre, B, Ronsseray S, Coen D (1993) Maternal repression of the P-element promoter in the germline of *Drosophila melanogaster*: a model for the P cytotype. Genetics 135: 149–160

Lindsley DL, Zimm GG (1992) The Genome of *Drosophila melanogaster*, 1. Academic, San Diego

McDonald JF (1993) Transposable elements and evolution. Kluwer Academic, London

Meister GA, Grialiatti TA (1993) Rapid spread of a P-element/Adh gene construct through experimental populations of *Drosophila melanogaster*. Genome 36: 1169–1175

Miller WJ, Hagemann S, Reiter E, Prinsker W (1992) P-element–homologous sequences are tandemly repeated in the genome of *Drosophila guanche*. Proc Natl Acad Sci USA 89: 4018–4022

Misra S, Rio DC (1990) Cytotype control of *Drosophila* P-element transposition: the 66–kD protein is a repressor of transposase activity. Cell 62: 269–284

Misra S, Buratowski RM, Ohkawa T, Rio DC (1993) Cytotype control of *Drosophila melanogaster* P-element transposition: genomic position determines maternal repression. Genetics 135: 785–800

Mullins MC, Rio DC, Rubin GM (1989) *Cis*-acting DNA sequence requirements for P-element transposition. Genes Dev 3: 729–738

Nassif NA, Engels WR (1993) DNA homology requirements for mitotic gap repair in *Drosophila*. Proc Natl Acad Sci USA 90: 1262–1266

Nassif NA, Penney J, Pal S, Engels WR, Gloor GB (1994) Efficient copying of nonhomologous sequences from ectopic sites via P-element-induced gap repair. Mol Cell Biol 14: 1613–1625

Niki Y (1986) Germline autonomous sterility of P-M dysgenic hybrids and their application to germline transfers in *Drosophila melanogaster*. Dev Biol 113: 255–258

Niki Y, Chigusa SI (1986) Developmental analysis of the gonadal sterility of P-M hybrid dysgenesis in *Drosophila melanogaster*. Jpn J Genet 61: 147–156

O'Brochta DA, Gomez SP, Handler AM (1991) P-element excision in *Drosophila melanogaster* and related drosophilids. Mol Gen Genet 225: 387–394

O'Hare K, Rubin GM (1983) Structure of P transposable elements and their sites of insertion and excision in the *Drosophila melanogaster* genome. Cell 34: 25–35

O'Hare K, Driver A, McGrath S, Johnson-Schlitz DM (1992) Distribution and structure of cloned P elements from the *Drosophila melanogaster* P strain π_2. Genet Res Camb 60: 33–41

O'Kane CJ, Gehring WJ (1987) Detection in situ of genomic regulatory elements in *Drosophila*. Proc Natl Acad Sci USA 84: 9123–9127

Ochman H, Gerber AS, Hart DL (1988) Genetic applications of an inverse polymerase chain reaction. Genetics 120: 621–623

Papoulas O, McCall K, Bender W (1994) Targeted gene conversion at the birthorax complex, 35th annual *drosophila* research conference, Chicago, p 218

Paques F, Wegnez M (1993) Deletions and amplifications of tandemly arranged ribosomal 5S genes internal to a P-element occur at a high rate in a dysgenic context. Genetics 135: 469–476

Paricio N, Perez-Alonso M, Martinez-Sebastian MJ, de Frutos R (1991) P sequences of *Drosophila subobscura* lack exon 3 and may encode a 66 kd repressor-like protein. Nucleic Acids Res 19: 6713–6718

Perkins HD, Howells AJ (1992) Genomic sequences with homology to the P-element of *Drosophila melanogaster* occur in the blowfly *Lucilia cuprina*. Proc Natl Acad Sci USA 89: 10753–10757

Pirrotta V (1988) Vectors for P-mediated transformation in *Drosophila*. Biotechnology 10: 437–456

Preston CR, Engels WR (1989) Spread of P transposable elements in inbred lines of *Drosophila melanogaster*. In: Cohn W, K Moldave (eds) Spread of P transposable elements in inbred lines of *Drosophila melanogaster*, (vol. 36) Academic, San Diego, pp 71–85

Rasmusson KE, Raymond JD, Simmons MJ (1993) Repression of hybrid dysgenesis in *Drosophila melanogaster* by individual naturally occurring P-elements. Genetics. 133: 605–622
Raymond JD, Ojala TA, White J, Simmons MJ (1991) Inheritance of P-element regulation in *Drosophila melanogaster*. Genet Res 57: 227–234
Rio DC (1990) Molecular mechanisms regulating *Drosophila* P-element transposition. Annu Rev Genet 24: 543–578
Rio DC, Rubin GM (1988) Identification and purification of a *Drosophila* protein that binds to the terminal 31-base–pair repeats of the P transposable element. Proc Natl Acad Sci USA 85: 8929–8933
Rio DC, Laski FA, Rubin GM (1986) Identification and immunochemical analysis of biologically active *Drosophila* P-element transposase. Cell 44: 21–32
Robertson HM (1993) The mariner tarnsposable element is widespread in insects (see comments). Nature 362: 241–245
Robertson HM, Engels WR (1989) Modified P elements that mimic the P cytotype in *Drosophila melanogaster*. Genetics 123: 815–823
Robertson HM, Preston CR, Phillis RW, Johnson-Schlitz D, Benz WK, Engels WR (1988) A stable genomic source of P-element transposase in *Drosophila melanogaster*. Genetics 118: 461-470
Roiha H, Rubin GM, O'Hare K (1988) P-element insertions and rearrangements at the *singed* locus of *Drosophila melanogaster*. Genetics 119: 75–83
Ronsseray S, Lehmann M, Anxolabéhère D (1991) The maternally inherited regulation of P elements in *Drosophila melanogaster* can be elicited by two P copies at cytological site 1A on the X chromosome. Genetics 129: 501–512
Ronsseray S, Lemaitre B, Coen D (1993) Maternal inheritance of P cytotype in *Drosophila melanogaster*: "pre- P cytotype" is strictly extra-chromosomally transmitted. Mol Gen Genet 241: 115–123
Rubin GM, Spradling AC (1982) Genetic transformation of *Drosophila* with transposable element vectors. Science 218: 348–353
Rubin GM, Kidwell MG, Bingham PM (1982) The molecular basis of P–M hybrid dysgenesis: The nature of induced mutations. Cell 29: 987–994
Salz HK, Cline TW, Schedl P (1987) Functional changes associated with structural alterations induced by mobilization of a P-element inserted in the *Sex-lethal* gene of *Drosophila*. Genetics 117: 221–231
Searles LL, Jokerst RS, Bingham PM, Voelker RA, Greenleaf AL (1982) Molecular cloning of sequences from a *Drosophila RNA polymerase II* locus by P-element transposon tagging. Cell 31: 585–592
Sentry JW, Kaiser K (1994) Application of inverse PCR to site-selected mutagenesis of *Drosophila*. Nucleic Acids Res 22: 3429–3430
Serano TL, Cheung HK, Frank LH, Cohen RS (1994) P-element transformation vectors for studying *Drosophila melanogaster* oogenesis and early embryogenesis. Gene 138: 181–186
Siebel CW, Rio DC (1990) Regulated splicing of the *Drosophila* P transposable element third intron in vitro: somatic repression. Science 248: 1200–1208
Simmons MJ, Raymond JD, Johnson N, Fahey T (1984) A comparison of mutation rates for specific loci and chromosome regions in dysgenic hybrid males of *Drosophila melanogaster*. Genetics 106: 85–94
Simonelig M, Anxolabéhère D (1991) A P-element of *Scaptomyza pallida* is active in *Drosophila melanogaster*. Proc Natl Acad Sci USA 88: 6102–6106
Spicer GS (1988) Molecular evolution among some *Drosophila* species groups as indicated by two-dimensional electrophoresis. J Mol Evol 27: 250–260
Spradling AC (1986) P-element-mediated transformation. In: Roberts DB (ed) P-element-mediated transformation. IRL Press, Oxford, pp. 175–197
Spradling AC, Rubin GM (1982) Transposition of cloned P elements into *Drosophila* germ line chromosomes. Science 218: 341–347
Staveley BE, Hodgetts RB, O'Keefe SL, Bell JB (1994) Targeting of an enhancer trap to vestigial. Dev Biol 165: 290–293
Sturtevant AH (1921) The North American species of *Drosophila*. Carnegie Institute Washington publication, vol 301
Sved JA, Eggleston WB, Engels WR (1990) Germline and somatic recombination induced by in vitro modified P elements in *Drosophila melanogaster*. Genetics 124: 331–337
Sved JA, Blackman LM, Gilchrist AS, Engels WR (1991) High levels of recombination induced by homologous P elements *Drosophila melanogaster*. Mol Gen Genet 225: 443–447
Syvanen M (1984) The evolutionary implications of mobile genetic elements. Annu Rev Genet 18: 271–293

Takasu-Ishikawa E, Yoshihara M, Hotta Y (1992) Extra sequences found at P-element excision sites in *Drosophila melanogaster.* Mol Gen Genet 232: 17–23

Tower J, Karpen GH, Craig N, Spradling AC (1993) Preferential tarnsposition in *Drosophila* P elements to nearby chromosomal sites. Genetics 133: 347–359

Tseng JC, Zollman S, Chain AC, Laski FA (1991) Splicing of the *Drosophila* P-element ORF2–ORF3 intron is inhibited in a human cell extract. Mech Dev 35: 65–72

Tsubota S, Schedl P (1986) Hybrid dysgenesis-induced revertants of insertions at the 5' end of the *rudimentary* gene in *Drosophila melanogaster*: transposon-induced control mutations. Genetics 114: 165–182

Tsubota S, Ashburner M, Schedl P (1985) P-element-induced control mutations at the *r* gene of *Drosophila melanogaster.* Mol Cell Biol 5: 2567–2574

Wei G, Oliver B, Mahowald AP (1991) Gonadal dysgenesis reveals sexual dimorphism in the embryonic germline of *Drosophila* [published erratum appears in Genetics (1992) 130 (1): 235] Genetics 129: 203–210

Williams JA, Bell JB (1988) Molecular organization of the vestigial region in *Drosophila melanogaster.* EMBO J 7: 1355–1363

Williams JA, Pappu SS, Bell JB (1988) Suppressible P-element alleles of the *vestigial* locus in *Drosophila melanogaster.* Mol Gen Genet 212: 370–374

Xu T, Rubin GM (1993) Analysis of genetic mosaics in developing and adult *Drosophila* tissues. Development 117: 1223–1237

The Tc1/*mariner* Transposon Family

R.H.A. Plasterk

1 Introduction

In many animals the main cause of mutations is transposon insertion. This is true, e.g., for strains of the nematode species *C. elegans*. It is not true for humans, where only relatively few cases have been reported of germline mutations caused by new transposon insertions, and where base-pair substitutions, frameshifts, and errors in replication of nucleotide repeats are more common (Dombroski et al. 1991, 1993; Caskey et al. 1992). *Caenorhabditis elegans* is a free-living nematode that can be found in the soil anywhere in the world. All *C. elegans* strains analyzed to date contain several copies of the transposable element Tc1 (*T*ransposon *C. elegans* number *1*) (Emmons et al. 1980, 1983; Liao

Netherlands Cancer Institute, Division of Molecular Biology, Plesmanlaan 121, NL-1066 CX Amsterdam, The Netherlands

et al. 1983). Insertion of Tc1 is the main cause of gene inactivation in the strain Bergerac (MOERMAN and WATERSTON 1984; EIDE and ANDERSON 1985). Since discovery of the Tc1 element, related elements have been found in the same species, and elements discovered in other species were also found to be homologous to Tc1. The best-described example is the *mariner* element, discovered in *Drosophila mauritiana* (JACOBSON et al. 1986; HARTL 1989). At present it seems that members of the Tc1/*mariner* transposon family are found in virtually all animal phyla: vertebrates (HENIKOFF 1992; HEIERHORST et al. 1992; GOODIER and DAVIDSON 1994; RADICE et al. 1994), nematodes (COLLINS et al. 1989; ABAD et al. 1991; PRASAD et al. 1991; SEDENSKY et al. 1994), arthropods (ROBERTSON 1993, 1994; HARRIS et al. 1988; HENIKOFF and PLASTERK 1988; BRIERLEY and POTTER 1985; HARTL 1989; HARRIS et al. 1990; GARCIA-FERNANDEZ et al. 1993; ROBERTSON et al. 1992; CAIZZI et al. 1993; FRANZ and SAVAKIS 1991; BIGOT et al. 1994; FRANZ et al. 1994; BREZINSKY et al. 1990), planarians (CAPY et al. 1994), ciliates (TAUSTA and KLOBUTCHER 1989; WILLIAMS et al. 1993), and even fungi (DABOUSSI et al. 1992). A recent alignment of the elements is in ROBERTSON (1995).

This review will focus primarily on the mechanism of transposition of these elements. Other aspects will not be covered at length, and readers are referred to the reviews mentioned: very good reviews of the knowledge on Tc1 accumulated up to 1989 are in MOERMAN and WATERSTON (1989) and HERMAN and SHAW (1987); evolutionary considerations are described in KIDWELL (1993), GARCIA-FERNANDEZ et al. (1993), HURST et al. (1992), MARUYAMA and HARTL (1993) and RADICE et al. (1994); the discovery of Tc1 is described in ANDERSON et al. (1992); the use of Tc1 for reverse genetics in PLASTERK (1992); and the *mariner* element is reviewed in HARTL (1989). Other transposons of *C. elegans* (such as the Tc2, Tc4, and Tc5 elements that also have long terminal inverted repeats), are beyond the scope of this review (LEVITT and EMMONS 1989; YUAN et al. 1991; RUVOLVO et al. 1992; LI and SHAW 1993; COLLINS and ANDERSON 1994).

2 The Structure of Tc1-like Elements: Inverted Repeats

In the remainder of this chapter I will focus predominantly on the two related members of the Tc1/*mariner* transposon family that have been studied in most detail: Tc1 and Tc3 of *C. elegans*. The structures of Tc1 and Tc3 are shown in Fig. 1 (ROSENZWEIG et al. 1983; COLLINS et al. 1989). The elements have inverted repeats that are almost perfect and contain one gene. The correct exon-intron boundaries of Tc1A were not immediately apparent from the sequence (ROSENZWEIG et al. 1983; ABAD et al. 1993; SCHUKKINK and PLASTERK 1990) and could not be determined experimentally due to the relatively high transposon copy number, but with the aid of alignment with a *C. briggsae* homolog (PRASAD et al. 1991) it was possible to make a guess that was confirmed experimentally (VOS et al. 1993; VAN LUENEN et al. 1993). Note that the names of the encoded

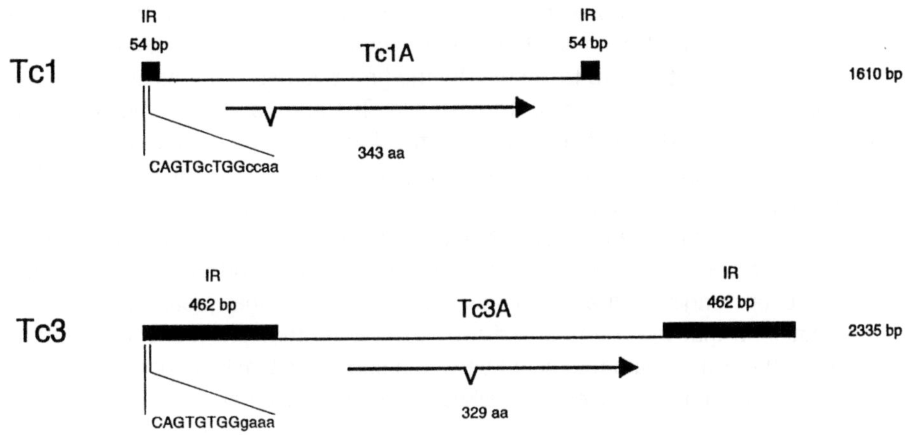

Fig. 1. Structure of the Tc1 and Tc3 elements. *Black blocks* indicate the terminal inverted repeats, *arrows* the intron-exon structure of the *Tc1A* and *Tc3A* genes

transposases follow the nomenclature for transposases of other transposable elements, rather than the nomenclature for genes in the nematode *C. elegans*.

Like several other transposable elements and repetitive DNA sequences, the Tc elements have virtually perfect inverted repeats. The function of these inverted repeats is not clear. If they had no function, one would expect the sequences to have drifted apart in evolution. One explanation may be that, somehow, the left end knows what the right end does, and that at some step in the element's life cycle base pairing is needed between the two ends. The success of the transposon replication might then depend on the capacity for (almost) perfect base pairing between these sequences, which would provide the selective pressure to keep them identical. Given what is known about the mechanism of transposition of other transposable elements, and of Tc1/Tc3 (see below), base pairing between strands from opposite transposon ends seems rather exotic.

A second possible explanation also depends on probing of similarity by base pairing, but now in mRNA. The explanation is suggested by the observation made by RUSHFORTH et al. (1993) that in some cases Tc1 sequences can be reasonably efficiently removed from exons in mRNA by aberrant splicing. They noticed that insertions in exons of the *myo-3* gene had no apparent phenotype. When they subsequently isolateo deletion derivatives of these insertion alleles they found that these had a clear phenotype, and they had to conclude that the insertions had not inactivated gene function. They were able to demonstrate the presence of an encoded protein in these insertion mutant animals. Amplification and sequence analysis of the mRNA revealed that the transposon is spliced out in an imprecise fashion. The splice sites do not resemble the consensus sequence (T.E. Blumenthal, personal communication). An explanation of this unusual splicing phenomenon may be that the inverted repeats of Tc1 snap back in RNA and thus force the splice apparatus to excise the transposon, using the best splice sequence around. This would result in a strong selective pressure: like any successful parasite, Tc1 would do well to minimize its damage to the host; this could be achieved by removal of Tc1 sequences from mRNAs of host genes.

A question that is hard to answer by the first two explanations above is why the terminal inverted repeats of, for instance, Tc1 and Tc3 are of such different length (54 versus 462 bp). No explanation that involves a function for the inverted repeats can easily account for these length differences: what would be the selective pressure to keep 462 bp identical for Tc3 if 54 bp suffice for Tc1?

There is a third explanation, i.e., that the virtually perfect inverted repeats serve no function for either transposon replication or host survival and are mere witnesses to the recent origin of these transposable elements. What pleads against this explanation is that the members of the Tc1/*mariner* family of transposons have terminal inverted repeats that share very little sequence similarity between family members but are nevertheless virtually perfect within each element.

3 The Transposase Proteins of Tc1 and Tc3

Figure 1 shows the Tc1A and Tc3A genes. The coding sequences are interrupted by small introns, at different positions in the two transposase genes. That these genes encode transposases is concluded primarily from the following experiment: the Bristol N2 strain contains approximately 15 copies of the Tc3 element (Collins et al. 1989), but no transposition of Tc3 has been observed in the germline of Bristol N2 (Collins et al. 1989) or in the soma (Van Luenen et al. 1993). However, when Tc3 transposase expression is induced with a heat-shock promoter in a transgenic strain, then the endogenous Tc3 elements move (Van Luenen et al. 1993). This leads to two conclusions: Tc3A is needed for Tc3 transposition, and it is probably the only factor lacking in the Bristol strain. Does TcA mobilize Tc elements through direct action, or does it induce activation of some other factor? Strong support for a direct role of TcA comes from the observation that Tc1A and Tc3A bind specifically to regions within the transposon close to transposon DNA ends (see below).

A second argument for a direct role in catalysis comes from the alignment of TcA to transposases of other elements. The integrases of retroelements and transposases of bacterial IS elements show sequence similarity, most notably in a motif referred to as the DD(35) E motif (Kulkosky et al. 1992; Fayet et al. 1990). Mutational analysis has shown that these residues are essential for transposase function. Recently, Doak et al. (1994) published a weak but potentially significant alignment between these transposases and those of the Tc1/*mariner* family. To test this hypothesis, Van Luenen et al. (1994) recently made point mutations in these DDE residues and in some other D and E positions in Tc3A and tested their action in vivo. They found that mutations in the DDE motif, but not in two other positions, completely inactivated the ability of Tc3A to mediate Tc3 transposition. This does not prove the significance of the observed similarity with retrovival integrases and IS transposases, but it at least supports it. Obviously, these experiments furthur argue for the fact that TcA is the protein directly mediating transposition, and for all practical purposes we may consider TcA the transposase protein.

To mediate transposition, TcA would need to bind transposon ends and probably also target DNA. Analysis of truncated transposases produced in *Escherichia coli* reveals that Tc1A and Tc3A each contain two DNA-binding domains: an N-terminal sequence-specific DNA-binding domain (amino acids 1–69 for Tc1A and 1–65 for Tc3A) and a second general DNA-binding domain (amino acids 71–207 for Tc1A and 98–192 for Tc3A). The general DNA-binding domain was recognized by South Western blotting (SCHUKKINK and PLASTERK 1990; VOS et al. 1993). It overlaps with the area of the DD(35)E motif described above (see Fig. 2). The binding of the N-terminal transposase domains to transposon DNA termini has been studied in some detail: bandshift, DNase-I footprinting, and methylation interference experiments reveal that the transposase can bind to a region of approximately 20 bp close to the transposon termini. Tc3 transposase specifically binds a second sequence in each terminal inverted repeat (see Fig. 3) (COLLOMS et al. 1994). Interestingly, the terminal 4 nucleotides, the only ones conserved between Tc1 and Tc3 ends and largely conserved in the whole

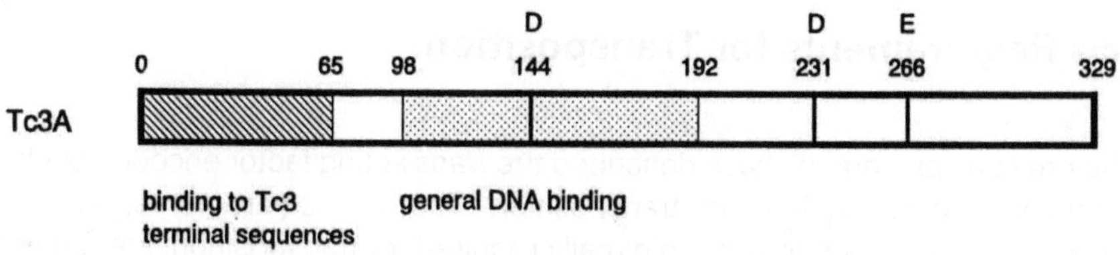

Fig. 2. Linear representation of the Tc3A protein. As described in the text, *shaded areas* represent DNA-binding domains. The N-terminal 65 amino acids can bind specifically to the termini of the transposon; the general DNA-binding domain shows no specificity in its DNA binding. The two aspartic acids and one glutamic acid of the "*DDE*" motif are indicated

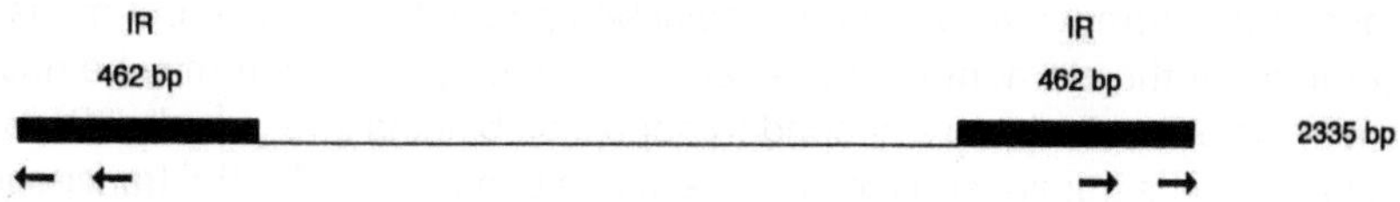


```
Tc3 end          1 TACAGTGTGGGAAAGTTCTATAGGACCCCCCCTAATTTGAAG 44
                   : :::::::  ::: :::::::::::::::::  :   :::
Tc3 internal   176 TCCAGTGTG..AAACTTCTATAGGACCCCCCGTTTTTTTTCA 217
```

Fig. 3. Tc3A-binding sites in the Tc3 DNA. *Arrows* indicate the binding sites for Tc3A. The two sites are shown aligned, with *dots* indicating identity. It is clear that the two binding sites are of similar sequence

Tc1/*mariner* family, are not necessary for specific binding and are not completely protected against DNAse I (Vos and Plasterk 1994). As might be expected on the basis of the dissimilarity between Tc1 and Tc3 transposon ends, the Tc1A transposase does not bind Tc3 DNA, and vice versa. This provides sufficient explanation why Tc1A overproduction in vivo does not mobilize Tc3 elements, and vice versa (Vos et al. 1993; Van Luenen et al.1993).

The properties of purified recombinant Tc1A transposase have recently been studied in some detail (Vos and Plasterk 1994). Interaction of different deletion derivatives of Tc1A with different mutant transposon end sequences revealed that the N-terminal DNA-binding domain is bipartite. The region of amino acids 1–68 of Tc1A specifically binds to a region 12–26 bp from the Tc1 ends, whereas additional specificity for recognition of base pairs 8–13 determined by the Tc1A region between amino acids 68 and 142. The purified transposase exhibits endonuclease activity and can cut a Tc1 transposon 5' end precisely at the position where it is thought to cut in vivo. Double-strand DNA cutting has not yet been observed, which may indicate that other factors are required for the complete excision reaction. A point mutation in the DDE motif abolishes the endonucleolytic activity.

4 *cis* Requirements for Transposition

In the previous paragraph I have described the *trans*-acting factor encoded by the transposon that is required for transposition. The *cis* requirements, the sequences at the transposon ends minimally required for transposition, are not yet fully defined. To date, the only analysis has been done for the Tc3 element (Van Luenen et al., in preparation).

Deletion analysis of Tc3 transposon ends has not been carried to the limits, but it is already clear that the 462 bp terminal inverted repeat sequences are, for the most part, not necessary for transposition. Deletions from the inside of the transposon that remove all but the terminal 94 bp result in reduced but measurable jumping of the element (Van Luenen et al., in preparation). It may be noted that these deletions remove a second transposase binding site virtually identical in sequence to the transposon binding site at positions 5–25 from the transposon terminus (Fig. 3). This second transposase binding site maps approximately 180 bp from the transposon end.

Whereas the transposase binding sites are relatively nonconserved between several members of the Tc1/*mariner* transposon family, the terminal 4 bp are strongly conserved (Henikoff 1992; Radice et al. 1994). All elements integrate into the target sequence TA, which is duplicated upon integration (see below). Therefore, all of these elements are flanked by TA sequences, and one can ask to what extent these are essential for subsequent excision and furthur transposition of the element. To address this question, VAN LUENEN et al. (1994) recently mutated the flanking TA sequences of Tc3 into GC sequences. They found that this did not affect transposition of the element: the element still excised and transposed at apparently normal frequencies and reintegrated into the target sequence TA.

5 Excision of Tc Elements

Induction of Tc3 transposition is accompanied by the appearance of extrachromosomal linear copies of Tc3 DNA (VAN LUENEN et al. 1993). Linear extrachromosomal Tc1 has also been observed (RUAN and EMMONS 1984). These are likely intermediates in the transposition reaction. The structure of the ends of the DNA of excised Tc3 has recently been determined (VAN LUENEN et al. 1994). It is shown in Fig. 4. The structure of the excised element indicates where these double-strand breaks occur: at the 3' ends of the transposon DNA a nick is made precisely between the transposon and the flanking TA sequence, leaving a 3'OH at the transposon ends, and at the 5' ends the cuts are made two nucleotides inside of the transposon end, with the result that two nucleotides of each transposon end are left at the donor site. In agreement with this, it has been found that the most common footprints (MOERMAN and WATERSTON 1989), presumably resulting from repair of the double-strand break left after Tc excision, are of the sequence TACATA; these can now easily be explained by the structure of the donor site after transposon excision. Similar footprints have been seen for the *mariner* element, where the preferred footprints seems to be TACCATA (BRYAN et al. 1990). This would suggest that the excision of *mariner* proceeds with a three nucleotide stagger. No information is available on the structure of the ends of excised linear *mariner* DNA to confirm this hypothesis.

Although circular excised versions of Tc1 have been detected (RUAN and EMMONS 1984; ROSE and SNUTCH 1984; RADICE and EMMONS 1993), it is likely (though not certain) that these are by products of the transposition reaction rather than intermediates.

The biochemical evidence for double-strand breaks initiating Tc transposition is in agreement with previous genetic experiments showing that transposon loss is associated with double-strand break repair at the donor locus (PLASTERK 1991; PLASTERK and GROENEN 1992). Precise loss of the Tc1 transposon depends upon the presence of a wild-type sequence on the homologous chromosome, as has been found for the P element of *Drosophila* (ENGLES et al. 1990). Homozygous transposon alleles therefore revert only infrequently, and if they revert, the reversion is hardly ever precise, but instead shows characteristic transposon "footprints" (see Fig. 5). Apparently the repair process is not always precise, and a genetic selection that recognizes the rare cases of imperfect DSB repair will visualize

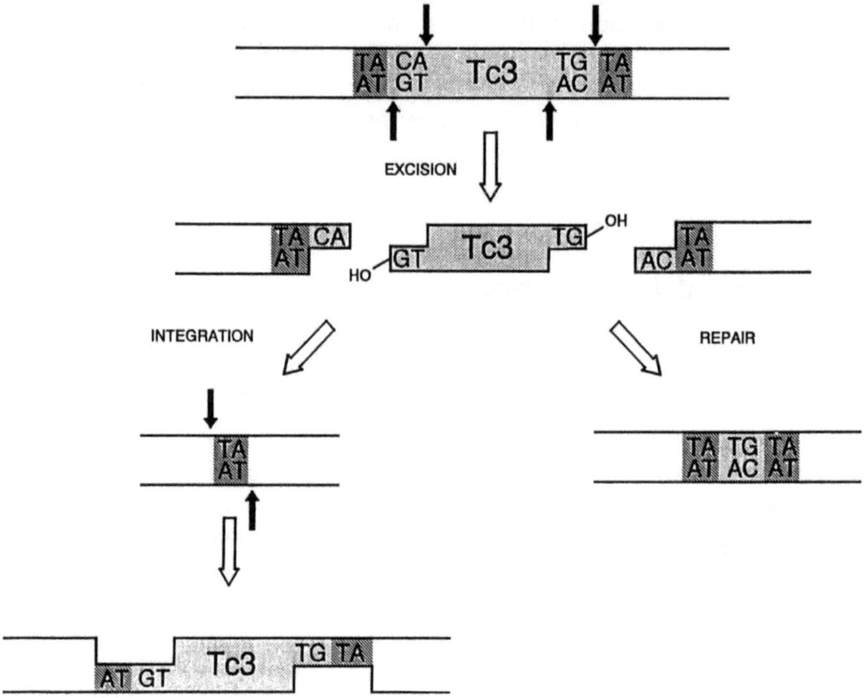

Fig. 4. Model for Tc3/Tc1 transposition. It is based primarily on experiments with Tc3 but probably applies to Tc1 and to the other members of the Tc1/*mariner* family. The TA base pairs flanking the element are *shaded dark*, the element itself is *shaded*, and its terminal base pairs are *drawn*. *Black arrows* are DNA breaks. The excised transposon can reintegrate, presumably by single nucleophilic attacks on phosphodiester bonds flanking TA base pairs. The resulting integrated element is flanked by four nucleotide single-strand gaps that can be repaired subsequently. One common possibility for repair of the broken donor DNA is drawn here, but others are also found. In this case, two nucleotides from one of the 3' protruding ends are removed, the ends are joined, and gaps and breaks are repaired. See text and literature for more detailed discussion. (From VAN LUENEN et al. 1994)

such footprints. It was shown (PLASTERK and GROENEN 1992) that also an ectopic transgenic template could be used for DSB repair. Note that this provides a way of introducing specific point mutations into the *C. elegans* genome.

Interestingly, the most common Tc1 footprint, an insertion of 4 bp compared with the sequence before Tc1 integration, can be part of a sequence with reasonable resemblance to the consensus 5' splice site in *C. elegans* (CARR and ANDERSON 1994). One may speculate that a Tc1 "visitation" thus helps the host organism to new splicing possibilities. The long-term selective advantage for the host is impossible to evaluate; it would be interesting to find natural pedigrees of strains without a transposon at a locus and strains that have undergone a Tc1 visitation and gained a new functional 5' splice site.

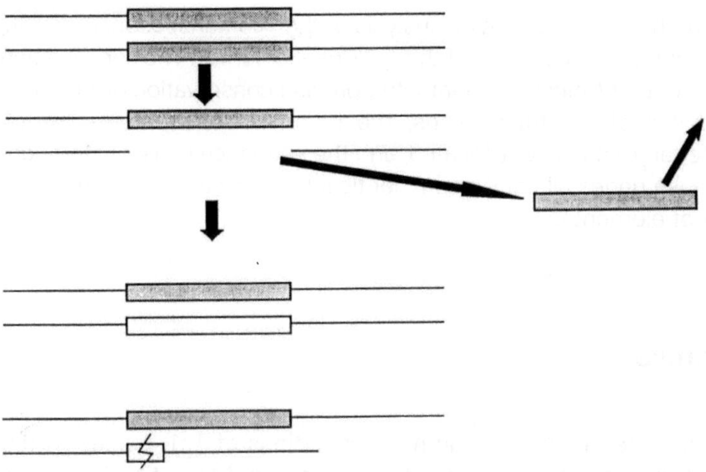

Fig. 5. Double-strand break repair following Tc1 excision. The two homologs of a dipoid cell are shown, each carrying the same Tc1 allele. One of the Tc1 elements excises, and can reintegrate somewhere else in the genome. The donor site is left with a double-strand break. One of the possibilities for repair is templated break repair, using the homologous chromosome as template. If that occurs, a new copy of the transposon is copied into the broken chromosome (*Open box*). In rare cases the repair may be imprecise, resulting in a "footprint" (*Small box*). If one selects for loss of the transposon, then all cases of complete repair will go unnoticed and only (some of) the footprints will be found. For further discussion see the text and the literature mentioned

6 A Model for Tc Transposition

The Tc1 element, and all family members analyzed to date, integrate into the sequence TA (first noticed by ROSENZWEIG et al. (1983)). Initially it was not clear whether the TA sequences were duplicated upon integration; an alternative explanation is that the inverted repeats of the Tc1 transposon have A and T at the terminal positions, so that a blunt cut at the target TA base pairs, followed by transposon insertion, would result in TA sequences at each transposon end. The experiment mentioned above, wherein transposition of a donor transposon flanked by GC sequences was monitored, disproved this alternative explanation and therefore leads to the conclusion that the TA sequences at the transposon ends are indeed the result of target duplication. This is in agreement with the structure of the excised Tc3 element (VAN LUENEN et al. 1994). Figure 4 shows a model for Tc3 transposition that incorporates the features described above. The element is excised by staggered double-strand breaks at the transposon ends, the excised transposon DNA with attached proteins finds a new target, and integration occurs through nucleophilic attacks of the 3' OH groups onto staggered phosphodiester bonds flanking the target TA base pairs. The primary product of this strand transfer has four nucleotide single-strand regions at each transposon end, which are thought to be repaired by host enzymes.

It is likely that this model, based on analysis of the Tc3 transposon, is equally applicable to Tc1 and other members of the Tc1/*mariner* family. The conservation of the overall structure of these elements, the precise conservation of the most terminal nucleotides of the transposons, the common requirements for TA sequences at the target that are duplicated, and the sequence similarity throughout the transposase genes all point to an identical transposition mechanism for this large family of elements.

7 Target Choice

The analysis of a limited number of germline insertions of Tc1 has led to the proposition of a consensus sequence for Tc1 integration (Mori et al. 1988; Eide and Anderson 1988). Our recent analysis (Van Luenen and Plasterk 1994) of the target choice within a small (approximately 1000 bp) area of the genome has lent further support to the notion that not all TA sequences are equally hot targets. One TA sequence may be hit 20 times more often than another that is only a few base pairs removed. Hot and cold sites do not seem to be regionally clustered or regularly spaced. The target choice of the related elements Tc1 and Tc3 is different, but the target choice seems to be the same for both orientations of the transposon. Alignment of the hot sites and the cold sites does not lead to a clear consensus sequence that has obvious predictive value, but, on the other hand, it seems likely that the primary sequence of the positions immediately flanking the target TA are the major determinants of target choice. Mutational analysis of hot sites and cold sites may be required to define what makes a hot site hot. A possible explanation for both the absolute requirement for a TA at the integration site and further preference for a subset of all potential TA targets is that the transposase needs to recognize TA target sequences for strand transfer and also interacts with some flanking base pairs.

8 Regulation of Tc/*mariner* Transposition

The regulation of Tc1/*mariner* transposition has not been studied extensively. Nevertheless, the data obtained in different experimental systems point to a common regulation of transposition.

All natural isolates of *C. elegans* contain several copies of the Tc1 and Tc3 transposon. In several strains, such as the Bristol N2 strain, no jumping of these elements in the germline can be detected, although somatic excision of Tc1 is frequent (Emmons and Yesner 1984; Emmons et al. 1986), as is somatic jumping (Vos et al. 1993). In other natural isolates, however, transposition of Tc1 occurs frequently in the germline (Moerman and Waterston 1984; Eide and Anderson 1985). The difference in transposition frequency is not explained simply by the copy numbers of Tc elements, since there is no correlation between Tc1 copy number and transpositional activity (Moerman and Waterston 1989). Transposition proficiency can be attributed to genetic factors, so-called mutators or *mut* genes. Some of these were mapped genetically (Mori et al. 1988, 1990); for example, *mut-4, mut-5,* and *mut-6* were found to be mobile. It has been suggested that these genes may represent copies of the Tc1 element itself that are transcribed at the right moment to produce transposase that can activate Tc1 jumping in the germline (Mori et al. 1988). The reason that these copies are germline mutators, whereas other Tc1 elements are not, might lie in rare but apparently important base-pair substitutions that have been found in different copies of Tc1 (Rose et al. 1985; Plasterk 1987); alternatively, the chromosomal position may be important for establishment of the expression pattern that makes a Tc1 element a mutator. The latter possibility seems the most plausible, but the former has not formally been excluded. The mutator of *mariner* has been referred to as the*Mos* factor (Bryan and Hartl 1988; Medhora et al. 1988). This factor has been cloned and found to be a copy of *mariner*, and also in this case it seems most likely that this *mariner* is the *Mos* factor because of its chromosomal location.

It has been reported that occasionally a Bristol N2 strain can turn into a mutator strain (Babity et al. 1990). It is conceivable that a mutation has occured within one of the nonautonomous transposons that made it a mutator, but the more likely explanation seems to be that the absence of transposase expression in the germline of Bristol N2 is not absolute, so that very rarely Tc1 elements in Bristol may jump. If one of these elements lands in a region where it can be expressed in the germline, then it is "promoted" to be a mutator. Mapping of the *mut-5* gene (Mori et al. 1988) has identified three copies of Tc1 that co-segregate with *mut-5*, and are therefore candidates for the mutator. Furthermore, three-factor crosses show that only one of these three, number 40, consistently co-segregated with *mut-5* (A.-M. Gruters and R.H.A. Plasterk, unpublished). Attempts to prove that Tc1 number 40 is a Tc1 mutator were unsuccessful: transgenic derivatives of Bristol N2 containing the complete Tc1 number 40 element plus 1 kbp of flanking DNA did not exhibit mutator activity (A.-M. Gruters and R.H.A. Plasterk, unpublished). These experiments are inconclusive, however, since more flanking DNA or the precise chromosomal position of this element may be essential for proper expression in the germline.

The *mut* genes mentioned above should not be confused with another factor affecting Tc1 and Tc3 transposition: the *mut-2* gene (Collins et al. 1987). The *mut-2* mutation was generated as the result of mutagenic treatment of a Bergerac strain with EMS and was initially recognized as a factor that enhanced reversion of a Tc1 insertion mutant of the *unc-54* gene. It was subsequently found to enhance transposition as well as excision of Tc1, and also of Tc3 and other transposons. The *mut-2* gene was mapped to chromosome 1 (Finney 1987). It is unlikely that *mut-2* is also a copy of Tc1, since it activates Tc1 as well as Tc3 transposition, whereas studies with Tc1A and Tc3A transposase in transgenic

Bristol N2 strains have shown that the transposase of one element cannot activate jumping of the other (VAN LUENEN et al. 1993). Also the *mut-2* mutation was isolated as an enhancer of Tc1 activity in a strain that was already mutator positive. Attempts in our laboratory to generate new mutators in a Bristol N2 strain by mutagenic treatment are ongoing. Furthur genetic and molecular analysis will be required to resolve the basis of regulation of Tc1 and its relatives and clarify the role of the *mut-2* gene in activation of different transposons.

As mentioned above, in the Bristol N2 strain of *C. elegans* Tc1 is active only in the soma, not in the germline (EMMONS and YESNER 1984). Is this a profound feature that reflects the importance for the organism to keep its germline intact? Is it related to chromosome diminution in the nematode *Ascaris* (TOBLER 1986), which involves extensive rearrangements of somatic DNA? The implication of the soma/germline difference for Tc1 activity should not be overinterpreted. Since the discovery of the phenomenon, it has become clear that there are also strains in which Tc1 is active in both cell types. Also there are transposons where the difference goes just the other way, absence of transposition in the soma (P element of *Drosophila*, ENGELS 1989). The experiment by VAN LUENEN et al. (1993), who showed that induced expression of Tc3 transposase in somatic cells activates Tc3 jumping, suggests that the soma/germline difference of Tc1 activity in some *C. elegans* strains is caused by differential expression of the transposase. That, in itself, is an interesting enough issue: not much is known about the signal sequences that are required for gene expression in the germline of *C. elegans*.

9 Tc1 as Tool

Tc1 is the only member of the Tc1/*mariner* that has been used extensively as a tool for genetic analysis. I will briefly mention three past and present applications of Tc1, and four possible future applications.

9.1 Identification and Isolation of Genes

The frequency of gene inactivation by Tc1 insertion is variable in different strains and for different genes, but in mutator strains such as Bergerac (RW7000) and "high hopper" strains such as TR679 for many genes one finds approximately one new mutation per thousand meioses. After several crosses with the low Tc1 copy number strain Bristol N2, Southern blot analysis is done to identify a possible copy of Tc1 that co-segregates with the mutant allele. The transposon-tagged gene can then be cloned. This strategy was follwed for the cloning of *unc-22* (MOERMAN et al. 1986), *lin-12* (GREENWALD 1985) and many genes since.

9.2 Gene Mapping

In those cases where no Tc1 allele of a gene of interest was available or easily obtained, Tc1 has been used to map genes. The high Tc1 copy number strains contain up to 500 copies of Tc1, approximately 1/200 kbp. These can be used to align the genetic and physical map in areas of interest and have been used to map and finally clone genes (for example, *lin-14*, Ruvkun et al. 1989).

More recently, a variant of this approach has become available, thanks to PCR technology. Williams et al. (1992) characterized Tc1 alleles at several positions in each of the chromosomes of a high Tc1 copy number strain and generated PCR primers that, in combination with Tc1-specific primers, can be used to visualize these insertions. They found that it was possible to look at five different Tc1 alleles in a single PCR. Thus each of these Tc1 alleles can be viewed as a 'conditional visible' mutation and can be used to map any new gene quickly. This is particularly important since, given the simple body plan and limited behavior of *C. elegans*, it is usually not possible to look at more than two or three visible mutations independently within single animals. Genes are mapped as follows: a mutation generated in a low-copy-number Tc1 strain is crossed into a high-copy-number Tc1 strain that contains the identified Tc1 alleles. Then, from heterozygous parents, homozygous mutants are picked individually and typed by PCR. Any Tc1 allele that is genetically linked to the gene of interest will be under-represented in homozygous mutants. When, after an initial experiment, the chromosome that contains the mutation, has been identified a subsequent experiment is done using Tc1 alleles that are specific for that chromosome to map the gene within that chromosome. In principle, this method can be furthur refined now that more Tc1 alleles have become available. (R. Korswagen and RHAP, manuscript in preparation).

9.3 Target-selected Gene Inactivation

The method of target-selected gene inactivation was initially developed for the P element of *Drosophila* (Ballinger and Benzer 1989; Kaiser and Goodwin 1990). Its usefulness is somewhat limited there by the logistics of sibling selection on very large pools of flies, particularly since the target choice of the P element seems far from random, so that many genes are frequently hit. The method has been applied to Tc1 of *C. elegans* (Rushforth et al. 1993; Plasterk 1992) and has been used to isolate transposon insertion alleles of many genes. The search of Tc1 alleles has been much facilitated by the generation of a frozen transposon insertion bank (Zwaal et al. 1993).

Many Tc1 insertions do not inactivate gene function, either because they are in introns and are therefore removed from mRNA, or because they are removed from exons by aberrant splicing (Rushforth et al. 1993). Fortunately, the phenomenon of imprecise DSB repair (described above) can be used: among the progeny of the animals containing a Tc1 allele, one can detect worms that have lost that Tc1 element plus a few kilobase pairs of flanking DNA (Kiff et al. 1988; Zwaal et al. 1993). These deletion derivatives usually have the gene of interest fully inactivated.

10 Perspectives

Some potential future applications of Tc1 for genetic analysis are the following:

10.1 Site-directed Mutagenesis

It has been shown (Plasterk and Groenen 1992) that a Tc1 insertion can be used to introduce specific point mutations into the genome. A strain is isolated containing a Tc1 insertion in the region of interest, either by forward genetics (as described in Sect. 9.1 above) or by reverse genetics (as described in Sect. 9.3). This strain is then micro injected with a plasmid construct that contains the region of interest, altered at a specific position in vitro. Once a transgenic line has been established, one can screen the progeny for worms that have lost the transposon allele and have used the transgene as template in double-strand break repair. The only case thus far where such an experiment has been reported (Plasterk and Groenen 1992) involved a visible allele (of the *unc-22* gene) and a transgene that contained several point mutations which left the coding capacity intact. Thus phenotypic analysis could be used to screen for *unc-22* revertants, approximately half of which were found to contain point mutations derived from the transgene. The method has not yet been used in a fully reverse genetic approach where site-directed point mutants were identified by DNA analysis (e.g., PCR), but in principle, such an approach seems possible. It may find an application in cases where specific point mutants have been found to be interesting in gene homologs of other species (transdominant mutations, thermosensitive alleles, etc.) and where one wants to introduce those into the proper chromosomal context, rather than into a transgenic ectopic version of the gene.

10.2 Enhancer Trapping

The method of enhancer trapping by screening of large numbers of transposon insertion alleles with a transgenic transposon marked with a monitor gene such as *lac Z* has been used extensively in *Drosophila* genetics (Kaiser 1993; Grossniklaus et al. 1989). The method has not yet been applied to *C. elegans*, largely because attempts to let transgenic Tc1 elements integrate into the nematode germline have failed. As discussed above, we have succeeded in demonstrating jumping from transgenic arrays into the chromosome; however, these experiments were done using PCR and probably visualized jumping in somatic cells (Van Luenen et al. 1993). They do demonstrate, however, that it is possible for a transposon to jump from an extrachromosomal transgenic array into a chromosome, and it therefore seems likely that the main obstacle to transposon-mediated transformation in the germline lies in the choice of the proper selectable markers. A positive-negative selection, such as developed by Sprunger (1992) would/probably be helpful.

10.3 Transgenesis of Other Species

The wide spread of the Tc1/*mariner* transposon family suggests that whether host factors may be required for transposition, they may be ubiquitous. Therefore, these transposons are probably well suited for introduction of foreign DNA into organisms that are not very well defined genetically. In contrast, the P element of *Drosophila* transposes only in a limited number of fly species. It has been shown that the *mariner* transposon from *Drosophila mauritiana* can be used successfully to transform *Drosophila melanogaster* (Lidholm et al. 1993). The coming years will probably witness many attempts to use Tc1/*mariner* transposons for the generation of transgenic insects of economic and/or medical importance. The discovery of Tc1 homologs in zebra fish (Radice et al. 1994) has raised the interest in application of either the fish homologs or Tc1 itself for the generation of transgenic zebra fish.

10.4 Global Reverse Genetics

Two applications mentioned above, gene mapping using mapped Tc1 alleles and target-selected gene inactivation by Tc1, may merge into an efforts to shotgun the entire *C. elegans* genome with Tc1 insertions. The purpose of such an effort would be to isolate many thousands of Tc1 insertion alleles, each of those defined by a short track of the DNA sequence immediately flanking the insertion site. Such an effort seems justified, given the efforts spent on the determination of the entire genome sequence (Waterston et al. 1992; Sulston et al. 1992; Wilson et al. 1994).

Acknowledgments. This review was written with the advice of Drs. Chris Vos, Scott Emmons, Phil Anderson, and Henri van Luenen and with the secretarial support of Miss Nicole Immink. I am grateful for their help.

References

Abad P, Quiles C, Tares S, Piotte C, Castagnone-Sereno P, Abadon M, Dalmasso A (1991) Sequences homologous to Tc(s) transposable elements of *Caenorhabditis elegans* are widely distributed in the phylum nematoda J Mol Evol 33: 251-258

Abad P, Cerutti M, Pauron D, Quiles C, Palin B, Devauchelle G, Dalmasso A (1993) Expression and biochemical characterization of the DNA-binding activity of TcA, the putative transposase of *Caenorhabditis elegans* transposable element Tc1. Biochem Biophys Res Commun 192: 1445–1452

Anderson P, Emmons SW, Moerman DG (1992) Discovery of Tc1 in the nematode *Caenorhabditis elegans*. In: Fedoroff N, Bostein D (eds) The dynamic genome: Barbara McClintock's ideas in the century of genetics. Cold Spring Harbor Laboratory Press, Cold Spring Harbor, New York, pp 319–333

Babity JM, Starr TVB, Rose AM (1990) Tc1 transposition and mutator activity in a Bristol strain of *Caenorhabditis elegans*. Mol Gen Genet 222: 65–70

Ballinger DG, Benzer S (1989) Targeted gene mutations in *Drosophila*. Proc Natl Acad Sci USA 86: 9402–9406

Bigot Y, Hamelin M, Capy P, Periquet G (1994) *Mariner*-like elements in hymenopteran species: insertion site and distribution. Proc Natl Acad Sci USA 91: 3408–3412

Brezinsky L, Wang GV, Humphreys T, Hunt J (1990) The transposable element Uhu from Hawaiian *Drosophila*–member of the widely dispersed class of Tc1-like transposons. Nucleic Acids Res 18: 2053–2059

Brierley HL, Potter SS (1985) Distinct characteristics of 100p sequences of two *Drosophila* foldback transposable elements Nucleic Acids Res 13: 485

Bryan GJ, Hartl DL (1988) Maternally inherited transposon excision in *Drosophila simulans*. Science 240: 215–217

Bryan GJ, Garza D, Hartl D (1990) Insertion and excision of the transposable element mariner in *Drosophila*. Genetics 125: 103–114

Caizzi R, Caggese C, Pimpinelli S (1993) *Bari-1*, a new transposon-like family in *Drosophila melanogaster* with a unique heterochromatic organization. Genetics 133: 335–345

Capy P, Anxolabéhère D, Langin T (1994) The strange phylogenies of transposable elements: are horizontal transfers the only explanation? Trends Genet 10: 7–12

Carr B, Anderson P (1994) Imprecise excision of the *Caenorhabditis elegans* transposon Tc1 creates functional 5' splice sites. Mol Cell Biol 14: 3426–3433

Caskey CT, Pizzuti A, Fu YH, Fenwick RG jr, Nelson DL (1992) Triplet repeat mutations in human disease. Science 256: 784–789

Collins J, Anderson P (1994) The Tc5 family of transposable elements in *Caenorhabditis elegans*. Genetics 137: 771–781

Collins J, Saari B, Anderson P (1987) Activation of a transposable element in the germline but not the soma of *Caenorhabditis elegans*. Nature 328: 726–728

Collins J, Forbes E, Anderson P (1989) The Tc3 family of transposable elements in *Caenorhabditis elegans*. Genetics 121: 47–55

Colloms SD, van Luemen HGAM, Plasterk RHA (1994) DNA binding activities of the Caenorhabditis elegans Tc3 transposase. Nucl Acids Res 25: 5548–5554

Daboussi M, Langin T, Brygoo Y (1992) Fotl, a new family of fungal transposable elements. Mol Gen Genet 232: 12–16

Doak TG, Doerder FP, Jahn CL, Herrick G (1994) A proposed superfamily of transposase-related genes: new members in transposase-like elements of ciliated protozoa and a common "D35E" motif Proc Natl Acad Sci USA 91: 942–946

Dombroski BA, Mathias SL, Nanthakumar E, Scott AF, Kazazian HHJ (1991). Isolation of an active human transposable element. Science 254: 1805–1808

Dombroski BA, Scott AF, Kazazian HHJ (1993) Two additional potential retrotransposons isolated from a human L1 subfamily that contains an active retrotransposable element. Proc Natl Acad Sci USA 90: 6513–6517

Eide D, Anderson P (1985) Transposition of Tc1 in the nematode *Caenorhabditis elegans*. Proc Natl Acad Sci USA 82: 1756–1760

Eide D, Anderson P (1988) Insertion and excision of *Caenorhabditis elegans* transposable element Tc1. Mol Cell Biol 8: 737–746

Emmons SW, Yesner L (1984) High-frequency excision of transposable element Tc1 in the nematode *Caenorhabditis elegans* is limited to somatic cells. Cell 36: 599–605
Emmons SW, Rosenzweig B, Hirsch D (1980) Arrangement of repeated sequences in the DNA of the nematode *Caenorhabditis elegans*. J Mol Biol 144: 481–500
Emmons SW, Yesner L, Ruan K, Katzenberg D (1983) Evidence for a transposon in *Caenorhabditis elegans*. Cell 32: 55–65
Emmons SW, Roberts S, Ruan K (1986) Evidence in a nematode for regulation of transposon excision by tissue-specific factors. Mol Gen Genet 202: 415–419
Engels WR (1989) P elements in *Drosophila melanogaster*. In: Berg DE, Howe MM (eds) Mobile DNA. American Society for Microbiology, Washington DC, pp 437–484
Engels WR, Johnson-Schlitz DM, Eggleston WB, Sved J (1990) High-frequency P element loss in *Drosophila* is homolog independent. Cell 62: 515–525
Fayet O, Ramond P, Polard P, Prére MF, Chandler M (1990) Functional similarities between retroviruses and the IS3 family of bacterial insertion sequences? Mol Microbiol 4: 1771–1777
Finney M (1987) The genetics and molecular biology of *unc-86*, a *C. elegans* cell lineage gene. PhD thesis, Massachussetts Institute of Technology, Cambridge MA
Franz G, Savakis C (1991) *Minos*, a new transposable element from *Drosophila hydei*, is a member of the Tc1-like family of transposons. Nucleic Acids Res 19: 6646
Franz G, Loukeris TG, Dialektaki G, Thompson CRL, Savakis C (1994) Mobile Minos elements from *Drosophila hydei* encode a two-exon transposase with similarity to the *paired* DNA-binding domain. Proc Natl Acad Sci 91: 4746–4750
Garcia-Fernandez J, Marfany G, Baguna J, Salo E (1993) Infiltration of *mariner* elements. Nature 364: 109–110
Goodier JL, Davidson WS (1994) Tc1 transposon-like sequences are widely distributed in salmonids. J Mol Biol 241: 26–34
Greenwald IS (1985) Lin-12. A nematode homeotic gene is homologous to a set of mammalian proteins that includes epidermal growth factor. Cell 43: 583–590
Grossniklaus U, Bellen HJ, Wilson C, Gehring WJ (1989) P-element-mediated enhancer detection applied to the study of oogenesis in *Drosophila*. Development 107: 189–200
Harris LJ, Baillie DL, Rose AM (1988) Sequence identity between an inverted repeat family of transposable elements in *Drosophila* and *Caenorhabditis*. Nucleic Acids Res 16: 5991–5998
Harris LJ, Prasad S, Rose AM (1990) Isolation and sequence analysis of *Caenorhabditis briggsae* repetitive elements related to the *Caenorhabditis elegans* transposon Tc1. J Mol Evol 30: 359–369
Hartl DL (1989) Transposable element *mariner* in Drosophila species. In: Berg DE, Howe MM (eds) Mobile DNA. American Society for Microbiology, Washington DC, pp 531–536
Heierhorst J, Lederis K, Richter D (1992) Presence of a member of the Tc1-like transposon family from nematodes and *Drosophila* within the vasotocin gene of a primitive vertebrate, the Pacific hagfish *Eptatretus stouti*. Proc Natl Acad Sci USA 89: 6798–6802
Henikoff S (1992) Detection of *Caenorhabditis* transposon homologs in diverse organisms. New Biol 4: 382–388
Henikoff S, Plasterk RHA (1988) Related transposons in *C. elegans* and *D. melanogaster*. Nucleic Acids Res 16: 6234
Herman RK, Shaw JE (1987) The transposable genetic element Tc1 in the nematode *Caenorhabditis elegans*. Trends Genet 3: 222–225
Hurst GDD, Hurst LD, Majerus MEN (1992) Selfish genes move sideways. Nature 356: 659–660
Jacobson JW, Medhora MM, Hartl DL (1986) Molecular structure of a somatically unstable transposable element in *Drosophila*. Proc Natl Acad Sci USA 83: 8684–8688
Kaiser K (1993) Second-generation enhancer traps. Curr Biol 3: 560–562
Kaiser K, Goodwin SF (1990) "Site-selected" transposon mutageneis of *Drosophila*. Proc Natl Acad Sci USA 87: 1686–1690
Kidwell MG (1993) Voyage of an ancient mariner. Nature 362: 202
Kiff JE, Moerman DG, Schriefer LA, Waterston R H (1988) Transposon-induced deletions in unc-22 of *C. elegans* associates with almost normal gene activity. Nature 331: 631–633
Kulkosky J, Jones KS, Katz RA, Mack JPG, Skalka AM (1992) Residues critical for retroviral integrative recombination in a region that is highly conserved among retroviral/retrotransposon integrases and bacterial insertion sequence transposases. Mol Cell Biol 12: 2331–2338
Levitt A, Emmons SW (1989) The Tc2 transposon in *Caenorhabditis elegans*. Proc Natl Acad Sci USA 86: 3232–3236
Li W, Shaw JE (1993) A variant Tc4 transposable element in the nematode *C. elegans* could encode a novel protein. Nucleic Acids Res 21: 59–67

Liao LW, Rosenzweig B, Hirsh D (1983) Analysis of a transposable element in *Caenorhabditis elegans*. Proc Natl Acad Sci USA 80: 3585–3589
Lidholm DA, Lohe AR, Hartl DL (1993) The transposable element *mariner* mediates germline transformation in *Drosophila melanogaster*. Genetics 134: 859–868
Maruyama K, Hartl DL (1993) Evidence for interspecific transfer of the transposable element *mariner* between *Drosophila* and *Zaprionus*. J Mol Evol 33: 514–524
Medhora MM, MacPeek AH, Hartl DL (1988) Excision of the *Drosophila* transposable element mariner: identification and characterization of the Mos factor. EMBO J 7: 2185–2189
Moerman DG, Waterston RH (1984) Spontaneous unstable unc-22 IV mutations in *C. elegans* var bergerac. Genetics 108: 859–877
Moerman DG, Waterston RH (1989) Mobile elements in *Caenorhabditis elegans* and other nematodes. In: Berg DE , Howe MM (eds) Mobile DNA. American Society for Microbiology, Washington DC, pp 537–556
Moerman DG, Benian GM, Waterston RH (1986) Molecular cloning of the muscle gene unc-22 in *Caenorhabditis elegans* by Tc1 transposon tagging. Proc Natl Acad Sci USA 83: 2579–2583
Mori I, Benian GM, Moerman DG, Waterston RH (1988a) Transposable element Tc1 of *Caenorhabditis elegans* recognizes specific target sequences for integration. Proc Natl Acad Sci USA 85: 861–864
Mori I, Moerman DG, Waterston RH (1988b) Analysis of a mutator activity necessary for germline transposition and excision of Tc1 transposable elements in *Caenorhabditis elegans*. Genetics 120: 397–407
Mori I, Moerman DG, Waterston RH (1990) Interstrain crosses enhance excision of Tc1 transposable elements in *Caenorhabditis elegans*. Mol Gen Genet 220: 251–255
Plasterk RHA (1987) Differences between Tc1 elements from the *Caenorhabditis elegans* strain Bergerac. Nucleic Acids Res 15: 10050
Plasterk RHA (1991) The origin of footprints of the Tc1 transposon of *Caenorhabditis elegans*. EMBO J 10: 1919–1925
Plasterk RHA (1992) Reverse genetics of *Caenorhabditis elegans*. Bioessays 14: 629–633
Plasterk RHA, Groenen JTM (1992) Targeted alterations of the *Caenorhabditis elegans* genome by transgene instructed DNA double-strand break repair following Tc1 excision. EMBO J 11: 287–290
Prasad SS, Harris LJ, Baillie DL, Rose AM (1991) Evolutionarily conserved regions in *Caenorhabditis* transposable elements deduced by sequence comparison. Genome 34: 6–12
Radice AD, Emmons SW (1993) Extrachromosal circular copies of the transposon Tc1. Nucleic Acids Res 21 2663–2667
Radice AD, Bugaj B, Fitch DHA, Emmons SW (1994) Widespread occurrence of the Tc1 transposon family: Tc1-like transposons from teleost fish. Mol Gen Genet 244: 606–612
Robertson HM (1993) The mariner transposable element is widespread in insects. Nature 362: 241–245
Robertson HM (1995) The Tc1-mariner superfamily of transposons in animals. J Insect Physiol 44: 99–105
Robertson HM, Lampe DJ, Macleod EG (1992) A mariner transposable element from a lacewing. Nucleic Acids Res 20: 6409
Rose AM, Snutch TP (1984) Isolation of the closed circular form of the transposable element Tc1 of *Caenorhabditis elegans*. Nature 311: 485–486
Rose AM, Harris LJ, Mawji NR, Morris WJ (1985) Tc1 (Hin): a form of the transposable element Tc1 in *Caenorhabditis elegans*. Can J Biochem 63: 752–756
Rosenzweig B, Liao LW, Hirsh D (1983) Sequence of the *C. elegans* transposable element Tc1. Nucleic Acids Res 11: 4201–4210
Ruan K, Emmons SW (1984) Extrachromosomal copies of transposon Tc1 in the nematode *Caenorhabditis elegans*. Proc Natl Acad Sci USA 81: 4018–4022
Rushforth AM, Saari B, Anderson P (1993) Site-selected insertion of the transposon Tc1 into a *Caenorhabditis elegans* myosin light chain gene. Mol Cell Biol 13: 902–910
Ruvkun GB, Ambros V, Coulson A, Waterston RH, Sulston JE, Horvitz HR (1989) Molecular genetics of the *Caenorhabditis elegans* heterochronic gene *lin-14*. Genetics 121: 501–516
Ruvolvo V, Hill JE, Levitt A (1992) The Tc2 transposon of *Caenorhabditis elegans* has the structure of a self-regulated element. DNA cell Biol 11: 111–122
Schukkink RF, Plasterk RHA (1990) TcA the putative transposase of the *C. elegans* Tc1 transposon, has an N-terminal DNA binding domain. Nucleic Acids Res 18: 895–900
Sedensky MM, Hudson SJ, Everson B, Morgan PG (1994) Identification of a *mariner*-like repetitive sequence in *C. elegans*. Nucleic Acids Res 22: 1719–1723

Sprunger SA (1992) The *Caenorhabditis elegans* alkali myosin light chain gene. PhD thesis, University of Wisconsin, Madison

Sulston J, Du Z, Thomas K, Wilson R, Hiller L, Staden R, Halloran N, Green P, Thierry-Mieg-J, Qiu L, Dear S, Coulson A, Craxton M, Durbin R, Berks M, Metzstein M, Hawkins T, Ainscough R, Waterston R (1992) The *C. elegans* genome sequencing project: a beginning. Nature 356: 37–41

Tausta SL, Klobutcher LA (1989) Detection of circular forms of eliminated DNA during macronuclear development in *E. crassus*. Cell 59: 1019–1026

Tobler H (1986) The differentiation of germ and somatic cell lines in nematodes. In: Hennig W (ed) Germline-soma differentiation, results and problems in cell differentiation. Springer, Berlin Heidelberg New York, pp 1–69

Van Luenen HGAM, Plasterk RHA (1994) Target site choice of the related transposable elements Tc1 and Tc3 of *Caenorhabditis elegans*. Nucleic Acids Res 22: 262–269

Van Luenen HGAM, Colloms SD, Plasterk RHA (1993) Mobilization of quiet endogenous Tc3 transposons of *Caenorhabditis elegans* by forced expression of Tc3 transposase. EMBO J 12: 2513–2520

Van Luenen HGAM, Colloms SD, Plasterk RHA (1994) The mechanism of transposition of Tc3 in *Caenorhabditis elegans*. Cell 79: 293–301

Vos JC, Plasterk RHA (1994) Tc1 transposase of *Caenorhabditis elegans* is an endonuclease with a bipartite DNA-binding domain. EMBO J 13: 6125–6132

Vos JC, Van Luenen HGAM, Plasterk RHA (1993) Chacracterization of the *Caenorhabditis elegans* Tc1 transposase in vivo and in vitro. Genes Dev 7: 1244–1253

Waterston R, Martin C, Craxton M, Huynh C, Coulson A, Hiller L, Durbin R, Green P, Shownkeen R, Halloran N, Metzstein M, Hawkins T, Wilson R, Berks M, Thierry-Mieg J, Sulston J (1992) A survey of expressed genes in *Caenorhabditis elegans*. Nature Gen 1: 114–123

Williams BD, Schrank B, Huynh C, Shownkeen R, Waterston RH (1992) A genetic mapping system in *Caenorhabditis elegans* based on polymorphic sequence-tagged sites. Genetics 131: 609–624

Williams K, Doak TG, Herrick G (1993) Developmental precise excision of *Oxytrichatrifallax* telomere-bearing elements and formation of cirlces closed by a copy of the flanking target duplication. EMBO J 12: 4593–4601

Wilson R, Ainscough R, Anderson K, Baynes C, Berks M, Bonfield J, Burton J, Connell M, Copsey T, Cooper J, Coulson A, Craxton M, Dear S, Du Z, Durbin R, Favello A, Fraser A, Fulton L, Gardner A, Green P, Hawkins T, Hiller L, Jier M, Johnson L, Jones M, Kershaw J, Kirsten J, Laisster N, Latreille P, Lightning J, Lloyd C, Mortimore B, O'Callaghan M, Parsons JT, Percy C, Rifken L, Roopra A, Saunders D, Shownkeen R, Sims M, Smaldon N, Smith A, Smith M, Sonnhammer E, Staden R, Sulston J, Thierry-Mieg J, Thomas K, Vaudin M, Vaughan K, Waterston R, Watson A, Weinstock L, Wilkinson-Sproat J, Wohldman P (1994) 2.2 Mb of contiguous nucleotide sequence from chromosome III of *C. elegans*. Nature 368: 32–38

Yuan J, Finney M, Tsung N, Horvitz HR (1991) Tc4, a *Caenorhabditis elegans* transposable element with an unusual fold-back structure. Genetics 88: 3334–3338

Zwaal RR, Broeks A, van Meurs J, Groenen JTM, Plasterk RHA (1993) Target-selected gene inactivation in *caenorhabditis elegans* by using a frozen transposon insertion mutant bank. Proc Natl Acad Sci USA 90: 7431–7435

The *En/Spm* Transposable Element of Maize

A. GIERL

1 Introduction

Exceptional phenotypes can be generated in the flowers and seeds of plants due to variation in the patterns of gene expression at the somatic bud. Such examples have attracted attention since naturalists began to investigate nature. One of the first reports on color variegation in a plant was given in 1588 by Jacob Theodor of Bergzabern, a village close to Strasbourg. He described color variegation in kernels of *Zea mays,* a species that had just arrived in the Old World.

An explanation for these puzzling "exceptions to the rule" had to wait until Barbara MCCLINTOCK (1947, 1948) developed the concept of transposable elements as mobile genetic entities. She used the term "controlling elements", since they seemed capable of influencing expression when integrated into or nearby a given locus. In the past decade, numerous transposable element systems,

Lehrstuhl für Genetik, Technische Universität München, Lichtenbergstraße 4, D-85747 Garching, Germany

mostly from maize and snapdragon (*Antirrhinum majus*), have been studied molecularly. It was established that the variegated phenotypes are due to genetic instability of the elements. During the organism's development, the transposable element leaves its site of integration within a gene, and thus the function of that gene is restored in this cell as well as in its progeny. These excision events result in a mosaic tissue of mutant and revertant (wild-type) cells.

The *En/Spm* transposable element system from maize was independently identified by PETERSON (1953) and McCLINTOCK (1954); they named it *Enhancer (En)* and *Suppressor-mutator (Spm)*, respectively. Based on their detailed genetic analysis, several groups have investigated the molecular aspects of *En/Spm* transposition. In general, transposition is the result of an interaction of proteins (*trans*-acting factors) with the termini (*cis*-determinants) of the transposon. The focus of this review is on the *cis/trans* requirements for transposition and the regulation of this process. The evolution of *En/Spm* and its use for gene isolation are also briefly considered.

2 DNA Structure and Gene Expression of *En/Spm*

The *En/Spm* element is 8.3 kb in length, has 13-bp perfect terminal inverted repeats (TIRs), and causes a 3-bp duplication of target sequences upon insertion (PEREIRA et al. 1986). It encodes two products, which have been termed TNPA and TNPD (Fig. 1; GIERL et al. 1988; MASSON et al. 1989). TNPA is 67 kD and, although expressed at a relatively low level, is about 100 times more abundant than the 131-kD TNPD protein (GIERL et al. 1988). Both proteins are absolutely required for transposition (see below). The *tnpA* and *tnpD* transcripts are derived by alternative splicing from a precursor transcript that is initiated at position 209. In addition to these two processed transcripts, other splice products have been identified (MASSON et al. 1989); however, it remains unclear whether they have any function

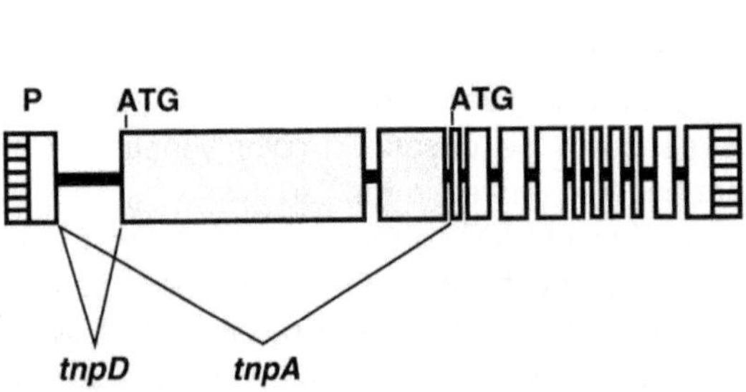

Fig. 1. Structural organization of *En/Spm*. The 8287-bp *En/Spm* element contains two genes: *tnpA* and *tnpD*. A single pre-mRNA is initiated at the promoter (*P*) and is differentially spliced into *tnpA* and *tnpD* mRNA. The exons encoding *TNPA* (*open boxes*) and *TNPD* protein (*shaded boxes*) are indicated. The first exon is common to both mRNAs; it is not translated and contains the GC-rich DCR region. About 200 bp of the 5' end and 300 bp of the 3'end (*hatched boxes*) represent the highly structured termini that are shown in more detail in Fig. 2

or are simply by-products of the mRNA maturation processes that have been detected with very sensitive methods.

The phenomenon of deviant splice products of *En/Spm* is seen more dramatically if dicotyledonous plant species are transformed with the autonomous element. While the mRNA pattern of *En/Spm* in *Arabidopsis thaliana* is relatively similar to that in maize, in *Nicotiana tabacum* and *Solanum tuberosum tnpD* mRNA is much more abundant than *tnpA* mRNA (CARDON et al. 1993b). This confirms the different processing requirements in dicotyledonous vs. monocotyledonous species (GOODALL and FILIPOWICZ 1991); however, it also reveals strong differences within the order of dicots.

Approximately 200 bp of the 5' end and 300 bp of the 3' end of *En/Spm* are highly structured, representing the *cis* determinants for excision. The single *En/Spm* promoter is contained within the structured 5' end of the element. In addition to the structurally intact, transposition-competent *En/Spm* element, numerous defective elements exist in maize. These elements represent internal deletion derivatives of *En/Spm* and transpose only in the presence of an active intact element. Because of this responsiveness, defective and active elements constitute a two-component system, consisting of the receptor (defective, *dSpm*) element and the autonomous *En/Spm* element.

3 *cis/trans* Requirements for Excision

In maize, the interaction of the element-encoded proteins of different transposable elements with their respective *cis* determinants is very specific. This was tested by making use of the fact that the maize elements exist in one of two forms: the fully intact autonomous element and defective elements (receptors) which have retained the *cis* determinants but do not encode functional proteins. Therefore, defective elements can transpose only in the presence of the autonomous elements. Family relationships can be determined by testing whether a particular autonomous element can trigger the excision of a variety of defective elements belonging to various families. Using this technique, separate transposable element systems have been defined in maize (for a detailed list see PETERSON 1987).

3.1 *cis* Determinants for Excision

Approximately 180 bp at the 5' end and 300 bp of the 3' end of *En/Spm* represent the *cis* determinants for excision (Fig. 2). Contained within these regions are reiterations of a 12-bp sequence motif that is recognized by the TNPA protein (see below), with six motifs present at the 5' end and eight at the 3' end. The 13-bp perfect TIRs of *En/Spm* are also *cis*-acting sequences absolutely required for excision. Deletion of the outermost nucleotide of the 13-bp TIR drastically

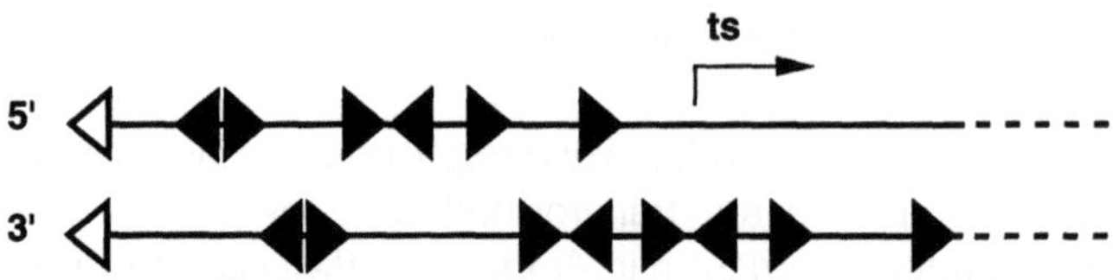

Fig. 2. *Cis* determinants for excision of *En/Spm*. The termini of *En/Spm* contain two different sequence motifs: the 13-bp terminal inverted repeats (TIRs, *open triangles*) and the 12-bp *TNPA*-binding sites (*filled triangles*). The 5' end has six and the 3' end has eight *TNPA*-binding motifs in different orientation. The tail-to-tail dimeric forms are closest to the TIRs. The transcription start site (*ts*) at position 209 is indicated

reduces the excision frequency (Schiefelbein et al. 1988). The TIR is not related in sequence to the 12-bp TNPA-binding motif. Partial deletion of the 12-bp motifs also correlates with decreased excision rates (Schwarz-Sommer et al. 1985, Schiefelbein et al. 1988). The extent of the deletion seems to be directly proportional to the amount of reduction. The entire deletion of the 12-bp TNPA binding sites at one end of *En/Spm* completely abolishes excision (Menssen et al. 1990). The 5' and the 3' ends of *En/Spm* are not interchangeable. An element with two 5' ends does not excise (Reinecke, personal communication).

A defective element has been isolated from the *brittle 1* locus (Sullivan et al. 1991) and the *C2* locus (Muszynski et al. 1993) of maize with slightly modified *cis* determinants. About 12% nucleotide changes have been detected within the ends that occur with and between the 12-bp TNPA-binding motifs. As a consequence, this element requires an additional function, termed mediator for excision (Muszynski et al. 1993). Mediator is encoded by a gene independent of *En/Spm* and seems to represent a helper function for TNPA, or may even replace TNPA with a functionally related TNPA-like product.

3.2 Functional Analysis of TNPA and TNPD in Transgenic Plants

The introduction of the autonomous *En/Spm* element into *Nicotiana tabacum* (Masson and Fedoroff 1989; Pereira and Saedler 1989), *Solanum tuberosum* (Frey et al. 1989), and *Arabidopsis thaliana* (Cardon et al. 1993a) showed that all mechanistic features of transposition are very similar, if not identical, to the behavior of *En/Spm* in maize. A 3-bp target site duplication is generated upon insertion, characteristic sequence alterations (footprints, see below) are left behind after excision, and about 50% of the transposition events occur to linked sites on the same chromosome (Cardon et al. 1993b). Preferential transposition into linked positions was detected earlier by genetic means (Novick and Peterson 1981).

Based on these results, it was possible to use transgenic tobacco in order to analyze the system in more detail. Three component systems were established (Frey et al. 1990; Masson et al. 1991; Cardon et al. 1993b) that allowed the definition of the two *En/Spm* functions necessary for transposition: TNPA and TNPD. These artificial reconstitution systems include excision reporter constructs consisting of *dSpm* elements inserted in the 5' untranslated leaders of various reporter genes. The other two components of the system, TNPA and TNPD, are expressed by fusing their cDNAs to plant-specific promoters. Only when all three components were combined in one plant was frequent somatic excision detected by restoration of the function of the reporter genes. No excision was detectable when either TNPA or TNPD was absent. Thereby, the absolute requirement and sufficiency of TNPA and TNPD for transposition of *En/Spm* was clearly established.

An artificial defective *Spm* element, named *dSpm-DHFR*, was constructed (CARDON et al. 1993b), which is marked by the insertion of a *dehydrofolate reductase* (*DHFR*) gene under the control of the *CaMV35S* promoter. The *DHFR* gene is a dominant selectable marker conferring resistance to methotrexate. The *dSpm-DHFR* element can be *trans* activated by TNPA and TNPD. Both excision and reintegration take place with characteristics similar to those for native *dSpm* elements. This indicates that internal *En/Spm* segments can be replaced by other DNA sequences without abolishing transposition. The excision frequency of *dSpm-DHFR* was reduced by a factor of 6 however. It was suggested (CARDON et al. 1993b) that the constitutive transcription of the *DHFR* gene driven by the relatively strong *CaMV35S* promoter might interfere with transposition.

The fate of the element after excision was also addressed using the marked *dSpm-DHFR*. It was found that about 50% of the excised elements reinserted (CARDON et al. 1993b). When reintegration of a normal *dSpm* element was analyzed by Southern hybridization, a somewhat lower frequency of loss was estimated.

Loss of the *En/Spm* element after excision was also demonstrated by an elegant genetic study of maize (DASH and PETERSON 1994) using various reporter alleles. It was concluded that loss results from transposition of *En/Spm* during chromosome replication. Transposition of En from a replicated segment of the chromosome to another site that has also undergone replication, followed by mitotic and/or meiotic segregation, explains these events. It was suggested (DASH and PETERSON 1994) that this mechanism may prevent a harmful build up of *En/Spm* copies in the genome and maintains a tolerable amount of transposition events. This study represents the first clear indication that, similar to *Ac* (GREENBLATT 1984), transposition of *En/Spm* might require DNA replication.

4 In Vitro Analysis of TNPA Function

The TNPA protein (621 amino acids) expressed in *Escherichia coli* (GIERL et al. 1988) and by in vitro translation (TRENTMANN et al. 1993) was used for DNA-binding studies. TNPA specifically binds to the 12-bp sequence motif that is reiterated several times in the subterminal regions of *En/Spm* (Fig. 2). Binding is reduced if the cytosine residues of CG dinucleotides and CNG trinucleotides within the motif are methylated (Gierl et al. 1988). The hemimethylated motif reduced complex stability by a factor of 5–10. The holomethylated form reduced binding even further.

The TNPA-binding motifs occur in three forms: as monomers and as closely spaced so-called head-to-head and tail-to-tail dimers. The dimeric forms are spaced by 2–3 bp. Oligonucleotides containing these motifs form three different complexes with TNPA (Trentmann et al. 1993). The simplest complex contains one TNPA molecule in specific contact with one 12-bp motif. This complex is detectable with the monomeric as well as with the dimeric forms of the motif. In the second complex, one TNPA is specifically bound to the motif and a second TNPA molecule is associated via protein-protein contacts of TNPA. This complex occurs with the monomeric and the head-to-head forms of the motif. A third complex is formed with the tail-to-tail motif. Here, two TNPA molecules are specifically in contact with the DNA and are additionally stabilized by protein interactions (Fig. 3). These findings, corroborated by DNA-protein and protein-protein cross-linking studies, indicate that TNPA contains two domains: a DNA-binding domain and a dimerization domain. These two domains have been coarsely defined by TNPA deletion derivatives. Amino acids 122–427 contain the DNA-binding domain, while the dimerization domain is contained in the region 428–542. The DNA-binding domain has no apparent homology to any sequences in the data bases, nor does it belong to the structurally defined classes of the

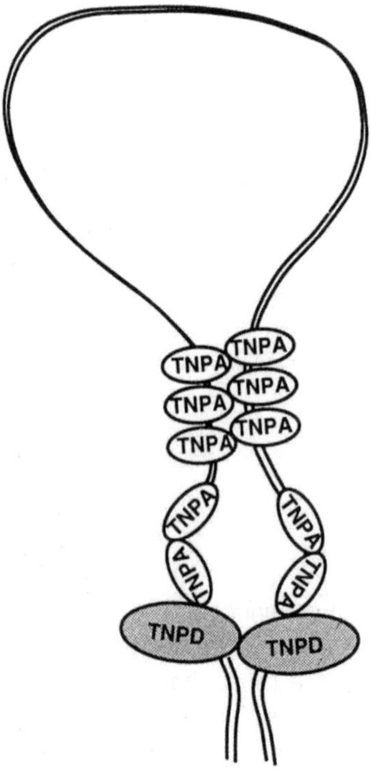

Fig. 3. Hypothetical structure of the transposition complex. *TNPA* is postulated to have a dual function: Binding of *TNPA* to the termini of *En/Spm* results in association of the ends and binding of *TNPA* to the outermost tail-to-tail motifs in activating recombination by DNA bending. *TNPD* is postulated to cleave at the elements' ends, thereby excising *En/Spm* from the chromosomal DNA

DNA-binding proteins such as the zinc finger, helix-turn-helix, or the basic domain of leucine zipper proteins.

Binding of TNPA results also in DNA bending (Gierl, unpublished). The bending of the DNA is twice as strong with the tail-to-tail motif compared with the other forms of the motif. DNA bending is often essential for DNA recombination events. This was studied in detail for integration host factor protein of *E. coli* (NASH 1990), which is required in lambda site-specific recombination to deform the DNA substrate into conformations active for recombination. Therefore, it may be speculated that deformation of ends of *En/Spm* by TNPA is necessary for excision of *En/Spm*. The fact that the tail-to-tail motifs are directly adjacent to the 13-bp TIRs, where recombination should occur, is further suggestive of this speculation.

5 Mechanism of Transposition

Transposition of *En/Spm* probably occurs via excision and integration. An essential requirement for such a mechanism is healing of the chromosome breaks that are generated during excision, thereby avoiding chromosomal loss. Joining of the breaks is most credible if the ends of the transposable element come into close physical proximity. Accordingly, two functions have been postulated (FREY et al. 1990), one that is required for association of the two ends of *En/Spm* and one that cuts close to the ends of the element. According to this hypothesis, a probable candidate for the joining functions would be the TNPA protein that binds to the subterminal binding motifs. The ability of TNPA to form dimers could lead to association of the two ends by cross-linking via protein-protein contacts. The asymmetric distribution of these motifs may facilitate correct positioning of the TIRs, thus forming the substrate for TNPD that would be involved in cutting at the ends and release of *En/Spm* (Fig. 3). Evidence in favor of this view is the finding that the TIRs of the so-called CACTA elements (see below) are nearly identical, and that these elements potentially encode a protein homologous to TNPD.

Almost at the beginning of the molecular analysis of *En/Spm*, the DNA analysis of revertant genes arising from excision events revealed that excision is rather imprecise (SCHWARZ-SOMMER et al. 1985) but still follows certain rules. As a consequence of excision, altered sequences, "footprints", are left behind at the site of excision. Therefore, gene function may be restored to varying degrees, ranging from zero to full activity. According to the model of SAEDLER and NEVERS (1985), the transposase recognizes the ends of the element and introduces staggered nicks at the ends of the target site duplication. Footprints result from the action of DNA repair enzymes (exonuclease, DNA polymerase, DNA ligase) involved in healing of the chromosome breaks, acting on the single-stranded fringes at the excision site. A similar staggered cut is formed during integration of *En/Spm* by the same endonuclease, causing the 3-bp TSD, whose length is specific for the *En/Spm* family.

The formation of footprints suggests a participation of cellular functions. However, it is not known to what extent TNPA and TNPD cooperate directly with cellular proteins in the transposition reaction. If they do these interactions would be highly conserved, because *En/Spm* transposes in several heterologous plant species.

6 The "Suppressor" Function of *En/Spm*

As described above, *dSpm* elements transpose if TNPA and TNPD are provided in *trans* by *En/Spm*. However, transposition may not be the only result of this interaction; it may also affect the activity of genes. McClintock (1954, 1961) observed both negative and positive *En/Spm*-mediated regulation of genes in which *dSpm* insertions reside. McClintock termed the ability of *En/Spm* to inhibit gene expression the suppressor function, and it is this type of control that is relatively well characterized in molecular terms.

Several suppressible alleles have been analyzed carrying a *dSpm* element within the transcribed regions (Schwarz-Sommer et al. 1985; Kim et al. 1987; Menssen et al. 1990). Usually, gene expression of these alleles is somewhat reduced due to the presence of *dSpm* but still detectable. The residual gene expression results from the fact that most of the *dSpm* sequences are removed by splicing from the pre-mRNA of the suppressible allele (Raboy et al. 1989; Menssen et al. 1990). However, this happens only in the absence of the autonomous *En/Spm* element. In its presence, the formation of the pre-mRNA is blocked by the suppressor function of *En/Spm*, and expression of the gene containing the *dSpm* element is thus abolished. The suppressor function resembles a negative regulatory circuit in which an *En/Spm*-encoded protein acts as a repressor. The repressor recognizes and binds to a defined *cis* element located within the *dSpm* elements. It was hypothesized (Gierl et al. 1985) that the bound protein sterically hinders progress of RNA polymerase through the gene, resulting in prematurely terminated transcripts at the site of *dSpm* insertion.

Transient expression studies of tobacco protoplasts (Grant et al. 1990, 1993) have identified TNPA as the only *En/Spm*-encoded protein that is necessary to suppress expression of a gene containing *dSpm* sequences. The tail-to-tail dimeric form of the TNPA-binding motif (see above) was identified as the minimal *cis*-acting sequence. Only this sequence seems to form a complex with TNPA that is stable enough to block progression of RNA polymerase through a gene.

Since the components of the suppressor system have been identified, it can now be used to control gene expression in transgenic plants (Grant et al. 1993). Suppressible genes could be created by insertion of the *cis* element in a position at which the *cis* element itself would not inhibit gene expression, for example between the transcription and translation start sites. The modified gene would then be normally active. However, expression could be suppressed if the modified gene were introduced by crossing into a genetic background in which TNPA is expressed.

The ability of *dSpm* elements to function as introns is due to the fact that they carry splice sites close to their termini. Splicing does not precisely remove all *dSpm* sequences from pre-mRNA; it usually leaves behind 12 bp of element sequences (MENSSEN et al. 1990). This process correlates with the size of *dSpm* elements, which is most effective if the elements are smaller than 2.2 kb. This feature probably mitigates the impact of transposable element insertion on gene expression.

7 Regulation of Transposition

Since transposition is tightly linked to mutation, it might be expected to be regulated to a level that is not deleterious to the cell. This is especially important in a system like *En/Spm*, where there are about 50–100 copies of *dSpm* elements present in the genome that could be potentially activated by *En/Spm* and could thus cause a burst of mutations within one cell cycle. The activity of the *En/Spm* system is regulated by element-specific mechanisms and by cellular functions.

7.1 Epigenetic Regulation of *En/Spm* Activity

The *En/Spm* system can reversibly and heritably change from an active to an inactive state (MCCLINTOCK 1958, 1971). Inactivity is correlated with an increased level of methylation of cytosine residues. Two regions adjacent to the transcription start site (position 209) have been identified as sites of altered C-methylation in maize (BANKS et al. 1988) and in transgenic tobacco (SCHLÄPPI et al. 1993), the upstream control region (UCR), containing the TNPA-binding sites, and the downstream control region (DCR), consisting of a 350-bp, GC-rich region. The DCR region is required for de novo methylation (SCHLÄPPI et al. 1994) and seems to initially attract the cellular methylation machinery. Methylation of the UCR and DCR regions seems to be a cellular mechanism to prevent high levels of *En/Spm* activity. It not only renders the *En/Spm* promoter inactive; methylated *dSpm* elements are also reduced in excision frequency when *trans* activated by an active *En/Spm* element (BANKS and FEDOROFF 1989). Therefore, methylated ends seem to inhibit excision of the element, probably because TNPA binding is hampered (GIERL et al. 1988).

Reactivation of an inactive element is usually a slow process requiring several plant generations. However, this process can be shortened by introduction of an active element (FEDOROFF 1989). Various lines of evidence suggest that TNPA is the element-encoded function responsible for this phenomenon.

7.2 Element-specific Regulation of Activity

Several element-specific properties seem to prevent high gene expression, which might be deleterious for the organism. The *En/Spm* promoter is relatively weak and is not stimulated by enhancers (RAINA et al. 1993), which seems to ensure that *En/Spm* expression is independent of the chromosomal location. Additionally, the single pre-mRNA is differentially spliced, yielding very little TNPD mRNA, which probably makes TNPD protein rate-limiting for transposition.

The finding that TNPA binds to sites in the UCR was the reason to hypothesize about an autoregulatory function of TNPA (GIERL et al. 1988), in addition to its requirement for transposition. Recent results (SCHLÄPPI et al. 1994) have clarified the part of TNPA in autoregulation of the *En/Spm* promoter. The effect of TNPA on the *En/Spm* promoter was analyzed in transgenic tobacco, using promoter luciferase reporter constructs. TNPA stimulates the activity of the inactive, methylated promoter. Activation is associated with reduced methylation of the UCR and DCR regions. On the other hand, TNPA acts as a repressor of the active unmethylated promoter. For repression the DNA-binding and the dimerization domains (see above) of TNPA are necessary, while for activation also the very carboxy-terminal sequences are required. Therefore, TNPA seems to be part of a sophisticated regulatory mechanism to ensure "tolerable" levels of transcription, by functioning both as a positive and negative regulator.

A different type of element-specific regulation was detected when the inhibitory effect of a particular *dSpm* element on transposition was found (CUYPERS et al. 1988). This element expresses an aberrant polypeptide (TNPR) that represents a fusion of TNPD and TNPA sequences. TNPR probably represses transposition by competitive inhibition of TNPA and/or TNPD function. Since there are many different *dSpm* elements distributed in the maize genome, it is feasible that products encoded by some of these elements could modulate transposition.

8 Evolution of the CACTA Transposable Element Family

Ac and *En/Spm* seem to be prototypes of two families of elements whose members are widely distributed in plants. In the case of *Ac*, remote relatives are found even in the animal kingdom (Kunze, this volume).

En/Spm was grouped into the so-called CACTA family (BONAS et al. 1984), based on the property that these elements produce a 3-bp target site duplication upon insertion and share nearly identical 13-bp TIRs. CACTA elements have been isolated from maize, snapdragon, soybean, pea, and Japanese morning glory (Table 1). Although there is a considerable size variation, the overall structural organization of the known autonomous elements of the CACTA family is strikingly similar. Among the other CACTA systems, the analysis of *Tam1* has contributed most of the molecular data. A comparison of *En/Spm* with *Tam1* will therefore best reveal the principle features of the CACTA family. Nevertheless, the properties described below also seem to hold true for *Tgm1* (RHODES and VODKIN 1988) and *Tpn1* (INAGAKI et al. 1994).

Tam1, like *En/Spm*, has highly structured ends and contains a single promoter close to the 5'-end. Two genes, termed *tnp 1* and *tnp 2*, are arranged in tandem and expressed by alternative splicing of a single pre-mRNA, yielding drastically different amounts of the respective mRNAs (NACKEN et al. 1991). The putative TNP2 protein is about 45% identical to TNPD at the amino acid level. An open reading frame identified in *Tgm1* (RHODES and VODKIN 1988) shares a similar amount of homology. The conservation of these polypeptides may indicate a common function. Since the 13-bp TIRs are also conserved, it was speculated (FREY et al. 1990) that TNPD interacts with the 13-bp TIRs and, in fact, may represent the endonucleolytic function which cleaves at the element's termini.

In contrast, TNPA of *En/Spm* and TNP1 of *Tam1* share no homology, nor was any homology detected with sequences of other CACTA elements. Therefore, the second gene products seem to be unrelated to each other. Interestingly, however, TNPA and TNP1 appear to be functionally equivalent. TNP1 binds to a 9-bp sequence motif that is reiterated several times in the termini of *Tam1*. This motif is not related to the 12-bp TNPA-binding sequence, however; it also occurs in monomeric and dimeric forms. By in vitro binding assays it was shown (TRENTMANN et al. 1993) that TNP1 forms three different complexes with these sequences in a manner very similar to TNPA (Fig. 3): complexes that contain one or two TNP1 molecules per oligonucleotide and a special complex with the tail-to-tail motif in which both TNP1 molecules seem to be in contact with the DNA. As in the case of *En/Spm*, the tail-to-tail forms of the TNP1-binding motif are close to the TIRs of *Tam1*.

In conclusion, the CACTA elements seem to require two proteins for transposition that interact with their respective *cis* determinants at the element's ends.

Table 1. The CACTA transposable element family

Element[a]	TIR	TSD	Size	Species	Reference
En/Spm	CACTACAAGAAAA	3	8287	maize	PEREIRA et al. (1986)
MPI1	CACTACCGGAATT	3	9 kb	maize	WEYDEMANN et al. (1987)
Tam1	CACTACAACAAAA	3	15164	snapdragon	NACKEN et al. (1991)
Tam2	CACTACAACAAAA	3	5187	snapdragon	KREBBERS et al. (1987)
Tam4	CACTACAAAAAAA	3	4329	snapdragon	LUO et al. (1991)
Tam7	CACTACAACAAAA	3	7 kb	snapdragon	[d]
Tam8	CACTACAACAAAA	3	3 kb	snapdragon	[d]
Tam9	CACTACAACAAAA	3	5.5 kb	snapdragon	[d]
Tgm	CACTATTAGAAAA	3	1.6-12 kb	soybean	RHODES and VOKIN (1988)
Pis1	CACTACGCCAAA	3	2.5 kb	pea	SHIRSAT (1988)
Tpn1	CACTACAAGAAAA[c]	3	6.4 kb	Japanese morning glory	INAGAKI et al. (1994)

TIR, Terminal inverted repeat; TSD, target size duplication upon insertion, in bp.
[a]Autonomous elements are in bold type.
[b]Sizez in kb refer to elements that have not completely sequenced.
[c]Tpnl has a 28-bp TIR, only the first 13 are indicated.
[d]Z.S. Schwarz-Sommer, personal communication.

One function (TNPD) and the 13-bp TIRs are evolutionarily conserved. The other function (TNPA) is specific for each element, as are its binding motifs that make up the subterminal repetitive regions flanking the 13-bp TIRs. Although not related evolutionarily, TNPA and its analogous seem to exert a similar mechanistic function. It is possible that the CACTA elements evolved independently in each species. The creation of a CACTA element requires the combination of two proteins and their DNA target sequences. The two proteins could have been derived from cellular functions that interact with DNA. The endonucleolytic function (TNPD) may be conserved, because there is a limited pool of such cellular proteins suitable to create a transposon. The other function (TNPA) is not conserved and is species specific. This could indicate that there are several cellular candidates that can potentially fulfill TNPA function. TNPA and TNP1 both contain DNA-binding and dimerization domains. Therefore, it is tempting to speculate that these proteins have evolved from the pool of cellular transcription factors; indeed, as described above, TNPA has the capacity to act as a transcription regulator.

9 Transposons as Tools for the Isolation of Genes and Determination of Gene Function

Once a transposable element has been isolated molecularly, it can be used as a probe to clone genes that are mutated by insertion of this element. In addition to *En/Spm*, other transposable elements that have been characterized in maize and snapdragon (for review, see Gierl and Saedler 1992) have been used successfully for gene isolation. The first step of this procedure is called transposon tagging. This means that the mutation caused by transposon insertion is identified in a genetic screen. Subsequently, the gene can be isolated by cloning the DNA sequences flanking the transposable element insertion. The technique requires no prior knowledge concerning the nature of the product of the tagged gene; it depends merely upon the expression of a mutant phenotype. In this way, genes involved in development, pathogen resistance, and biochemical pathways have been isolated.

The finding that *Ac* and *En/Spm* transpose in many heterologous transgenic plants has permitted extention of the transposon tagging technique to other plant species in which no suitable elements have been identified (for review, see Balcells et al. 1991; Haring et al. 1991; Shimamoto 1994). The progress made in this area, especially for *Ac* but also for *En/Spm*, has led to the isolation of genes, for example a flower-color gene from the petunia (Chuck et al. 1993), *Cf-9* of tomato, conferring resistance to a fungal pathogen (Jones et al. 1994), and a male sterility gene from *Arabidopsis* (Aarts et al. 1993).

A fairly recent development, first employed for *Drosophila,* is the attempted use of transposons in order to define the function of plant genes for which only

DNA sequence data are known. A large population is required in which statistically every gene is impaired by a transposon insertion. The members of the population are grouped into batches of 100–500 individuals and the heterozygous mutant alleles for a particular gene are subsequently identified using a PCR-based strategy involving primers derived from the transposon and, for example, a cDNA sequence. Once a putative mutant has been identified within a batch, and later as an individual, the mutant phenotype generated by the original element insertion can be uncovered by selfing. Although this procedure involves numerous DNA isolations, the progress made in automatization of standard molecular techniques will probably help to establish this approach as a useful tool for genome analysis.

References

Aarts MG, Dirkse WG, Stiekema WJ, Pereira A (1993) Transposon tagging of a male sterility gene in *Arabidopsis*. Nature 363: 715–717

Balcells I, Swinburne J, Coupland G (1991) Transposons as tools for the isolation of plant genes. Trends Biotech 9: 31–37

Banks JA, Fedoroff N (1989) Patterns of developmental and heritable change in methylation of the suppressor-mutator transposable element. Dev Genet 10: 425–437

Banks JA, Masson P, Fedoroff N (1988) Molecular mechanisms in the developmental regulation of the maize suppressor-mutator transposable element. Genes Dev 2: 1364–1380

Bonas U, Sommer H, Saedler H (1984) The 17-kb *Tam1* element of *Antirrhinum majus* induces a 3-bp duplication upon integration into the chalcon synthase gene. EMBO J 3: 1015–1019

Cardon GH, Frey M, Saedler H, Gierl A (1993a) Mobility of the maize transposable element *En/Spm* in *Arabidopsis thaliana*. Plant J 3: 773–784

Cardon GH, Frey M, Saedler H, Gierl A (1993b) Definition and characterization of an artificial *En/Spm*-based transposon tagging system in transgenic tobacco. Plant Mol Biol 23: 157–178

Chuck G, Robbins TP, Nijjar C, Ralston E, Courtney-Gutterson N, Dooner HK (1993) Tagging and cloning of a petunia flower color gene with the maize transposable element activator. Plant Cell 5: 371–378

Cuypers H, Dash S, Peterson PA, Saedler H, Gierl A (1988) The defective En–I102 element encodes a product reducing the mutability of the *En/Spm* transposable element system of *Zea mays*. EMBO J 7: 2953–2960

Dash S, Peterson PA (1994) Frequent loss of the *En* transposable element after excision and its relation to chromosome replication in maize (*Zea mays* L.). Genetics 136: 653–671

Fedoroff N (1989) Maize transposable elements and development. Cell 56: 181–191

Frey M, Tavantzis SM, Saedler H (1989) The maize En–1/Spm element transposes in potato. Mol Gen Genet 217: 172–177

Frey M, Reinecke J, Grant S, Saedler H, Gierl A (1990) Excision of the *En/Spm* transposable element of *Zea mays* requires two element-encoded proteins. EMBO J 9: 4037–4044

Gierl A, Saedler H (1992) Plant-transposable elements and gene tagging. Plant Mol Biol 19: 39–49

Gierl A, Schwarz-Sommer ZS, Saedler H (1985) Molecular interactions between the components of the En-I transposable element system of *Zea mays*. EMBO J 4: 579–583

Gierl A, Lütticke S, Saedler H (1988) TnpA product encoded by the transposable element En-1 of *Zea mays* in a DNA-binding protein. EMBO J 7: 4045–4053

Goodall GJ, Filipowicz W (1991) Different effects of intron nucleotide composition and secondary structure on pre-mRNA splicing in monocot and dicot plants. EMBO J 10: 2635–2644

Grant SR, Gierl A, Saedler H (1990) *En/Spm* encoded tnpA protein requires a specific target sequence for suppression. EMBO J 9: 2029–2035

Grant SR, Hardenack S, Trentmann S, Saedler H (1993) Functional *cis*-element sequence requirements for suppression of gene expression by the TNPA protein of the *Zea mays* transposon *En/Spm*. Mol Gen Genet 241: 153–160

Greenblatt IM (1984) A chromosomal replication pattern deduced from pericarp phenotypes resulting from movements of the transposable element, modulator, in maize. Genetics 108: 471–485
Haring MA, Caius, MT, Rommens H, Nijkamp HJJ, Hille J (1991) The use of transgenic plants to understand transposition mechanisms and to develop transposon tagging strategies. Plant Mol Biol 16: 449–461
Inagaki Y, Hisatomi Y, Suzuki T, Kasahara K, Iida S (1994) Isolation of a suppressor-mutator/enhancer-like transposable element, *Tpn1*, from Japanese morning glory bearing variegated flowers. Plant Cell 6: 375–383
Jones DA, Thomas CM, Hammond-Kosack KE, Balint-Kurti PJ, Jones JDG (1994) Isolation of the tomato Cf-9 gene for resistance to *Cladosporium fulvum* by transposon tagging. Science 266: 789–793
Kim HY, Schiefelbein JW, Raboy V, Furtek DB, Nelson OE (1987) RNA splicing permits expression of a maize gene with a defective suppressor-mutator transposable element in an exon. Proc Natl Acad Sci USA 84: 5863–5867
Krebbers E, Hehl R, Piotriwiak R, Lönning EE, Sommer H, Saedler H (1987) Molecular analysis of paramutant plants of *Antirrhinum majus* and the involvement of transposable elements. Mol Gen Genet 209: 499–507
Luo D, Coen ES, Doyle S, Carpenter R (1991) Pigmentation mutants produced by transposon mutagenesis in *Antirrhinum majus*. Plant J 1: 59–69
Masson P, Fedoroff N (1989) Mobility of the maize suppressor-mutator element in transgenic tobacco cells. Proc Natl Acad Sci USA 86: 2219–2223
Masson P, Rutherford G, Banks J, Fedoroff N (1989) Essential large transcripts of the maize Spm transposable element are generated by alternative splicing. Cell 58: 755–765
Masson P, Strem M, Fedoroff N (1991) The tnpA and tnpD gene products of the Spm element are required for transposition in tobacco. Plant Cell 33: 73–85
McClintock B (1947) Cytogenetic studies of maize and *Neurospora*. Carnegie Inst Washington Yearb 46: 146–152
McClintock B (1948) Mutable loci in maize. Carnegie Inst Washington Yearb 47: 155–169
McClintock B (1954) Mutations in maize and chromosomal aberrations in *Neurospora*. Carnegie Inst Washington Yearb 53: 254–260
McClintock B (1958) The suppressor-mutator system of control of gene action in maize. Carnegie Inst Washington Yearb 57: 415–429
McClintock B (1961) Further studies on the suppressor-mutator system of control of gene action in maize. Carnegie Inst Washington Yearb 60: 469–476
McClintock B (1971) The contribution of one component of a control system to versatility of gene expression. Carnegie Inst Washington Yearb 70: 5–17
Menssen A, Höhmann S, Martin W, Schnable PS, Peterson PA, Saedler H, Gierl A (1990) The *En/Spm* transposable element of *Zea mays* contains splice sites at the termini generating a novel intron from a dSpm element. EMBO J 9: 3051–3057
Muszynski MG, Gierl A, Peterson PA (1993) Genetic and molecular analysis of a three-component transposable-element system in maize. Mol Gen Genet 237: 105–112
Nacken WKF, Piotrowiak R, Saedler H, Sommer H (1991) The transposable element *Tam1* from *Antirrhinum majus* shows structural homology to the maize transposon *En/Spm* and has no sequence specificity of insertion. Mol Gen Genet 228: 201–208
Nash HA (1990) Bending and supercoiling of DNA at the attachment site of bacteriophage lambda. Trends Biochem Sci 15: 222–227
Pereira A, Saedler H (1989) Transcriptional behaviour of the maize *En/Spm* element in transgenic tobacco. EMBO J 8: 1315–1321
Pereira A, Cuypers H, Gierl A, Schwarz-Sommer ZS, Saedler H (1986) Molecular analysis of the *En/Spm* transposable element system of *Zea mays*. EMBO J 5: 835–841
Novick, Peterson (1981) Transposition of the Enhancer controlling element system of maize.
Peterson PA (1953) A mutable pale green locus in maize. Genetics 45: 115–133
Peterson PA (1987) Mobile elements in plants. CRC Crit Rev Plant Sci 6: 105–208
Raboy V, Kim HY, Schiefelbein JW, Nelson OE (1989) Deletions in a dSpm insert in a maize bronze-1 allele alter RNA processing and gene expression. Genetics 122: 695–703
Raina R, Cook D, Fedoroff N (1993) Maize Spm transposable element has an enhancer-insensitive promoter. Proc Natl Acad Sci USA 90: 6355–6359
Rhodes PR, Vodkin LO (1988) Organization of the Tgm family of transposable elements in soybean. Genetics 120: 597–604

Saedler H, Nevers P (1985) Transposition in plants: a molecular model. EMBO J 4: 585–590

Schiefelbein JW, Raboy V, Kim HY, Nelson OE (1988) Molecular characterization of suppressor-mutator (Spm)-induced mutations at the bronze-1 locus in maize: the bz-m 13 alleles. In: Nelson OE (ed) Proceedings international symposium on plant transposable elements. Plenum, New York, pp 261–278

Schläppi M, Smith D Fedoroff N (1993) TnpA *trans*-activates methylated maize suppressor-mutator transposable elements in transgenic tobacco. Genetics 133: 1009–1021

Schläppi M, Raina R Fedoroff N (1994) Epigenetic regulation of the maize Spm transposable element: novel activation of a methylated promotor by TnpA. Cell 77: 427–437

Schwarz-Sommer ZS, Gierl A, Berntgen R, Saedler H (1985) Sequence comparison of 'states' of al-ml. EMBO J 4: 2439–2443

Shimamoto K (1994) Gene expression in transgenic monocots. Curr Opin Biotech 5: 158–162

Shirsat AH (1988) A transposon-like structure in the 5' flanking sequence of a legumin gene from *Pisum sativum*. Mol Gen Genet 212: 129–133

Sullivan TD, Strelow L1, Illingworth CA, Phillips RL, Nelson OE (1991) Analysis of maize brittle-1 alleles and a defective suppressor-mutator-induced mutable allele. Plant Cell 3: 1337–1348

Trentmann SF, Saedler H, Gierl A (1993) The transposable element *En/Spm*-encoded TNPA protein contains a DNA binding and a dimerization domain. Mol Gen Genet 238: 201–208

Weydemann U, Wienand U, Niesbach-Klösgen U, Peterson PA, Saedler H (1987) Cloning of the transposable element Mpil from c2-m3. Maize Genet Coop News1 62: 48

[illegible]

The Maize Transposable Element *Activator (Ac)*

R. Kunze

Institut für Genetik, Universitat zu Köln, Weyertal 121, D-50931 Köln, Germany
Present address: Institut für Genetik, Universität München, Maria-Ward-Str. 1a, D-80638 München, Germany

1 Introduction

Almost 50 years after the discovery of the autonomous transposable element *Activator (Ac)* and its nonautonomous relatives, termed *Dissociation (Ds)*, in maize by Barbara McClintock, we have learned that transposons occur in many organisms; they may even be ubiquitous inhabitants of both prokaryotes and eukaryotes. Meanwhile, there are more than ten DNA-based transposable element "families" known in maize which consist of a small number of autonomous and a much larger number of inactive, nonautonomous elements. Based on the structure of an *Ac*-like sequence in pearl millet, it was calculated that the *Ac/Ds* elements have existed in the grasses for at least 25 million years (MacRae et al. 1994).

Activator (Ac), Enhancer/Suppressor-mutator (En/Spm), and Mutator (Mu) are the genetically and molecularly best characterized autonomous DNA-based transposons in maize. In contrast to the retrotransposons, which replicate through an RNA intermediate, these elements can excise nonreplicatively from one locus and reinsert somewhere else in the genome. *Ac, En/Spm*, and *Mu* and their nonautonomous derivatives have been successfully used for transposon tagging in maize (for review see Walbot 1992), and meanwhile, *Ac* and *En/Spm* have also proven useful for gene tagging in heterologous plants in which no active endogenous transposons were identified.

The maize-transposable elements including *Ac* and the pertinent nonautonomous *Ds* elements have been comprehensively reviewed (Döring and Starlinger 1986; Fedoroff 1983, 1989; Gierl et al. 1989). Therefore, the emphasis of this review is on recent progress in the molecular analysis of the transposition reaction and its regulation.

2 The Structure of *Ac* and *Ds* Elements

2.1 The DNA Structure of *Ac*

Ac is a simply structured and rather small transposable element (Fig. 1). It is 4565 bp long and has 11-bp terminal inverted repeats (IRs), whose outermost nucleotides are not complementary (Pohlman et al. 1984a, b; Müller-Neumann et al. 1984; Dooner et al. 1988). With respect to its size, its structural organization, and the sequence of the IRs, *Ac* has similarities to the *Tam3* element from *Antirrhinum majus* (Hehl et al. 1991) and the *hobo* and *P* elements from *Drosophila melanogaster* (Streck et al. 1986; O'Hare and Rubin 1983).

The nucleotide composition of *Ac* is highly biased. The G+C contents of the terminal 240 bp at the left end (or 5'-end) and the right end (or 3'-end) are 45% and 40%, respectively. The 5'-end contains 26 and the 3'-end 24 CpG dinucleotides; however, both ends have only one GpC dinucleotide. In contrast, the G+C content of the long untranslated leader is 68%, without a bias in CpG versus

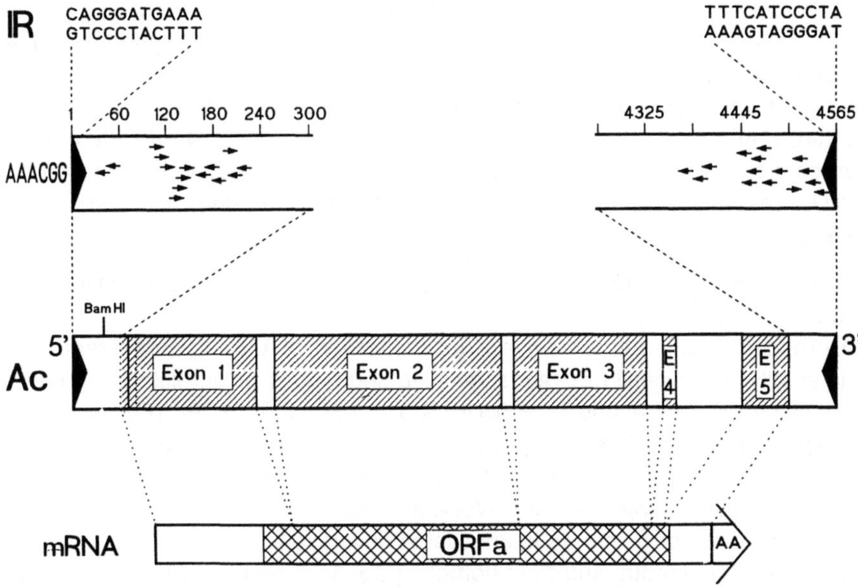

Fig. 1. The *Ac* element is 4565 bp in length and has 11-bp imperfect terminal inverted repeats whose sequences are shown in the *upper line* (*IR*). The distribution of AAACGGs and closely related motifs, the TPase-binding sites, within the subterminal *Ac* regions is indicated by *arrows* underneath (AAACGG). The left *Ac* end containing the unique *Bam*HI site at position 181 is designated the 5'-end. Transcription of *Ac* is initiated at multiple sites between positions 280 and 380. The 3.5-kb mRNA consists of 5 exons. The 5'- and 3'-untranslated regions are indicated as *stippled boxes; ORFa* is the 2421-nucleotide-long TPase open reading frame

GpC distribution. Within the coding region, which contains 38% G+C, CpG dinucleotides are clearly under-represented (Kunze et al. 1988). This uneven sequence composition in different segments of *Ac* reflects the different functions of these segments and presumably different roles of DNA methylation in them (see Sects. 6.3 and 7.6).

The prevalence of CpG dinucleotides in the ends of *Ac* reminds one of CpG islands (Bird 1986; Antequera and Bird 1988; Gardiner-Garden and Frommer 1992). It was suggested that CpG dinucleotides are evolutionarily conserved in DNA regions, like certain promoters which are frequently bound by proteins and in which C-methylation may have a regulatory effect. The numerous CpG motifs in the ends of *Ac* could also indicate a (permanent?) protection of these sequences by proteins. In fact, many of them are situated within the recognition sites for the transposase protein (see Sect. 6.2).

2.2 The Nonautonomous *Ds* Elements

Ds elements are the nonautonomous members of the *Ac/Ds* transposable element family (McClintock 1948). Hybridization experiments indicate that from

30 to more than 100 copies of *Ac*-related sequences occur in the maize genome (GEISER et al. 1982; FEDOROFF et al. 1983; THERES et al. 1987). Among them are the *Ds* elements which can be mobilized in *trans* by *Ac*, but also immobile transposon fragments which have lost essential sequences required in *cis* for transposition (R. Kunze, unpublished results). The sequence analysis of several different *Ds* elements allowed an initial definition of the *trans*- and *cis*-acting functions of *Ac* and *Ds*. Based on their structures, *Ds* elements can be divided into three distinct classes:

1. The simple *Ds* elements are deletion derivatives of *Ac* which have lost internal, *trans*-acting sequences.
2. Composite *Ds* elements have retained the termini of *Ac* to variable extents, but internally they consist of rearranged *Ac* sequences or unrelated segments (reviewed by DÖRING and STARLINGER 1986 and FEDOROFF 1989). The composite *wxB4::Ds* element has only 259 bp of the 5'-end and 317 bp of the 3'-end of uninterrupted *Ac* termini, suggesting that these sequences are sufficient in *cis* for transposition (VARAGONA and WESSLER 1990).
3. The *Ds*1 class of elements is unusual in several aspects. *Ds*1 elements are only 0.4 kb in length, and their homology to *Ac* is restricted to 13 bp of the 5'-end and 26 bp of the 3'-end of *Ac* (SUTTON et al. 1984; GERLACH et al. 1987). The *Ds*1 internal sequences are very AT rich (78%) and not homologous to *Ac*. In simple *Ds* elements, *Ac* termini as short as those in *Ds*1 are not sufficient for transposition (COUPLAND et al. 1989). Yet, *Ds*1 elements are unequivocally responsive to *As*, as the capability of *Ac* to mobilize *Ds*1 elements has been verified in transgenic tobacco plants (LASSNER et al. 1989) and in agroinfected maize plants (SHEN and HOHN 1992). However, *Ds*1 can also be mobilized by the *Ubiquitous* (*Uq*) transposable element in maize, which does not mobilize *Ds* elements of the other two classes (PISABARRO et al. 1991; CALDWELL and PETERSON 1992).

2.3 The *Ac2* Element Is Presumably a Weak Variant of *Ac*

The *Ac*2 element is an autonomous member of the *Ac/Ds* transposable element system, whose transactivation properties differ markedly from those of *Ac* (RHOADES and DEMPSEY 1983). In contrast to *Ac*, one copy of the *Ac*2 element does not transactivate any *Ds* elements. Two copies induce infrequent excisions of the composite *Ds*2 element in the *bz2-m* allele(THERES et al. 1987), whereas mobilization of other *Ds* elements requires four copies of *Ac2*. The timing of *Ac2*-induced mutations is always late in endosperm development, and no inverse dose effect, which is characteristic for *Ac* (see Sect. 7.3), has been observed. It is therefore assumed that *Ac2* is a weak variant of *Ac* (DEMPSEY 1993). Interestingly, the activity of *Ac2* seems to be temperature sensitive, a phenomenon which has not been observed with *Ac*. Exposure of the developing maize ear containing the *Ac2*-*bz2-m* system to elevated temperatures stimulates a great increase in excision frequency of the *Ds2* element (OSTERMAN 1991). This effect is opposite to the temperature response of the *Tam* elements in *Antirrhinum majus*, whose transposition frequencies at 25°C are 10–1000 times lower than at 15°C (for review, see COEN et al. 1989).

3 Transposition Substrate Requirements of *Ac* and *Ds*

3.1 The Target Site Duplication Is Not Required for Excision

Like most other transposable elements, *As* and *Ds* create during integration short target site duplications which subsequently flank the element directly. These duplications are usually retained upon excision ("footprints"), but excision frequently is imprecise and associated with nucleotide additions, deletions, or inversions at their junction (SUTTON et al. 1984; WECK et al. 1984; POHLMAN et al. 1984a). Two slightly different transposition models propose 8-bp staggered cuts at the insertion site to explain the generation of the target site duplication and cuts staggered between 1bp and 8 bp in both copies of the duplication in conjunction with exonucleolytic and DNA repair processes to account for the diversity of excision products (PEACOCK et al. 1984; SAEDLER and NEVERS 1985; COEN et al. 1986; for review, see GIERL and SAEDLER 1989).

In several instances it was shown that the target site duplication is not required for the excision reaction. The *bz-m2::Ac* allele gave rise to the *bz-s: 2114(Ac)* allele, in which an adjacent deletion has removed one of the 8-bp direct repeats. Nevertheless, the *Ac* in this allele transposes with a frequency similar to that in the progenitor allele (DOONER et al. 1988). Another example is the *Ac* element in the *P-vv* allele, which is not flanked by a duplication but transposes frequently (LECHELT et al. 1989; PETERSON 1990). The target site duplication is also dispensable for the excision of *Ds* elements. A 2-kb *Ds* can transpose from a 3-kb "double-*Ds*" structure (DÖRING et al.1989), and the *Ds*1 element can excise from the maize streak virus genome in the absence of a flanking duplication (SHEN et al. 1992). As the flanking sequences have apparently no influence on the excision reaction, the 8-bp duplication created during intergration is obviously not involved in any sequence-specific protein/DNA interaction during the excision reaction, nor is the presence of any directly repeated motif required.

3.2 The Terminal Inverted Repeats Are Essential for Transposition

The correct sequence of both IRs is an essential determinant for transposition triggered by the *Ac* TPase, as a replacement of the four terminal bases of the 3'-IR of *Ac* (HEHL and BAKER 1989) or of the five terminal bases of the 5'-IR (HEALY et al. 1993) as well as the replacement of the 11 bp *Ac*-IRs by the related 12 bp IRs from the *Antirrhinum majus Tam3* element (S. Chatterjee and R. Kunze, unpublished results) immobilizes the element completely (Table 1).

On the other hand, the outermost nucleotide of the 11-bp IRs may be mismatched, as is inferred by the observations that:

1. The outermost nucleotides of *Ac* elements are not complementary.
2. The outermost nucleotides of the *Ds*1 element in the *bz-wm* allele are mismatched as in *Ac*, but with the cytosine residue at its 3'-end.
3. The *Ds*1-01 element has 11-bp IRs with mismatched outermost nucleotides. However, the capability of this element to transpose has not yet been shown.

Apparently, certain mismatches inside the IRs are also tolerated. The fourth nucleotide in the 5'-IR of the *Ds*1 element in the *ruq-st* allele is replaced by adenine. The same substitution is also found in the 3'-IR of a *Ds*1 element from the teosinte species *Zea perennis* (Table 1). It may be of interest that the IRs of *Tam3* and the *Drosophila melanogaster* transposon *hobo* also carry an adenine at the corresponding position (Table 2).

Table 2 shows an alignment of the IRs of a number of transposable elements from diverse eukaryotes. According to their IR sequences and lengths of target site duplications, the elements can be classified into an *Ac*-like group, whose members create 8-bp duplications, and a "CACTA family", whose members terminate with a CACTA motif and create 3-bp duplications. The *Tat*1 and the *Jordan* elements differ from both groups, in that they are the only elements which do not terminate with CA or TA. Some IRs of *Ac*-like elements are conspicuously homologous to the *Ac/Ds* IRs, whereas the central sequences of these elements have no obvious homologies. The transposase proteins of only four of these elements are known (*Ac, Tam3, hobo, P*), and the amino acid sequences of the *Ac, Tam3,* and *hobo* TPases are over about 600 amino acids co-linear and remarkably well conserved (Fig. 4) (Calvi et al. 1991; Feldmar and Kunze 1991). Thus, it is conceivable that some, if not all, of these transposable elements have a common evolutionary origin and were horizontally transmitted between the plant and animal kingdoms (Calvi et al. 1991), perhaps in a mode similar to that

Table 1. Alignment of the inverted repeats of different *Ac* and *Ds* isolates. Small letters indicate deviations from the IR sequences of *Ds* elements. In the column "TP" is noted whether the element can transpose in the presence of an active *Ac* element or not

Element (allele)	5'-IR	3'-IR	TP	Reference
Ds (sh-m5933)	TAGGGATGAAA ..	TTTCATCCCTA	+	Döring et al. (1984)
Ac (wx-m7::Ac)	cAGGGATGAAA ...	TTTCATCCCTa	+	Müller-Neumann et al. (1984)
Ds1 (bz-wm)	TAGGGATGAAA ...	TTTCATCCCTg	+	Schiefelbein et al. (1988)
Ds1-01	gAGGGATGAAA ...	TTTCATCCCTg	?	Gerlach et al. (1987)
Ds1 (ruq-st)	TAGaGATGAAA ..	TTTCATCCCTA	+	Pisabarro et al. (1991)
Ds1 (Zea perennis)	TAGGGATGAAA ...	TTTCATCtCTA	+	MacRae and Clegg (1992)
Ac-18	cAGGGATGAAA	TTTCATCtgag	–	Hehl and Baker (1989)
Ac3	gAGctATGAAA ...	TTTCATCCCTA	–	Healy et al. (1993)
Tam3	TAaaGATGtgAA	TTcaCATCttTA	(–)	Hehl et al. (1991) (S. Chatterjee, unpublished)

Table 2. Alignment of the inverted repeat sequences of transposable elements from different organisms. If the two IRs are not perfect both are shown.

Element	5'/3'	- IR sequence	IR (bp)	TSD (bp)	Species	Reference
Ac/Ds	5'	CAGGG.ATGAAA				
	3'	TAGGG.ATGAAA	11	8	Maize	MÜLLER-NEUMANN et al. (1984)
Bg		CAGGG	5	8	Maize	HARTINGS et al. (1991)
rDt	5'	CAGtGttTtAAAtc				
	3'	CAatGtATtAAAtc	14	8	Maize	BROWN et al. (1989)
Tag1		CAatG.tTttcAcgc..	22	8	*Arabidopsis*	TSAY et al. (1993)
Tam3		TAaaG.ATGtgAa	12	8	*Antirrhinum*	HEHL et al. (1991)
Tpc1	5'	TAGGG..TGtAAa				
	3'	TAGGG.cTGtAAa	11/12	8	Parsley	HERRMANN et al. (1988)
Ips-r		TAGGGg.TGgcAa	12	8	*Pisum sativum*	BHATTACHARYYA et al. (1990)
Tst1	5'	CAGGGg.cGtAt				
		CAGaGg.cGtAt	11	8	Potato	KÖSTER-TÖPFER et al. (1990)
dTnp1	5'	CAGtGc cGgctcaa				
	3'	CAaGGg cGgctcaa	14	8	*N. plumbaginif.*	MEYER et al. (1994)
dTph1		CAGGGg.cGgAgc	12	8	*Petunia*	GERATS et al. (1990)
Gulliver		CAGGGgtcGtAtctt	15	8	*Chlamydomonas*	FERRIS (1989)
hobo		CAGaG.Aactgca	12	8	*Drosophila*	STRECK et al. (1986)
P		CA.tG.ATGAAAtaa..	31	8	*Drosophila*	O'HARE and RUBIN (1983)
1723	5'	TAGGG.ATGtAgcga..				
	3'	TAGaG.ATGtcgcgg..	16/18	8	*Xenopus laevis*	KAY and DAWID (1983)
Tat1		tgtGG.ATGtcgga	13	5	*Arabidopsis*	PELEMAN et al. (1991)
En/Spm		CActacAaGAAAa	13	3	Maize	PEREIRA et al. (1995)
Tpn1		CActacAaGAAAaatg..	28	3	Japanese morning glory	INAGAKI et al. (1994)
Tam1		CActacAacAAAa	13	3	*Antirrhinum*	BONAS et al. (1984)
Tgm1		CActattaGAAAa	13	3	Soybean	RHODES and VODKIN (1988)
Pis1		CActacgccAAA	12	3	*Pisum sativum*	SHIRSAT (1988)
Jordan	5'	CcctatggcAta				
	3'	CccaatggcAta	12	3	*Volvox*	MILLER et al. (1993)
Tc1		CAGtG.cTGgccaaa	54	2	*C. elegans*	ROSENZWEIG et al. (1983)
pogo		CAGtataattcgctt..	21	2	*Drosophila*	TUDOR et al. (1992)

IR, Inverted repeat sequence; *TSD*, target site duplication

suggested for the horizontal transfer of *P* and *mariner* elements between insects (HOUCK et al. 1991; ROBERTSON 1993).

3.3 Subterminal Sequences Are Involved in Transposition

The observation that *Ac* can autonomously transpose in transgenic tobacco plants (BAKER et al. 1986, 1987) opened the way for the fine analysis of the *Ac* sequences required in *cis* and in *trans* for transposition. By analyzing the effects of different internal deletions on the excision frequency, COUPLAND et al. have shown that (a) the untranslated leader sequence is neither in *cis* nor in *trans* required for the transposition function; (b) deletions of part of the *Ac* open reading frame mutate the autonomous *Ac* into a nonautonomous *Ds* element; and (c) 238 or more nucleotides at the 5'-end and 209 or more nucleotides at the 3'-end of the transposable element are required for wild-type excision frequencies. At either end of *Ac* internal deletions extending farther towards the termini result in a gradual reduction in excision frequency. The element is immobilized when 116 bp or less at the 5'-end or less than 102 bp at the 3'-end are retained (COUPLAND et al. 1988, 1989).

Obviously, in addition to the IRs, about 240 bp subterminal nucleotides of *Ac* are essential for transposition. In fact, these sequences are retained in all known simple and composite *Ds* elements, whereas they are lacking in the *Ds*1 family. Since *Ds*1 can also be mobilized by *Uq*, which does not act on the other *Ds* elements (CALDWELL and PETERSON 1992), it seems possible that the transposition mechanism of *Ds*1 elements differs from that of *Ac* and the other *Ds* classes.

A very similar structural organization of the *cis*-acting sequences is found in the *En/Spm* family of maize transposons. The ends of *En/Spm* consist of perfect 13-bp IRs and repetitive 12-bp motifs, differently arranged in each end. Partial deletions of these motifs in one end result in reduced transposition and complete deletion in immobilization (for review see GIERL et al. 1989; FEDOROFF 1989).

3.4 A Left and a Right Transposon End Are Obligatory for Transposition

Although the 5'- and 3'-subterminal sequences of *Ac* and of simple and composite *Ds* elements have a similar bias in CpG versus GpC dinucleotides, their sequences differ. These differences seem to reflect a functional difference, because elements having either two 5'-ends or two 3'-ends neither transpose (COUPLAND et al. 1989) nor cause chromosome breakage (ENGLISH et at. 1993).

The conclusion from these results is that the determinants for transposability of *Ac* and *Ds* are a combination of two nearly perfect IRs of defined sequence with two differently structured but similarly composed subterminal regions. The *Ac* TPase – though able to bind to each single transposon end – obviously does not induce double-strand breaks at a single transposon end, whereas DNA cleavage and coordinate religation of the donor DNA is induced when a left and a right transposon end come in contact (presumably mediated by the TPase). This lack of activity at a single transposon end could be explained if the TPase bound to one end cleaved exclusively at the other end.

3.5 The Chromosomal Chromatin Structure Is Not Required for Transposition

In the plant genome, the transposable elements are embedded within the highly organized chromatin structure. Yet a location on a chromosome is not required for transposition, as *Ac* and *Ds* elements can efficiently excise from a wheat dwarf virus replicon after naked DNA transfection into maize, wheat, and rice protoplasts (LAUFS et al. 1990), and also from a maize streak virus vector after agroinfection into maize plants (SHEN and HOHN 1992). Excision of *Ac* and *Ds* can also happen from plasmids transfected into *Petunia* (HOUBA-HÉRIN et al. 1990) and *Nicotiana plumbaginifolia* protoplasts. In the latter species, it was shown that the *Ds* can also reintegrate into the plant genome (HOUBA-HÉRIN et al. 1994). Therefore, it appears that a chromosomal environment is not necessary for *Ac* or *Ds* excision, although it cannot be excluded that the DNA introduced into the plant cells aquires a chromatin-like structure before the element transposes.

4 Characteristics of the Transposition Reaction

4.1 *Ac* transposes in a Nonreplicative Manner and During Replication

The phenotype of the unstable *P-vv* allele, kernels with variegated red-striped pericarp, is caused by the insertion of an *Ac* element into the *P* gene. When *Ac* transposes from the *P-vv* allele, about 80% of the time so-called twin-sector mutations are generated, i.e., contiguous sectors of fully revertant, red pericarp and light, variegated pericarp indicative of an increased *Ac* copy number (GREENBLATT 1974). Genetic analyses of the reversion patterns and cloning and sequencing of *Ac* target sites in mitotic daughter cell lineages of twin sectors led to the following conclusions:

1. *Ac* transposes by a nonreplicative mechanism. The element is physically excised from the donor site and reinserts in a new position (Fig. 2) (GREENBLATT and BRINK 1962).
2. The element transposes primarily, if not exclusively, during or shortly after replication. Accordingly, during the other phases of the cell cycle, either the transposon is in a state which is not recognized by the transposition machinery or there is no active transposase present (GREENBLATT 1984; CHEN et al. 1987).
3. *Ac* transposes from only one of two daughter chromosomes. This means that the two daughter *Acs* have to be physically distinct after replication, and the transposition apparatus including the transposase can distinguish between them (FEDOROFF 1989; CHEN et al. 1992).
4. *Ac* and *Ds* elements reinsert into both replicated (Fig. 2a) and unreplicated target sites (Fig. 2b). Thus, the transposition machinery apparently is insensitive to replication-dependent alterations of the state of the target DNA (i.e., methylation) (CHEN et al. 1992).

However, it should be noted that the efficient excision of *Ac* and *Ds* from plasmids which supposedly do not replicate in plant cells might indicate that DNA replication is not an absolute requirement for the excision reaction (HOUBA-HÉRIN et al. 1990, 1994).

4.2 *Ac* Transposes Preferentially to Physically Linked Target Sites

In maize, *Ac* has a pronounced preference for short-range transpositions. In one study, after transpositions of *Ac* away from the *P-vv* allele, 61% of reinserted *Acs* remained closely linked to *P*. However, no reinsertions within 4 cM proximal of the *P* gene were found, whereas 23 reinserted *Ac* elements mapped within the same distance distal to *P* (GREENBLATT 1984). The asymmetrical distribution around the donor site is rather unusual, because in several other studies the directions of transpositions were random, and *Ac* transposition from *P-vv* into a site located 15 kb proximal to *P* was recently described (CHEN et al. 1992). In the case of transposition of *Ac* from the *bz–m2::Ac* allele, about 60% of the elements reintegrate within 5 cM distance on either side of bz (DOONER and BELACHEW 1989). Over 250 *Ac* reinsertions into the *P* gene were randomly distributed around the donor site (MORENO et al. 1992). Schwartz describes four *Ac* transpositions 0.1–1.1 map units away from the *wx-m9::Ac* allele (SCHWARTZ 1989b). Furthermore, frequent *Ac* and *Ds* transpositions in a variety of genes over distances ranging from only 6 bases to several kilobases were isolated (DOWE et al. 1990;

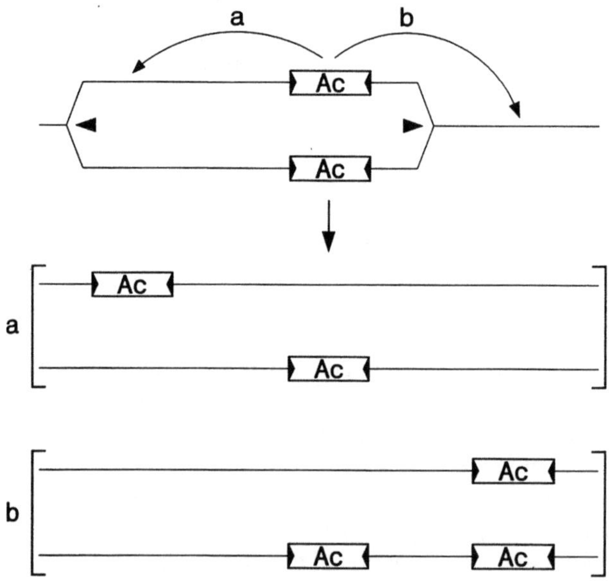

Fig. 2. *Ac* transposes shortly after replication **(a)** into replicated DNA, or **(b)** into unreplicated target DNA, which results in an increased copy number of *Ac*. Only one of the two freshly replicated *Ac* elements is capable of transposition (the *shaded Ac* does not transpose), and reinsertion occurs frequently into genetically linked sites

PETERSON 1990; GROTEWOLD et al. 1991; ATHMA et al. 1992; WEIL et al. 1992; CHEN et al. 1992). This high frequency of short-range transpositions can be exploited to generate large numbers of mutations within a single gene, a strategy which was termed "cyclic mutagenesis" (BRINK and WILLIAMS 1973; KERMICLE et al. 1989; ATHMA et al. 1992; WEIL et al. 1992; MORENO et al. 1992; ALLEMAN and KERMICLE 1993).

Different models are suggested to account for the high frequency of nearby transpositions, which have in common that the *Ac* donor and recipient site are somehow physically connected during transposition (GREENBLATT 1984; SCHWARTZ 1989b; ROBBINS et al. 1989). A physical connection between donor and target sites is also suggested by the observation that *Ac* transpositions to genetically unlinked sites are not random. Ten of 37 *Ac* transpositions from the *bz-m2::Ac* allele on the short arm of chromosome 9 to genetically unlinked sites mapped on the long arm of the same chromosome, and in the remaining 27 events some chromosomes are preferred against others (DOONER et al. 1994).

These results imply that it might be of advantage to initiate transposon tagging experiments with an *Ac* or *Ds* element closely linked to the desired target locus. Indeed, this strategy has been successfully used in maize (DELONG et al. 1993). Of course, preferential transposition of *Ac* to closely linked sites would also be extremely useful for transposon tagging in heterologous plants (see also Sect. 8), since in some important plant species a collection of transgenic lines was generated which carry *Ac* or *Ds* insertions in many different genomic positions. JONES and colleagues (1990) have found that in one transgenic tobacco line, 11 of 14 independent *Ac* transpositions were to very closely linked sites. However, short-range transposition is not a universal mechanism in tobacco. The preference for nearby transpositions varies considerably in different transgenic lines. In four of six independent transformants a majority of the linked transposed *Acs* are clustered within 5 cM of the donor site, whereas in two of them they are more dispersed. Possibly, the preferred transposition range is a function of the chromatin structure in the vicinity of the donor site (DOONER and BELACHEW 1991).

In *Arabidopsis, Ds* elements transpose with a frequency (68%) similar to that in maize and tobacco into linked sites. In four independent transformants the distribution of transposed elements varied around the different donor sites, but no tight clusters such as those seen in maize and tobacco were observed (BANCROFT and DEAN 1993; KELLER et al. 1993b).

In the tomato the transposition pattern of *Ac* and *Ds* elements is more complicated. In several cases it was observed that the introduced autonomous *Ac* transposed without preference for short-range transpositions to unlinked and linked sites as well. Yet the transposed *Acs* were arranged in small dispersed clusters of linked insertions (YODER 1990; OSBORNE et al. 1991; BELZILE and YODER 1992). Both linked and unlinked clusters of insertions were also obtained with *Ds* elements carrying either GUS, NPTII, or HPT as marker genes (ROMMENS et al. 1993; HEALY et al. 1993; KNAPP et al. 1994). The transposition patterns may be strikingly different in plants carrying *Ds* in two different, but very closely linked T-DNA insertion loci. Whereas in one family the element reinserted preferentially close to the primary insertion site, in the second family no such preference, but rather a small reinsertion cluster on another chromosome was observed (Knapp et al. 1994). The formation of reinsertion clusters in the tomato could be the consequence of an early primary transposition event to an unlinked site, followed by secondary transpositions to linked loci. Less likely is the existence of hot spots for integration.

4.3 The *Ac* 5'- and 3'-Ends in Direct Orientation Induce Chromosome Breakage

One of McClintocks original discoveries was a phenomenon she termed the "breakage-fusion-bridge cycle" (McClintock 1942). Chromosome breakage occurred frequently at a specific genetic locus, which she therefore called *Dissociation (Ds)* (McClintock 1946). She subsequently detected that the *Ds* locus could change its position in the genome, and that chromosome breakage at *Ds*, as well as transposition of *Ds*, requires the activity of a second locus, which she designated *Activator (Ac)* (McClintock 1947, 1948). Two classes of *Ds* elements exist which differ in their response to *Ac*. State-I *Ds* elements cause frequent chromosome breaks and rather rare reversions (excisions), whereas state-II *Ds* elements produce higher rates of reversions and barely detectable rates of chromosome breakage (McClintock 1949).

When the structures of the state-I *Ds* elements from the *sh-m5933* (Courage-Tebbe et al. 1983; Döring and Starlinger 1984; Döring et al. 1989) and the *sh-m6233* allele (Weck et al. 1984) were elucidated, it turned out that they both contain "double *Ds*" (d*Ds*) elements, consisting of a 2-kb *Ds* element (a simple, internally deleted *Ac* derivative) inserted in opposite orientation into the center of a second, identical 2-kb *Ds* (Fig. 3a) (for review, see Döring and Starlinger 1986; Fedoroff 1989). Another state-I *Ds* from the *sh-m6258* allele contains a similarly complex 3-kb *Ds* structure, i.e., a 2-kb *Ds* flanked by a 1-kb half-*Ds*. This 3-kb *Ds* consists essentially of 1 kb from the 5'-end of *Ac*, flanked on either side by 1 kb from the 3'-end in opposite orientation (Döring et al. 1990). The proposed models to explain chromosome breakage at d*Ds* elements suggest that breakage is a result of an aberrant transposition attempt involving the two transposon ends which are in direct orientation (Döring and Starlinger 1984; Döring et al. 1989). These have to be two different ends (one 5'- and one 3'-end), as the insertion of one transposable element into another in the same orientation gives rise exclusively to excisions but not to chromosome breaks (Fig. 3d) (Michel et al. 1994).

However, neither the inverted orientation of the complete *Ds* element relative to the half-*Ds* nor the close contact of the transposon segments is a necessary prerequisite for chromosome breakage. At the bronze locus a number of chromosome breaking structures were characterized which consist of a 3'-half of *Ac* and a complete *Ac* in direct orientation and separated by 0.05 cM (>25 kb) (Ralston et al. 1989), or pairs of either two very closely linked (<1 cM) *Ac* elements, or one *Ac* and one *Ds* element (Dooner and Belachew 1991). Together with the interval between them, the two elements constitute a

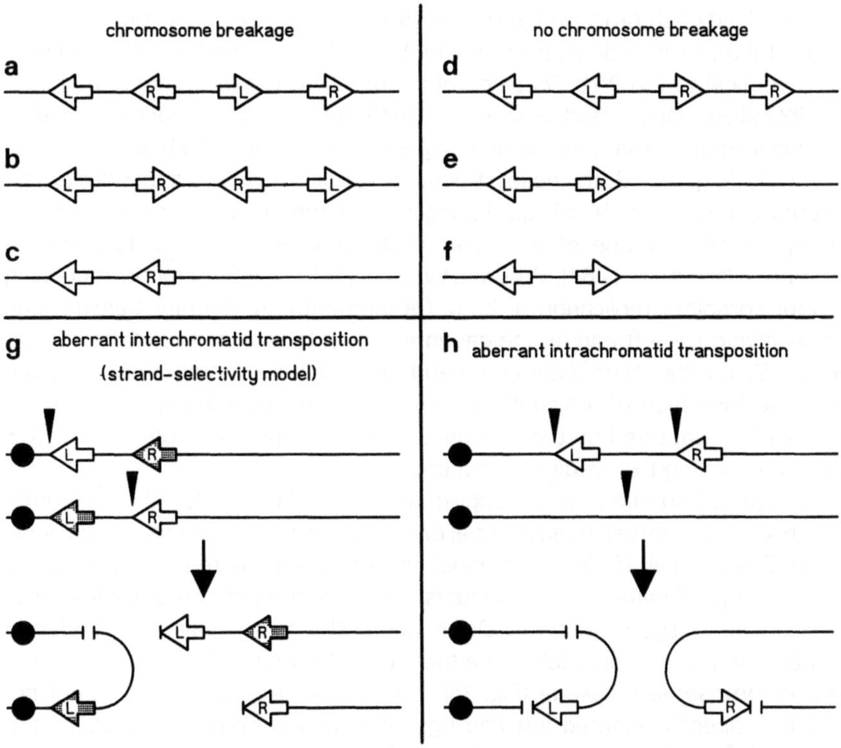

Fig. 3a–h. Chromosome breakage is a consequence of aberrant transposition attempts between two different element ends in direct orientation. *Arrows* marked "*L*" and "*R*" indicate the left end (5'-end) and the right end (3'-end) of *Ac*, respectively. **a–c** Chromosome breaking structures: **(a)** insertion of one *Ds* in opposite orientation into another *Ds* = double*Ds* structure (DÖRING et al. 1989); **(b)** two *Ds* elements in opposite orientation (WEIL and WESSLER 1993); **(c)** left and right *Ac* ends in direct orientation (ENGLISH et al. 1993). **d** Insertion of one *Ac* into another, equally oriented one (MICHEL et al. 1994). **e** Simple *Ac* or *Ds* element (MCCLINTOCK 1949). **f** Two left or right *Ac* ends in opposite orientation (ENGLISH et al. 1993). **g** The strand-selectivity model for chromosome breakage predicts that after replication of two directly oriented left and right transposon ends, one end on each daughter chromatid is transposition incompetent (*shaded arrows*). After DNA cleavage at the termini of the two competent ends (*filled arrowheads*) religation of the two "donor site" halves will lead to a dicentric and subsequently breaking chromosome if the centromere (*filled circle*) is located at the left hand side. **h** The aberrant intrachromatid transposition model suggests that both transposon ends on one chromatid are transposition competent. DNA cleavage and religation at the two "donor site" halves and the "acceptor site" on the sister chromatid will lead to the formation of one acentric and one dicentric chromosome and subsequent chromosome breakage

"macrotransposon". Chromosome breakage at these structures could be explained by excision of the partially replicated macrotransposon, followed by reinsertion in opposite orientation before replication of the internal sequences is complete. In contrast to the breakage model for the "double *Ds*" elements, the key feature here is not the recognition of a left and right transposon end in direct orientation, but rather the presence of left and right transposon ends separated by a stretch of chromosome that is in a different replication state from one or both transposon ends at the time of transposition (DOONER and BELACHEW 1991).

A third kind of chromosome breaking structure has been described (Weil and Wessler 1993). In two derivatives of the *Wx-m5* allele two 2-kb *Ds* elements [nearly identical to the 2-kb *Ds* elements comprising the d*Ds* structures in the *sh-m5933* allele (Döring and Starlinger 1984)] are inserted in opposite relative orientations in the *Waxy* gene and separated by only 1.5 kb and 0.45 kb, respectively (Fig. 3b). Both these structures cause *Ac* TPase-dependent high-frequency chromosome breakage. Taking into account the genetic evidence that after replication only one of each pair of daugther elements is transposition competent (Greenblatt 1984), Weil and Wessler (1993) invoke a model proposing that after complete replication of both *Ds* elements an aberrant transposition event involving the left end of one daughter *Ds* and the right end of the second daughter *Ds* on the other sister chromatid fuses the two sister chromatids and causes the formation of a dicentric chromosome, which subsequently breaks. This model is supported by the existence of a fusion product detected by PCR analysis and having the predicted sequence.

In a very different approach, English and co-workers analyzed the chromosome breaking properties of an artificial d*Ds* and several derivatives in transgenic tobacco (English et al. 1993). Preferential transposition of state-II *Ds* and chromosome breakage of state-I *Ds* elements is obviously not restricted to maize, but also occurs in transgenic tobacco. Dissection of the d*Ds* revealed that a left and a right *Ds* end in direct orientation are the minimal substrate for the *Ac* TPase to induce chromosome breakage (Fig. 3c). An aberrant transposition attempt between two directly oriented left and right transposon ends should cause the inversion of one half-transposon, and the existence of these predicted products was verified by PCR analysis. Four different constructs carrying a left and a right *Ds* end in direct orientation caused chromosome breakage, whereas seven other constructs containing complete *Ac* or *Ds* elements (Fig. 3e), or consisting of two inverted left or right *Ds* ends (Fig. 3f) or a single transposon end, respectively, did not induce breakage.

The PCR-detected footprints analyzed by Weil and Wessler and English and co-workers are consistent with two different models for chromosome breakage induced by aberrant transposition reactions. One model is basically similar to the one proposed by Döring and Starlinger (1984). It predicts that the two directly oriented transposon ends are cleaved on the same chromatid and get fused to an acceptor site on the sister chromatid (Fig. 3h). The other model takes into account the genetic and molecular evidence, which suggests that *Ac* and *Ds* transpose during or soon after replication and that only one of the two daughter elements is transposition competent (Greenblatt 1984; Chen et al. 1987, 1992). These phenomena could be explained by a strand selectivity of the transposition machinery (Fedoroff 1989), and in fact, the *Ac* TPase has a strand-specific DNA-binding capacity in vitro (Kunze and Starlinger 1989). According to the "strand selectivity model", the left transposon end is cleaved on one chromatid and the right end on the other chromatid, followed by interchromatid fusion (Fig. 3g).

5 Expression of *Ac*

5.1 Transcription of *Ac* in Maize

Northern blot analysis with various *Ac*-containing and *Ac*-free maize lines and cDNA cloning led to the identification of a 3.5-kb messenger RNA whose appearance is correlated with *Ac* activity. The primary transcript spans most of the *Ac* sequence, leaving only about 300 bp at the 5'-end and 264 bp at the 3'-end untranscribed (Fig. 1). Transcription begins at several sites within a 100-bp region, the most prominent initation site being 334 bp proximal of the 5'-end of *Ac*. Polyadenylation takes place 265 bp proximal to the 3'-end (Kunze et al. 1987; Finnegan et al. 1988).

The *Ac* promoter is lacking a CAAT and a TATA box and therefore is reminiscent of mammalian housekeeping gene promoters (Dynan 1986). Like these, the *Ac* promoter is weak and seems to be constitutively active. It gives rise to an average mRNA abundance of approximately 10 mRNA molecules per cell (Fußwinkel et al. 1991), which is low, but not low enough to explain the rareness of transposition events. The *Ac* promoter retains its basic properties also in transgenic tobacco plants, where similarly low transcript levels are expressed (Kunze et al. 1987; Hehl and Baker 1990) and the same multiple initiation sites are used (Finnegan et al. 1988; R.Kunze, unpublished results).

As the analysis of maize cDNA by PCR amplification yielded no alternatively processed *Ac* transcripts, it is likely that the 3.5-kb mRNA is the only *Ac* transcription product (C. Roßbach and P. Starlinger, unpublished). In this respect, *Ac* clearly differs from the *En/Spm* transposable element, which is transcribed into various alternatively spliced mRNAs (Pereira et al. 1986; Masson et al. 1989).

The first 600–700 nucleotides of the *Ac* mRNA constitute an untranslated leader sequence. To date, there are no unambiguous data regarding whether this unusually long leader sequence has a particular function. It is obviously dispensable for the transposition reaction, as in transgenic tobacco plants the deletion of nearly the whole leader sequence does not abolish autonomous excision of the *Ac* derivative (Coupland et al. 1988). In *Arabidopsis thaliana* the deletion of the leader sequence in an *Ac* derivative even leads to a significantly increased transactivation frequency of a *Ds* element (Bancroft et al. 1992). Thus, in *Arabidopsis* the untranslated leader plays an inhibitory role. In flax, however, the same deletion causes at best a marginal activity increase (Finnegan et al. 1993). A possible explanation for these differences could be that the leader sequence is responsible for misprocessing of the *Ac* mRNA or leads to destabilization in *Arabidopsis* (Grevelding et al. 1992), but not in other plants.

5.2 *Ac*-encoded Protein

The first ATG codon on the *Ac* mRNA opens a 2421-base open reading frame (ORFa) which encodes the transposase protein (TPase) (Kunze et al. 1987). In

Western blotting experiments, a protein with an apparent molecular weight of 112 kD was detected exclusively in nuclear endosperm extracts from maize lines containing an active *Ac* element (Fußwinkel et al. 1991). This protein presumably is the full-length ORFa protein, the TPase, because the expression of ORFa in transgenic tobacco plants (Coupland et al. 1988) and *Petunia* protoplasts (Houba-Hérin et al. 1990) leads to the appearance of a protein with the same electrophoretic mobility and induces the mobilization of nonautonomous *Ds* elements. With an estimated abundance in maize endoperm in the range of 10^3 molecules per genome equivalent, the TPase is a quite rare protein.

In addition to the 112-kD protein, small and variable amounts of a second ORFa-derived protein with an apparent molecular weight of approximately 70 kD were detected in endosperm extracts (Fußwinkel et al. 1991). Since no corresponding mRNA was identified, it is assumed that this protein species is a proteolytic product of the 112-kD TPase. It is not known whether the 70-kD protein has an influence on the transposition reaction.

By immunostaining the TPase protein in maize endosperm sections, a fraction of cells was found to contain rod-shaped TPase aggregates about 2 μm in length. It is not known if these aggregates have a function during transposition. However, as in transfected *Petunia* protoplasts the expression of increasing amounts of TPase leads to increasing amounts of TPase aggregates, but not to a corresponding increase in transposition frequency, it is suggested that the TPase aggregates are transpositionally inactive (Heinlein et al. 1994).

6 The *Ac* Transposase Protein

6.1 Structural Features of the Transposase Polypeptide

The "*Ac*" bar in Fig. 4 schematically shows the 807-amino acid TPase protein {TPase(1–807} with some structurally or functionally distinct segments highlighted. The N-terminal 200 amino acids of the *Ac* protein contain clusters of basic residues which constitute the nuclear localization signals (NLSs). Three peptides with NLS activity were identified, encompassing TPase residues 44–62, 159–178, and 174–206, respectively (Boehm et al. 1995) (Fig. 4). The nuclear targeting capability of the three NLSs is cumulative, as the nuclear import efficiency of a N-terminally truncated TPase(103–807) is reduced (Heinlein et al 1994).

For TPase activity, the N-terminal 102 residues including NLS(44–62) are dispensable (Li and Starlinger 1990). Surprisingly, in transfected *Petunia* protoplasts the N-terminally truncated TPase triggers even higher excision frequencies than the full-length TPase when expressed at low levels (Houba-Hérin et al. 1990; Becker et al. 1992). The mechanism by which the N-terminal 102 amino acids reduce the TPase activity in *Petunia* cells is not yet understood. However, as high TPase concentrations seem to have an autoinhibitory effect (Sect. 7.5), it is

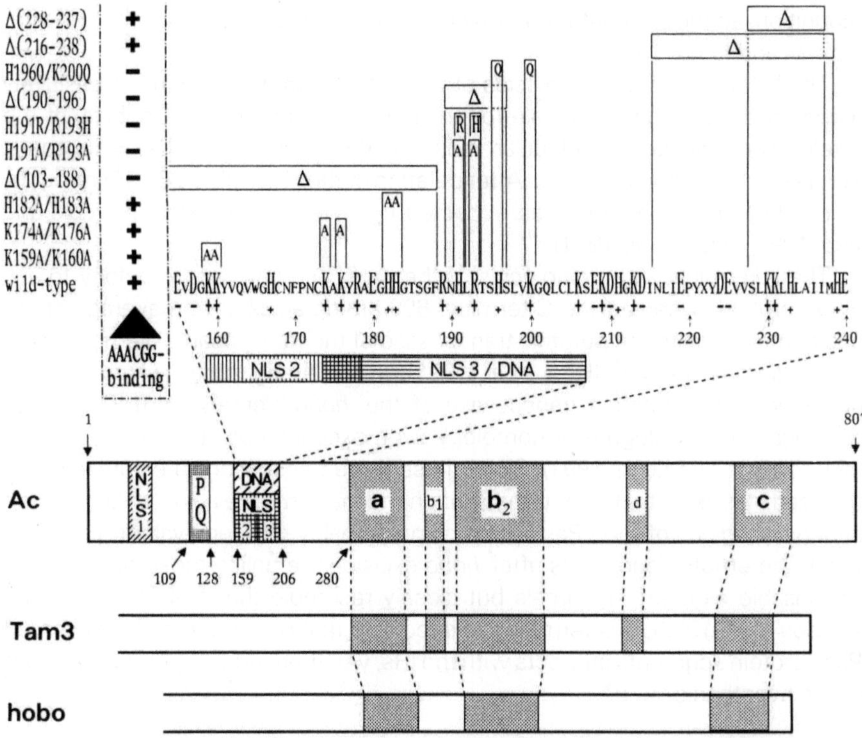

Fig. 4. Structure of the *Ac* transposase (TPase) protein. The horizontal bar "*Ac*" schematically shows the 807-amino acid TPase protein. "NLS" indicates the three nuclear localization signals. Residues 109–129 consist of a tenfold repeat of the dipeptide proline-glutamine. Residues 159–206 are the basic region which is shown enlarged above the *Ac bar* in single-letter code. The effects of several deletion and substitution mutations (shown above the wild-type TPase sequence) on in vitro DNA binding to the ends of *Ac* and to synthetic AAACGG repeats are indicated in the column *AAACGG binding. Bars at bottom* marked "Tam3" and "hobo" symbolize the putative transposase polypeptides of the *Tam3* and *hobo* transposable elements. Within the *Shaded segments,* 40–65% of the amino acids are identical in the three proteins

conceivable that the lower nuclear concentration of the TPase(103–807) protein causes the apparently higher activity.

Between residues 109 and 128, the *Ac* TPase consists of a tenfold repeat of a proline-glutamine dipeptide. This sequence is required for TPase activity (Kunze et al. 1993), but it is not involved in the DNA-binding reaction (Feldmar and Kunze 1991).

The DNA-binding domain of the TPase is located between residues 159 and 206 (Feldmar and Kunze 1991) and overlaps with two NLSs (Fig. 4). However, the two functions can be separated (Boehm et al 1995). The basic amino acids between residues 189 and 200 are presumably directly involved in recognition of the AAACGG motifs (see Sect. 6.2), as their substitution with alanine and the reciprocal exchange of His_{191} and Arg_{193} abolishes DNA binding completely. In contrast, the basic amino acids between 159 and 183 are not involved in DNA recognition, as their substitution by alanines does not affect DNA binding in vitro (FELDMAR and KUNZE 1991).

The N-terminal 200 amino acids of the TPase with their surplus of positively charged residues are responsible for an aberrant electrophoretic mobility of the protein. The calculated molecular mass of the TPase is 92kDa, whereas it migrates at 112 kDa in SDS gels. After deletion of the N-terminal 200 amino acids, the residual protein migrates as expected (FUßWINKEL et al. 1991; FELDMAR and KUNZE 1991; KUNZE et al. 1991).

The N-terminal 200 amino acids of the *Ac* TPase have no homology to any known protein, whereas the C-terminal 600 amino acids are on average about 30% homologous to the putative transposase of the transposable element *Tam3* from *Antirrhinum majus* (HEHL et al. 1991). In three segments, which are also conserved in the putative transposase of the *hobo* element from *Drosophila melanogaster,* the degree of homology even exceeds 65% (Fig. 4) (CALVI et al. 1991; FELDMAR and KUNZE 1991). Since these three transposons are very similarly organized, it is likely that their transposition mechanisms are also similar, and the conserved regions of the TPase proteins presumably have equivalent functions. In fact, the empty donor sites after *hobo* excision are unlike those of any other transposable element in animals but closely resemble those of *Ac* and *Tam3* (ATKINSON et al. 1993). It is tempting to speculate that one of the higly conserved TPase protein segments interacts with the IRs, yet, there is no experimental proof for this hypothesis.

6.2 Transposase Binds In Vitro to Subterminal, Repetitive Sequence Motifs

The DNA-binding properties of the *Ac* TPase were investigated in gel-shift assays with TPase protein overexpressed in the eukaryotic baculovirus system (HAUSER et al. 1988) or in *E. coli* cells (FELDMAR and KUNZE 1991). In these experiments, no binding of the TPase to the IRs of *Ac* was detected, but binding to those subterminal about 250 bp sequences which are required for transposition in vivo was (COUPLAND et al. 1989; KUNZE and STARLINGER 1989). Within these regions the TPase binds to repetitive AAACGGs or closely related motifs, but the formation of stable TPase/DNA complexes in vitro requires several AAACGG motifs on the target DNA. If the number of AAACGG motifs is reduced to less than about six, the stability of the complexes decreases drastically, which could indicate a cooperative interaction of TPase molecules on the DNA (KUNZE and STARLINGER 1989; FELDMAR and KUNZE 1991). These observations are consistent with the gradual reduction in element excision in vivo by deletions extending progressively into the TPase-binding region (COUPLAND et al. 1988, 1989).

It is remarkable that most of the asymmetric AAACGG TPase-binding motifs in the subterminal regions of *Ac* are directly repeated (Fig. 1). Accordingly, if individual TPase molecules interact with each other on the DNA, it seems possible that the TPase oligomers have not a closed structure, like many di- or tetrameric prokaryotic repressors or eukaryotic transcription factors, but rather an open structure with no defined size. Such an oligomerization property is proposed for the *Drosophila melanogaster* protein Zeste, whose DNA-binding properties are reminiscent of those of the *Ac* TPase (CHEN and PIRROTTA 1993).

The properties of the TPase/DNA complexes indicate that several TPase molecules have to bind to each end of *Ac* during transposition. However, it is not known whether each AAACGG motif has to be occupied by TPase simultaneously, or if only the occupation of a certain subset promotes transposition and other occupational patterns result in inhibition of transposition. If this were so, the likelihood of attaining the correct occupation must be rather low, due to the high number of individual binding sites in each *Ac* end. Such a mechanism could explain why – in spite of the proposed constitutive expression of about 10^3 TPase molecules per cell – *Ac* and *Ds* elements transpose in only a small fraction of cells, and why two transposable elements in the same cell usually do not transpose simultaneously (HEINLEIN and STARLINGER 1991).

6.3 DNA Binding of Transposase Is Methylation Sensitive

A significant fraction of the potential methylation sites (CpG and CpNpG) in the ends of *Ac* are part of the TPase-binding sites, and DNA-binding by the TPase is indeed strictly dependent on the methylation status of the target DNA. The affinity of the TPase to its target motifs increases if one or both cytosine residues on the one strand are replaced by 5-methylcytosine. In contrast, DNA binding is totally

Table 3. Relative efficiencies of TPase binding to target DNAs consisting of 18 tandem copies of the indicated hexamer motifs as determined by gel-shift assays (KUNZE et al. 1991)

Target DNA efficiency	TPase binding
5'-AAACGG-3' 3'-TTTGCC-5'	+
5'-AAAC GG-3' 3'-TTTG^{m}CC-5'	++
5'-AAACG G-3' 3'-TTTGC^{m}C-5'	++
5'-AAAC G G-3' 3'-TTTG^{m}C^{m}C-5'	+++
5'-AAA^{m}CGG-3' 3'-TTT GCC-5'	–
5'-AAA^{m}C GG-3' 3'-TTT G^{m}CC-5'	–
5'-AAA^{m}CG G-3' 3'-TTT GC^{m}C-5'	–
5'-AAA^{m}C G G-3' 3'-TTT G^{m}C^{m}C-5'	(+)

^{m}C,5-methylcytosine; (+), very weak binding that can be obscured by increasing the competitor-DNA concentration

inhibited by C-methylation on the other strand, irrespective of whether the cytosine residues on the complementary strand are methylated or not (Table 3) (Kunze and Starlinger 1989; Kunze et al. 1991). This differential recognition of the two hemimethylated forms of the binding site emphasizes the asymmetric nature of the TPase-DNA interaction.

It is intriguing to speculate that the selective recognition of only one hemimethylated DNA species by the TPase and inhibition of TPase binding to the other species is the mechanism that couples transposition to replication (see Sect. 4.1). Such a phenomenon is well known from the bacterial *IS*10 element, which transposes preferentially from one of the two differentially hemimethylated daughter molecules after replication (Roberts et al. 1985; for review see Kleckner 1990). According to this hypothesis, before replication the *Ac* element is holomethylated and stable. As the majority of AAACGG motifs in the 5'-end of *Ac* are inversely oriented to those in the 3'-end, after replication the two hemimethylated daughter elements would differ, in that one will preferentially bind TPase at its 5'-end but not at its 3'-end, whereas the opposite happens at the other daughter *Ac*.

Of course, a prerequisite for this scenario is that the TPase-binding sites in maize are methylated in vivo. As the asymmetric AAACGG motifs are not accessible by restriction analysis, this question can be approached only by genomic sequencing. However, the methylation states of the *Hpa*II restriction sites of an active (*wx-m9::Ac*) and a transiently inactivated *Ac* (*wx-m9::Ds-cy*) have been investigated. Three *Hpa*II sites at the right end of *Ac* which are located between TPase-binding motifs are always methylated (Schwartz and Dennis 1986). It will be interesting to see whether the methylation state of these *Hpa*II sites reflects the methylation state of the TPase-binding motifs in their vicinity. If so, the TPase-binding sites at the right end of *Ac* should always be methylated. Immediately after replication the two daughter elements were differentially hemimethylated, and this could account for the strand selectivity of transposition (Greenblatt 1984; Fedoroff 1989).

One *Hpa*II site in the left *Ac* end (nt 178) is situated in a 10-bp spacer between two AAACGG motifs. This site is unmethylated in the active *Ac* but methylated in the inactive phase. The *Bam*HI site (nt 181) which overlaps this *Hpa*II site and one AAACGG motif is cleavable, and hence unmethylated in the active *Ac* but uncleavable and presumably methylated in the inactive *wx-m9::Ds-cy* allele and the *wx-m9::Ds* allele as well (Schwartz 1989a). If an active *Ac* is crossed in, the *Bam*HI site of the *wx-m9::Ds* and, to a lesser degree, that of the *wx-m9::Ds-cy* become demethylated. This could indicate a protection of the *Bam*HI site by the TPase against methylation. Since reactivation of *wx-m9::Ds-cy* is greatly enhanced in the presence of an active *Ac* (Schwartz 1989a), and reactivation is correlated with the reappearance of the *Ac* mRNA (Kunze et al. 1988), it seems possible that the TPase-binding region and the promoter of *Ac* overlap, and that the promoter remains protected against inactivation by methylation by TPase bound to it. This would imply that the TPase is permanently bound to the left end of *Ac* but binds only after replication to the right end in only one of the two daughter elements.

Ott and co-workers have generated genomic sequences of the 5'-subterminal *Ac* regions in five transgenic tobacco plants (OTT et al. 1992). Only at one single C residue did they detect methylation, whereas at least 80% of the C residues at any other analyzed site (including nine AAACGG motifs) are unmethylated. Accordingly, methylation is unlikely to play a role during transposition of *Ac* in tobacco. As the TPase binds also to unmethylated AAACGG motifs, it is conceivable that *Ac* transposition in tobacco is – unlike in maize – not coupled to replication.

Surprisingly, de novo methylation of *Ac* in tobacco is generally very low. Of 87 cytosine residues in *Ac* analyzed with C-methylation-sensitive restriction enzymes, only three were partially methylated. Even three *Hpa*II sites that are always methylated in maize are unmethylated in tobacco (NELSEN-SALZ and DÖRING 1990).

6.4 The Transpositionally Active Transposase Is an Oligomeric Protein

Houba-Hérin et al. have developed a transient excision assay that enables a much quicker investigation of mutant derivatives of the TPase protein than is possible with transgenic plants (HOUBA-HÉRIN et al. 1990). A variety of deletion and two-amino acid insertion mutants were tested for their transpositional activity and their influence upon coexpression with the wild-type TPase. It turned out that a deletion of the PQ repeat (residues 109–128; Fig. 4), as well as of the carboxy-terminal 54 amino acids, inactivates the protein, indicating that these sequences have essential functions during transposition. The appearance of dominant, transposition-negative, and DNA-binding-deficient mutants indicates that the active TPase is an oligomeric protein (KUNZE et al. 1993). Furthermore, by complementation analysis with several inactive, but recessive TPase mutants, functionally distinct domains of the TPase were located on the polypeptide chain (R. Kunze, unpublished results).

7 Regulation of Transposition

7.1 The Transposase Concentration Is Not the Only Transposition-limiting Factor

In maize, the *Ac* element transposes usually only in a fraction of the cells, but the transposition frequencies vary over a wide range. The germinal excision frequencies of *Ac* or *Ds* at different loci may vary in the range of 0.01% to more than 15% (for review see DÖRING 1989), and the frequency of *Ds2*-element excision from the *bz2::Ds2* allele in the aleurone cells is in the range of 0.2–1% (LEVY and WALBOT

1990). The simplest explanation for these low transposition frequencies would be an extremely low TPase concentration, as described for the *IS*10 element (RALEIGH and KLECKNER 1986). In maize endosperm, the average abundance of the TPase was estimated in the range of 10^3 molecules per genome equivalent (FUßWINKEL et al. 1991), but it is not known whether these TPase molecules are evenly distributed among the cells. Indeed, the appearance of TPase aggregates in only a fraction of endosperm cells could be an indication for uneven expression (HEINLEIN et al. 1994). If the TPase concentration were the transposition frequency-limiting factor, one would expect a high coincidence of excisions of different *Ac* and *Ds* elements in the few cells having enough of the enzyme. Yet, as co-excision of *Ds* elements is rather the exception than the rule (HEINLEIN and STARLINGER 1991), the scarcity of transposition events cannot be explained easily by a lack of TPase. Rather, the probability of assembling a functional transposition complex might be low. Conceivably, the TPase protein stochastically binds to some of the multiple TPase-binding sites in the ends of the element, but only the occupation of a certain subset of them promotes the excision reaction.

7.2 Is the Transposition Frequency Modulated by (a) Host Factor(s)?

In analyzing variegation patterns in maize endosperm it becomes obvious that not only the frequency, but also the developmental timing of transposition is controlled (MCCLINTOCK 1951). The excision frequency of the *Ds*2 element from the *bz2::Ds*2 allele fluctuates between 0.2 and 1% during proliferation of the aleurone layer (LEVY and WALBOT 1990). These changes reflect either temporal fluctuations in TPase expression or the influence of a transposition-modulating host factor. It has been genetically shown that the two identical *Acs* in the *wx-m7::Ac* and the *wx-m9::Ac* alleles induce very different excision patterns of the *Ds* in the *bz-m2(DI)* allele, and that these distinct phenotypes are not due to the presence of a modifier gene somewhere in the maize genome (HEINLEIN 1995). Though this does not exclude the existence of a modulating host factor, it implies that subtle differences in TPase expression may result in spatially and temporally distinct transposition patterns.

7.3 *Ac* Can Have an Inverse or Positive Dosage Effect in Maize

McClintock has described a peculiar phenomenon in the regulation of *Ac* transposition, which is called the "dosage effect". In maize endosperm with two or three *Ac* elements in the genome, *Ac* and *Ds* excisions are developmentally delayed compared with plants with only one *Ac*, resulting in less and smaller revertant sectors on the kernels (MCCLINTOCK 1951). This has led to the conclusion that *Ac* senses its own copy number as well as the developmental stage of the endosperm, which means it interacts somehow with a cellular signal

(SCHWARTZ 1984). However, this signal does not interact equally with all *Ac* elements, as the dosage effects of different *Acs* are not uniform and may even change from negative to positive during endosperm development (HEINLEIN and STARLINGER 1991).

How could sensing of its own copy number be achieved? Direct interactions between *Ac* elements are unlikely, because the overall number of *Ac*-homologous sequences, including the *Ds* elements, is so high that one or two additional elements do not make much of a difference. The dosage effect is rather dependent on expression of the TPase protein, because the deletions in the *wx-m9::Ds* and *bz-m2(DI)* alleles, which destroy the TPase-coding region, mutate *Ac* into *Ds* elements that do not contribute to the dosage effect any more (DOONER et al. 1986), and even an *Ac* element which is transiently inactivated by methylation and not transcribed behaves like *Ds* and does not contribute to the dosage effect (MCCLINTOCK 1964, 1965; SCHWARTZ 1986; KUNZE et al. 1988).

Sensing its own copy number could be achieved by negative autoregulation at the transcriptional or translational level. However, this also seems unlikely, because in endosperm (KUNZE et al. 1987) and also in seedlings (BRUTNELL and DELLAPORTA 1994) the *Ac* mRNA concentration and the 112-kD TPase concentration as well (FUẞWINKEL et al. 1991) increase with *Ac* copy number. Therefore, it is more likely that the *Ac* dosage effect is manifested at a post-translational stage.

7.4 *Ac* Has a Positive Dosage Effect in Heterologous Plants

The dosage effect of *Ac* has also been investigated in transgenic plants. In tobacco, plants homozygous for *Ac* clearly show more frequent transposition than heterozygous plants, which suggests a positive dosage effect (JONES et al. 1989; HEHL and BAKER 1990). Surprisingly, the major transposition frequency-determining factor seems to be the copy number of the TPase source, rather than the number of mobile elements (JONES et al. 1991). Furthermore, *Ac* tends to transpose earlier in homozygous plants than in heterozygotes (KELLER et al. 1993a).

In *Arabidopsis thaliana, Ac* has an extremely low transposition frequency (SCHMIDT and WILLMITZER 1989). As transformants with several *Ac* elements tend to have a higher transposition frequency than single-*Ac* lines, the dosage effect of *Ac* in this plant also seems to be positive (KELLER et al. 1992; DEAN et al. 1992). By using a two-element system consisting of a stable TPase source and a *Ds* element, it was shown that – as in tobacco – increasing the copy number of the TPase source causes a higher transposition frequency, whereas altering the number of *Ds* elements has no effect (BANCROFT and DEAN 1993). This could indicate that the TPase concentration conditions the general TPase activity state, and that the distribution of the total TPase activity over an increasing number of elements reduces the probability of transposition for each single element.

In other heterologous plants *Ac* dosage effects have not yet been analyzed in great detail. In the tomato, no obvious reduction in transposition frequency was detectable when more than one *Ac* copy was present (BELZILE et al. 1989; YODER 1990). In *Nicotiana Plumbaginifolia* the transposition frequency increases with *Ac* copy number (MARION-POLL et al. 1993).

In conclusion, a delay in timing and decrease in frequency of *Ac* or *Ds* transpositions upon an increase in the copy number of *Ac* in the genome has so far not been observed in any heterologous plant. Since the inverse dosage effect in maize is presumably expressed at the post-translational level, it is conceivable that the TPase interacts specifically with a signal in maize but not in transgenic plants.

7.5 Autoinhibition of Transposition at High Transposase Expression Levels

Most, if not all, transposable elements underlie mechanisms which limit transposition frequencies to levels that are not detrimental to the host organism. These mechanisms may be either intrinsic to the element itself or based on interactions with a host factor.

The inverse dose effect of *Ac* is caused by such an autoregualatory mechanism. As the TPase protein concentration increases with *Ac* copy number, apparently an inverse relation exists between TPase concentration and transposition frequency. Though the inverse dose effect has not been described in transgenic plants, it was observed that raising the *Ac* transcription level in tobacco about 1000-fold does not necessarily lead to increased germinal transposition rates (SCOFIELD et al. 1992). Furthermore, in developing tobacco cotyledons, this high TPase-expression level triggers transpositions early during development but causes a complete loss of transposition events at later stages. Since this phenotype is dominant over different transposition patterns caused by low expression of TPase, it was proposed that TPase levels above a certain threshold inhibit the transposition reaction (SCOFIELD et al. 1993). Based on the observation that the TPase aggregates in vivo, the possibility is discussed that beyond this threshold TPase starts to aggregate and transpositions cease, because either the TPase aggregates actively inhibit transposition or all soluble TPase is sequestered into inactive aggregates (HEINLEIN et al. 1994).

7.6 *Ac* Can Be Reversibly Inactivated by Methylation

Ac can undergo a reversible change in activity, called "change in phase" by MCCLINTOCK. The inactivation of *Ac* is a somatic process (BRUTNELL and DELLAPORTA 1994). In the inactive phase, the element behaves like a *Ds* and does not contribute to the dosage effect (MCCLINTOCK 1964, 1965).

Three *Ac* elements in the inactive phase (*wx-m9::Ds-cy*, *wx-m7::inactive*, and *P-vv::I*) and a metastable *Ac* (*P-vv::Cy*) were molecularly analyzed. All ten *Hpa*II sites at the left end of *Ac* [one at nt 178 in the TPase-binding region and nine within the untranslated leader region (5'-UTR)] were found to be fully methylated in the inactive *wx-m9::Ds-cy* allele (SCHWARTZ and DENNIS 1986). Similarly, all nine *Eco*RII and *Sst*II sites in the 5'-UTR of the *wx-m7::inactive* allele are fully methylated (CHOMET et al. 1987), and most of the *Ava*I, *Nci*I, *Nru*I, and *Sst*II sites in the *P-vv::I* allele near the transcription start are methylated (BRUTNELL and DELLAPORTA 1994). In contrast, the corresponding active *Ac* elements are hypomethylated at these sites. In the metastable *P-vv::Cy* allele, an intermediate level of methylation was observed (BRUTNELL and DELLAPORTA 1994). Methylation of the *Ac* untranslated leader sequence and/or the promoter is accompanied by a dramatically reduced transcript level, and hence no or very little TPase is expressed (KUNZE et al. 1988), i.e., the *Ac*-mRNA level in the inactive *P-vv::I* allele is reduced to about 2% (BRUTNELL and DELLAPORTA 1994). This residual level of *Ac* message is not sufficient to induce transpositions. As the inactive *Ac* itself behaves like a *Ds* and does not contribute to the dosage effect, this effect is expressed at the level of the TPase protein.

In a usually small fraction of the kernels on an ear (<10%) the inactive element reverts spontaneously back to the active phase. Reappearance of TPase activity is accompanied by demethylation and expression of the 3.5-kb *Ac* transcript, but reversion to full activity apparently is a multistep process, because the revertant element remains partially methylated and the mRNA level is lower than normal (SCHWARTZ 1986; KUNZE et al. 1988; BRUTNELL and DELLAPORTA 1994).

A spontaneous appearance of transposable element activity in the progeny of plants lacking an active transposon was observed after the induction of chromosome breakage by different means (for review, see FEDOROFF 1983), called "genomic stress" by MCCLINTOCK (1984). Tissue culture conditions may also cause "genomic stress", as in some maize plants regenerated from material lacking *Ac* activity an active *Ac* element was detected (PESCHKE et al. 1987). This spontaneous appearance of *Ac* may in fact be a demethylation-associated reactivation of a silent element like the *wx-m9::Ds-cy* (BRETTELL and DENNIS 1991). Activation of transposable elements as a response to "genomic stress" may be a widespread phenomenon, as not only DNA-based transposons but also retransposons can be activated during tissue culture (HIROCHIKA 1993).

Methylation seems to be a widely used mechanism for reversible inactivation of transposable elements in plants. Like *Ac*, maize transposons *En/Spm* (BANKS and FEDOROFF 1989) and *Mutator* (CHANDLER and WALBOT 1986; BENNETZEN 1987), and presumably the *Antirrhinum majus Tam3* element also (MARTIN et al. 1989) can be inactivated by methylation.

Inactivation of *Ac* by methylation has not yet been unambiguosly observed in heterologous plants. In *Arabidopsis thaliana*, no correlation between *Ac* activity and methylation at the two *Pvu*II sites within *Ac* was observed: inactive and active *Ac* elements alike were unmethylated at both *Pvu*II sites, which are methylated in *wx-m7::inactive* (KELLER et al. 1992). However, in another case inactivated *Ac* elements in *Arabidopsis* were found to be methylated at the *Hpa*II sites in the 5'-UTR (VAN SLUYS et al. 1993).

8 *Ac* Transposes in Many Heterologous Plants

In maize, *Ac* has been successfully used for transposon tagging (for reviews, see Döring 1989; Balcells et al. 1991; Walbot 1992). Consequently, the suitability of *Ac* as a mutagen in heterologous plants was analyzed (Baker et al. 1986). Since *Ac* is apparently able to transpose in almost any plant (Table 4), it is extensively used for transposon tagging strategies in a variety of plants in which no endogenous transposable elements are known (for review see Balcells et al. 1991; Walbot 1992).

The transposition frequencies of *Ac* vary widely in different species, however (Haring et al. 1991). In the tomato, *Ac* usually is very active. It was estimated that about 30% of the progeny inherit a transposed *Ac* from their parent (Belzile et al. 1989). In tobacco *Ac* can also be very active, but the germinal excision frequencies in individual transformants vary over a wide range from almost 0 to 83%, the majority of plants showing between 2 and 5% germinal excisions (Hehl and Baker 1990; Jones et al. 1989, 1991; Rommens et al. 1992; Scofield et al. 1992). Even within a single plant, *Ac* excision frequency varies randomly from flower to flower, and transpositions occur at various times during flower development (Keller et al. 1993a). As mentioned before (see Sect. 7.5), replacing the weak *Ac* promoter in tobacco with the at least 1000-fold stronger CaMV 35S promoter causes only a slight increase of germinal excisions to an average of 6% (Scofield et al. 1992).

Table 4. Transposition of the maize element *Ac* in heterologous plants

Plant species	Family	Class	Transformation	Reference
Maize	Poaceae	Monocot	–	McClintock (1948)
Rice	Poaceae	Monocot	Transient	Laufs et al. (1990)
Rice	Poaceae	Monocot	Transgenic	Izawa et al. (1991)
Rice	Poaceae	Monocot	Transgenic	Jing-Liu et al. (1991)
Rice	Poaceae	Monocot	Transgenic	Murai et al. (1991)
Wheat	Poaceae	Monocot	Transient	Laufs et al. (1990)
Tobacco	Solanaceae	Dicot	Transgenic	Baker et al. (1986)
Nicotiana plumbaginifolia	Solanaceae	Dicot	Transgenic	Marion-Poll et al. (1993)
Tomato	Solanaceae	Dicot	Transgenic	Yoder et al. (1988)
Potato	Solanaceae	Dicot	Transgenic	Knapp et al. (1988)
Petunia	Solanaceae	Dicot	Transgenic	Haring et al. (1989)
Datura innoxia	Solanaceae	Dicot	Transgenic	Schmidt-Rogge et al. (1994)
Carrot	Umbelliferae	Dicot	Transgenic	Van Sluys et al. (1987)
Parsley	Umbelliferae	Dicot	Transient	R. Lütticke and R. Kunze (unpublished)
Arabidopsis	Brassicaceae	Dicot	Transgenic	Van Sluys et al. (1987)
Soybean	Fabaceae	Dicot	Transgenic	Zhou and Atherly (1990)
Flax	Linaceae	Dicot	Transgenic	Roberts et al. (1990)
Lettuce	Compositae	Dicot	Transgenic	Yang et al. (1993)

In contrast, the germinal transposition frequency of *Ac* is rather low in *Arabidopsis,* ranging from 0.07 to 5.7% with a peak between 0.2 and 2% (SCHMIDT and WILLMITZER 1989; DEAN et al. 1992). Surprisingly, 30–50% of transformed plants carrying one or more *Ac* elements showed no evidence of germinal transposition, although somatic *Ac* excision was detected by a variegated phenotype (DEAN et al. 1992; KELLER et al. 1992). In contrast to tobacco, the level of excision can be greatly increased in *Arabidopsis* by fusing stronger promoters like the 35S promoters to the TPase reading frame (GREVELDING et al. 1992; SWINBURNE et al. 1992; HONMA et al 1993). Interestingly, when TPase expression is driven by the 35S promoter, large somatic excision and reintegration sectors occur that frequently cover entire flowers, while this rarely occurs with the also strong *Arabidopsis* rbcS promoter or the weak *Ac* promoter (HONMA et al. 1993; LONG et al. 1993b). Low *Ac* transposition frequencies are also observed in flax (ELLIS et al. 1992), in *Nicotiana plumbaginifolia* (MARION-POLL et al. 1993), in lettuce (YANG et al. 1993), and in *Petunia* (HARING et al. 1989; ROBBINS et al. 1994). In the latter species the primary rate of transposition of *Ac* from the T-DNA is significantly below 1%. Remarkably, the secondary transposition frequency in progeny of primary germinal transpositions is enhanced to 10% or more. This increase in *Ac* activity is not accompanied by changes in the methylation state (ROBBINS et al. 1994).

Meanwhile, several genes have been isolated in heterologous plants by non-targeted tagging with *Ac.* A flower-color gene was isolated in *Petunia* (CHUCK et al. 1993), and in *Arabidopsis thaliana* two genes, one causing abnormal morphogenesis (BANCROFT et al. 1993) and the other an albino phenotype (LONG et al. 1993a), were tagged. Another gene causing a male sterile phenotype was tagged in *Arabidopsis* with the maize element *En/Spm* (AARTS et al. 1993). These first successes in heterologous plants demonstrate the general usefulness of transposon tagging. The first genes tagged on purpose will be reported soon, and it is expected that the progress made in understanding the transposition mechanism and its regulation will lead to a further improvement of the transposon tagging efficiency.

References

Aarts MG, Dirkse WG, Stiekema WJ, Pereira A (1993) Transposon tagging of a male sterility gene in *Arabidopsis.* Nature 363: 715–717

Alleman M, Kermicle JL (1993) Somatic variegation and germinal mutability reflect the position of transposable element *Dissociation* within the maize R-gene. Genetics 135: 189–203

Antequera F, Bird AP (1988) Unmethylated CpG islands associated with genes in higher plant DNA. EMBO J 7: 2295–2299

Athma P, Grotewold E, Peterson T (1992) Insertional mutagenesis of the maize *P* gene by intragenic transposition of *Ac.* Genetics 131: 199–209

Atkinson PW, Warren WD, O'Brochta DA (1993) The *hobo* transposable element of *Drosophila* can be cross-mobilized in houseflies and excises like the *Ac* element of maize. Proc Natl Acad Sci USA 90: 9693–9697

Baker B, Schell J, Lörz H, Fedoroff NV (1986) Transposition of the maize-controlling element "*Activator*" in tobacco. Proc Natl Acad Sci USA 83: 4844–4848
Baker B, Coupland G, Fedoroff NV, Starlinger P, Schell J (1987) Phenotypic assay for excision of the maize-controlling element *Ac* in tobacco. EMBO J 6: 1547–1554
Balcells L, Swinburne J, Coupland G (1991) Transposons as tools for the isolation of plant genes. TIBTECH 9: 31–37
Bancroft I, Dean C (1993) Transposition pattern of the maize element *Ds* in *Arabidopsis thaliana*. Genetics 134: 1221–1229
Bancroft I, Bhatt AM, Sjodin C, Scofield S, Jones JDG, Dean C (1992) Development of an efficient two-element transposon tagging system in *Arabidopsis thaliana*. Mol Gen Genet 233: 449–461
Bancroft I, Jones JDG, Dean C (1993) Heterologous transposon tagging of the DRL1 locus in *Arabidopsis*. Plant Cell 5: 631–638
Banks JA, Fedoroff N (1989) Patterns of developmental and heritable change in methylation of the suppressor-mutator transposable element. Dev Genetics 10: 425–437
Becker D, Lütticke R, Li M-g, Starlinger P (1992) Control of excision frequency of maize transposable element *Ds* in *Petunia* Protoplasts. Proc Natl Acad Sci USA 89: 5552–5556
Belzile F, Yoder JI (1992) Pattern of somatic transposition in a high copy *Ac* tomato line. Plant J 2: 173–179
Belzile F, Lassner MW, Tong Y, Khush R, Yoder JI (1989) Sexual transmission of transposed *Activator* elements in transgenic tomatoes. Genetics 123: 181–189
Bennetzen JL (1987) Covalent DNA modification and the regulation of *Mutator* element transposition in maize. Mol Gen Genet 208: 45–51
Bhattacharyya MK, Smith AM, Ellis T, Hedley C, Martin C (1990) The wrinkled-Seed character of pea described by Mendel is caused by a transposon-like insertion in a gene encoding starch-branching enzyme. Cell 60: 115–122
Bird AP (1986) CpG-rich islands and the function of DNA methylation. Nature 321: 209–213
Boehm U, Heinlein M, Behrens U, Kunze R (1995) One of three nuclear localization signals of maize *Activator (Ac)* transposase overlaps the DNA-binding domain. Plant J (in press)
Bonas U, Sommer H, Saedler H (1984) The 17-kb *Tam1* element of *Antirrhinum majus* induces a 3-bp duplication upon integration into the chalcone synthase gene. EMBO J 3: 1015–1019
Brettell RIS, Dennis ES (1991)Reactivation of silent *Ac* following tissue culture is associated with heritable alterations in its methylation pattern. Mol Gen Genet 229: 365–372
Brink RA, Williams E (1973) Mutable R-navajo alleles of cyclic origin in maize. Genetics 73: 273–296
Brown JJ, Mattes MG, O'Reilly C, Shepherd NS (1989) Molecular characterization of *rDt*, a maize transposon of the "*Dotted*" controlling element system. Mol Gen Genet 215: 239–244
Brutnell TP, Dellaporta SL (1994) Somatic inactivation and reactivation of *Ac* associated with changes in cytosine methylation and transposase expression. Genetics 138: 213–225
Caldwell EEO, Peterson PA (1992) The *Ac* and *Uq* transposable element systems in maize: interactions among components. Genetics 131: 723–731
Calvi BR, Hong TJ, Findley SD, Gelbart WM (1991) Evidence for a common evolutionary origin of inverted repeat transposons in *Drosophila* and plants: *hobo*, *Activator* and *Tam3*. Cell 66: 465–471
Chandler VL, Walbot V (1986) DNA modification of a maize transposable element correlates with loss of activity. Proc Natl Acad Sci USA 83: 1767–1771
Chen J, Greenblatt IM, Dellaporta SL (1987) Transposition of *Ac* from the *P* locus of maize into unreplicated chromosomal sites. Genetics 117: 109–116
Chen J, Greenblatt IM, Dellaporta SL (1992) Molecular analysis of *Ac* transposition and DNA replication. Genetics 130: 665–676
Chen JD, Pirrotta V (1993) Multimerization of the *Drosophila* zeste protein is required for efficient DNA binding. EMBO J 12: 2075–2083
Chomet PS, Wessler S, Dellaporta SL (1987) Inactivation of the maize transposable element *Activator* (*Ac*) is associated with its DNA modification. EMBO J 6: 295–302
Chuck G, Robbins T, Nijjar C, Ralston E, Courtney-Gutterson N, Dooner HK (1993) Tagging and cloning of a *Petunia* flower color gene with the maize transposable element *Activator*. Plant Cell 5: 371–378
Coen ES, Carpenter R, Martin C (1986) Transposable elements generate novel spatial patterns ofgene expression in *Antirrhinum majus*. Cell 47: 285–296
Coen ES, Robbins TP, Almeida J, Hudson A, Carpenter R (1989) Consequences and mechanism of transposition in *Antirrhinum majus*. In: Berg DE, Howe MM (eds) Mobile genetic elements. American Society for Microbiology, Washington DC, pp 413–436

Coupland G, Baker B, Schell J, Starlinger P (1988) Characterization of the maize transposable element *Ac* by internal deletions. EMBO J 7: 3653–3659

Coupland G, Plum C, Chatterjee S, Post A, Starlinger P (1989) Sequences near the termini are required for transposition of the maize transposon *Ac* in transgenic tobacco plants. Porc Natl Acad Sci USA 86: 9385–9388

Courage-Tebbe U, Döring H-P, Fedoroff NV, Starlinger P (1983)The controlling element *Ds* at the *Shrunken locus* in Zea mays: structure of the unstable *sh-m5933* allele and several revertants. Cell 34: 383–393

Dean C, sjodin C, Page T, Jones JDG, Lister C (1992) Behavior of the maize transposable element *Ac* in *Arabidopsis thaliana*. Plant J 2: 69–81

DeLong A, Calderon-Urrea A, Dellaporta SL (1993) Sex determination gene TASSELSEED2 of maize encodes a short-chain alcohol dehydrogenase required for stage-specific floral organ abortion. Cell 74: 757–768

Dempsey E (1993) The *Ac2-bz2m* mutable system of maize. Maydica 38: 151–161

Dooner HK, Belachew A (1989) Transposition pattern of the maize element *Ac* from the *bz-m2(Ac)* allele. Genetics 122: 447–457

Dooner HK, Belachew A (1991) Chromosome breakage by pairs of closely linked transposable elements of the *Ac-Ds* family in maize. Genetics 129: 855–862

Dooner HK, English J, Ralston EJ, Weck E (1986) A single genetic unit specifies two transposition functions in the maize element *Activator*. Science 234: 210–211

Dooner HK, English J, Ralston EJ (1988) The frequency of transposition of the maize element *Activator* is not affected by an adjacent deletion. Mol Gen Genet 211: 485–491

Dooner HK, Belachew A, Burgess D, Harding S, Ralston M, Ralston E (1994) Distribution of unlinked receptor sites for transposed *Ac* elements from the *bz-m2(Ac)* allele in maize. Genetics 136: 261–279

Döring H-P (1989) Tagging genes with maize transposable elements. An overview. Maydica 34: 73–88

Döring H-P, Starlinger P (1984) Barbara McClintock's controlling elements: now at the DNA level. Cell 39: 253–259

Döring H-P, Starlinger P (1986) Molecular genetics of transposable elements in plants. Annu Rev Genet 20: 175–200

Döring H-P, Freeling M, Hake S, Johns MA, Kunze R, Merckelbach A, Salamini F, Starlinger P (1984) A *Ds* mutation of the *Adh*1 gene in *Zea mays* L. Mol Gen Genet 193: 199–204

Döring H-P, Nelsen-Salz B, Garber R, Tillmann E (1989) Double *Ds* elements are involved in specific chromosome breakage. Mol Gen Genet 219: 299–305

Döring H-P, Pahl I, Durany M (1990) Chromosomal rearrangements caused by the aberrant transposition of double Ds elements are formed by *Ds-sequences* and adjacent non-*Ds* sequences. Mol Gen Genet 224: 40–48

Dowe MF jr, Roman GW, Klein AS (1990) Excision and transposition of two *Ds* transposons from the *bronze mutable 4* derivative 6856 allele of *Zea mays* L. Mol Gen Genet 221: 475–485

Dynan WS (1986) Promoters for housekeeping genes. Trends Genet August: 196–197

Ellis JG, Finnegan EJ, Lawrence GJ (1992) Developing a transposon tagging system to isolate rust-resistance genes from flax. Theor Appl Genet 85: 46–54

English J, Harrison K, Jones JDG (1993) A genetic analysis of DNA sequence requirements for *Dissociation* state-I activity in tobacco. Plant Cell 5: 501–514

Fedoroff N, Wessler S, Shure M (1983) Isolation of the transposable maize controlling elements *Ac* and *Ds*. Cell 35: 235–242

Fedoroff NV (1983) Controlling elements in maize. In: Shapiro JA (eds) Mobile genetic elements. Academic, New York, pp 159–221

Fedoroff NV (1983) Maize transposable elements. In: Berg DE, Howe MM (eds) Mobile DNA. American Society for Microbiology, Washington DC, PP 374–411

Feldmar S, Kunze R (1991) The ORFa protein, the putative transposase of maize transposable element *Ac*, has a basic DNA binding domain. EMBO J 10: 4003–4010

Ferris PJ (1989) Characterization of a *Chlamydomonas* transposon, *Gulliver*, resembling those in higher plants. Genetics 122: 363–377

Finnegan EJ, Taylor BH, Dennis ES, Peacock WJ (1988) Transcription of the maize transposable element *Ac* in maize seedlings and in transgenic tobacco. Mol Gen Genet 212: 505–509

Finnegan EJ, Lawrence GJ, Dennis ES, Ellis JG (1993) Behaviour of modified *Ac* elements in flax callus and regenerated plants. Plant Mol Biol 22: 625–633

Fußwinkel H, Schein S, Courage U, Starlinger P, Kunze R (1991) Detection and abundance of mRNA and protein encoded by transposable element *Activator (Ac)* in maize. Mol Gen Genet 225: 186–192

Gardiner-Garden M, Frommer M (1992) Significant CpG-rich regions in angiosperm genes. J Mol Evol 34: 231–245
Geiser M, Weck E, Döring H-P, Werr W, Courage-Tebbe U, Tillmann E, Starlinger P (1982) Genomic clones of a wild-type allele and a transposable element-induced mutant of the sucrose synthase gene of *Zea mays* L. EMBO J 1: 1455–1460
Gerats AGM, Huits H, Vrijlandt E, Marana C, Souer E, Beld M (1990) Molecular characterization of a nonautonomous transposable element (*dTph*1) of *petunia*. Plant Cell 2: 1121–1128
Gerlach WL, Dennis ES, Peacock WJ, Clegg MT (1987) The Ds1 controlling element family in maize and *Tripsacum*. J Mol Evol 26: 329–334
Gierl A, Saedler H (1989) The En/Spm transposable element of *Zea mays*. Plant Mol Biol 13: 261–266
Gierl A, Saedler H, Peterson PA (1989) Maize transposable elements. Annu Rev Genet 23: 71–85
Greenblatt IM (1974) Proximal-distal polarity of *Modulator* transposition upon leaving the *P* locus. Maize Genet Coop Newsl 48: 188–189
Greenblatt IM (1984) A chromosomal replication pattern deduced from pericarp phenotypes resulting from movements of the transposable element, *Modulator*, in maize. Genetics 108: 471–485
Greenblatt IM, Brink RA (1962) Twin mutations in medium variegated pericarp maize. Genetics 47: 489–501
Grevelding C, Becker D, Kunze R, von Menges A, Fantes V, Schell J, Masterson R (1992) High rates of *Ac/Ds* germinal transposition in *Arabidopsis* suitable for gene isolation by insertional mutagenesis. Proc Natl Acad Sci USA 89: 6085–6089
Grotewold E, Athma P, Paterson T (1991) A possible hot spot for *Ac* insertion in the maize P gene. Mol Gen Genet 230: 329–331
Haring MA, Gao J, Volbeda T, Rommens CMT, Nijkamp HJI, Hille J (1989) A comparative study of *Tam3* and *Ac* transposition in transgenic tobacco and petunia plants. Plant Mol Biol 13: 189–201
Haring MA, Teeuwen de Vroomen MJ, Nijkamp HJ, Hille J (1991) *Trans*-activation of an artifical *dTam3* transposable element in transgenic tobacco plants. Plant Mol Biol 16: 39–47
Hartings H, Spilmont C, Lazzaroni N, Rossi V, Salamini F, Thompson RD, Motto M (1991) Molecular analysis of the *Bg-rbg* transposable element system of *Zea mays* L. Mol Gen Genet 227: 91–96
Hauser C, Fußwinkel H, Li J, Oellig C, Kunze R, Müller-Neumann M, Heinlein M, Starlinger P, Doerfler W (1988) Overproduction of the protein encoded by the maize transposable element *Ac* in insect cells by a baculovirus vector. Mol Gen Genet 214: 373–378
Healy J, Corr C, Deyoung J, Baker B (1993) Linked and unlinked transposition of a genetically marked *Dissociation* element in transgenic tomato. Genetics 134: 571–584
Hehl R, Baker B (1989) Induced transposition of *Ds* by a stable *Ac* in crosses of transgenic tobacco plants. Mol Gen Genet 217: 53–59
Hehl R, Baker B (1990) Properties of the maize transposable element *Activator* in transgenic tobacco plants: a versatile inter-species genetic tool. Plant Cell 2: 709–721
Hehl R, Nacken W, Krause A, Saedler H, Sommer H (1991) Structural analysis of *Tam3*, a transposable element from *Antirrhinum majus*, reveals homologies to the *Ac* element from maize. Plant Mol Biol 16: 369–371
Heinlein M (1995) Variegation patterns caused by excision of the maize transposable element *Dissociation* (*Ds*) are autonomously regulated by allele-specific *Activator* (*Ac*) elements and are not due to *trans*-acting modifier genes. Mol Gen Genet 246: 1–9
Heinlein M, Starlinger P (1991) Variegation patterns caused by transposable element *Ac*. Maydica 36: 309–316
Heinlein M, Brattig T, Kunze R (1994) In vivo aggregation of maize *Activator (Ac)* transposase in nuclei of maize endosperm and *Petunia* protoplasts. Plant J 5: 705–714
Herrmann A, Schulz W, Hahlbrock K (1988) Two alleles of the single-copy chalcone synthase gene in parsley differ by a transposon-like element. Mol Gen Genet 212: 93–98
Hirochika H (1993) Activation of tobacco retrotransposons during tissue culture. EMBO J 12: 2521–2528
Honma MA, Baker BJ, Waddell CS (1993) High-frequency germinal transposition of *DsALS* in *Arabidopsis*. Proc Natl Acad Sci USA 90: 6242–6246
Houba-Hérin N, Becker D, Post A, Larondelle Y, Starlinger P (1990) Excision of a *Ds*-like maize transposable element (AcΔ) in a transient assay in *Petunia* is enhanced by a truncated coding region of the transposable element *Ac*. Mol Gen Genet 224: 17–23
Houba-Hérin N, Domin M, Pedron J (1994) Transposition of a *Ds* element from a plasmid into the plant genome in *Nicotiana plumbaginifolia* protoplast-derived cells. Plant J 6: 55–66
Houck MA, Clark JB, Peterson KR, Kidwell MG (1991) Possible horizontal transfer of *Drosophila* genes by the mite *Proctolaelaps regalis*. Science 253: 1125–1129

Inagaki Y, Hisatomi Y, Suzuki T, Kasahara K, Iida S (1994) Isolation of a *suppressor-mutator/enhancer*-like transposable element, *Tpn1*, from Japanese Morning Glory bearing variegated flowers. Plant Cell 6: 375–383

Izawa T, Miyazaki C, Yamamoto M, Terada R, Iida S, Shimamoto K (1991) Introduction and transposition of the maize transposable element *Ac* in rice (*Oryza sativa* L.). Mol Gen Genet 227: 391–396

Jing-Liu Z, Xiao-Ming L, Rui-Zhu C, Rui-Xin H, Meng-Min H (1991) Transposition of maize transposable element *Activator* in rice. Plant Sci 73: 191–198

Jones JDG, Carland FM, Maliga P, Dooner HK (1989) Visual detection of transposition of the Maize element *Activator (Ac)* in tobacco seedlings. Science 244: 204–207

Jones JDG, Carland FM, Lim E, Ralston EJ, Dooner HK (1990) Preferential transposition of the maize element *Activator* to linked chromosomal locations in tobacco. Plant Cell 2: 701–707

Jones JDG, Harper L, Carland FM, Ralston EJ, Dooner HK (1991) Reversion and altered variegation of an SPT: *Ac* allele in tobacco Maydica 36: 329–335

Kay BK, Dawid IB (1993) The 1723 element: a long, homogeneous, higly repeated DNA unit interspersed in the genome of *xenopus laevis*. J Mol Biol 170: 583–596

Keller J, Lim E, James DW, Dooner HK (1992) Germinal and somatic activity of the maize element *Activator (Ac)* in *Arabidopsis*. Genetics 131: 449–459

Keller J, Jones JDG, Harper E, Lim E, Carland F, Ralston EJ, Dooner HK (1993a) Effects of gene dosage and sequence modification on the frequency and timing of transposition of the maize element *Activator (Ac)* in tobacco. Plant Mol Biol 21: 157–170

Keller J, Lim E, Dooner HK (1993b) Preferential transposition of *Ac* to linked sites in *Arabidopsis*. Theor Appl Genet 86: 585–588

Kermicle JL, Alleman M, Dellaporta SL (1989) Sequential mutagenesis of a maize gene, using the transposable element *Dissociation*. Genome 31: 712–716

Kleckner N (1990) Regulating Tn10 and IS10 transposition. Genetics 124: 449–454

Knapp S, Coupland G, Uhrig H, Starlinger P, Salamini F (1988) Transposition of the maize transposable element *Ac* in *Solanum tuberosum*. Mol Gen Genet 213: 285–290

Knapp S, Larondelle Y, Roßberg M, Furtek D, Theres K (1994) Transgenic tomato lines containing *Ds* elements at defined genomic positions as tools for targeted transposon tagging. Mol Gen Genet 243: 666–673

Köster-Töpfer M, Frommer WB, Rocha-Sosa M, Willmitzer L (1990) Presence of a transposon-like element in the promoter region of an inactive patatin gene in *Solanum tuberosum* L. Plant Mol Biol 14: 239–247

Kunze R, Starlinger P (1989) The putative transposase of transposable element *Ac* from *Zea mays* L. interacts with subterminal sequences of Ac. EMBO J 8: 3177–3185

Kunze R, Stochaj U, Laufs J, Starlinger P (1987) Transcription of transposable element *Activator* (*Ac*) of *Zea mays* L. EMBO J 6: 1555–1563

Kunze R, Starlinger P, Schwartz D (1988) DNA methylation of the maize transposable element *Ac* interferes with its transcription. Mol Gen Genet 214: 325–327

Kunze R, Coupland G, Fußwinkel H, Feldmar S, Courage U, Schein S, Becker H-A, Chatterjee S, Li M-G, Starlinger P (1991) Structure and function of the maize transposable element *Activator* (*Ac*). Plant Mol Biol (285–298): NATO ASI series, life sciences

Kunze R, Behrens U, Courage-Franzkowiak U, Feldmar S, Kühn S, Lütticke R (1993) Dominant transposition-deficient mutants of maize *Activator* (*Ac*) transposase. Proc Nat Acad Sci USA 90: 7094–7098

Lassner MW, Palys JM, Yoder JI (1989) Genetic transactivation of *Dissociation* elements in transgenic tomato plants. Mol Gen Genet 218: 25–32

Laufs J, Wirtz U, Kammann M, Matzeit V, Schaefer S, Schell J, Czernilofsky AP, Baker B, Gronenborn B (1990) Wheat dwarf virus *Ac/Ds* vectors: expression and excision of transposable elements introduced into various cereals by a viral replicon. Proc Natl Acad Sci USA 87: 7752–7756

Lechelt C, Peterson T, Laird A, Chen J, Dellaporta SL, Dennis ES, Peacock WJ, Starlinger P (1989) Isolation and molecular analysis of the maize *P* locus. Mol Gen Genet 219: 225–234

Levy AA, Walbot V (1990) Regulation of the timing of transposable element excision during maize development. Science 248: 1534–1537

Li M-G, Starlinger P (1990) Mutational analysis of the N terminus of the protein of maize transposable element *Ac*. Proc Natl Acac Sci USA 87: 6044–6048

Long D, Martin M, Sundberg E, Swinburne J, Puangsomlee P, Coupland G (1993a) The maize transposable element system *Ac/Ds* as a mutagen in *Arabidopsis*: identification of an albino mutation induced by *Ds* insertion. Proc Natl Acad Sci USA 90: 10370–10274

Long D, Swinburne J, Martin M, Wilson K, Sundberg E, Lee K, Coupland G (1993b) Analysis of the frequency of inheritance of transposed *Ds* elements in *Arabidopsis* after activation by a CaMV 35 S promoter fusion to the *Ac* transposase gene. Mol Gen Genet 241: 627–636

MacRae AF, Clegg MT (1992) Evolution of *Ac* and *Ds*1 elements in select grasses (*Poaceae*). Genetica 86: 55–66

MacRae AF, Huttley GA, Clegg MT (1994) Molecular evolutionary characterization of an *Activator (Ac)*-like transposable element sequence from pearl millet (*Pennisetum glaucum*) (*Poacae*). Genetica 92: 77–89

Marion-Poll A, Martin E, Bonnefoy N, Pautot V (1993) Transposition of the maize autonomous element *Activator* in transgenic *Nicotiana plumbaginifolia* plants. Mol Gen Genet 238: 209–217

Martin C, Prescott A, Lister C, MacKay S (1989) Activity of the transposon *Tam3* in *Antirrhinum* and tobacco: possible role of DNA methylation. EMBO J 8: 997–1004

Masson P, Rutherford G, Banks JA, Fedoroff N (1989) Essential large transcripts of the maize *Spm* transposable element are generated by alternative splicing. Cell 58: 755–765

McClintock B (1942) The fusion of broken ends of chromosomes following nuclear fusion. Proc Natl Acad Sci USA 28: 458–463

McClintock B (1946) Maize genetics. Carnegie Inst Washington Year Book 45: 176–186

McClintock B (1947) Cytogenetic studies of maize and *Neurospora*. Carnegie Inst Washington Year Book 46: 146–152

McClintock B(1948) Mutable loci in maize. Carnegie Inst Washington Year Book 47: 155–169

McClintock B (1949) Mutable loci in maize. Carnegie Inst Washington Year Book 48: 142–154

McClintock B (1951) Chromosome organization and genic expression. Cold Spring Harb Symp Quant Biol 16: 13–47

McClintock B (1964) Aspects of gene regulation in maize. Carnegie Inst Washington Year Book 63: 592–602

McClintock B (1965) Components of action of the regulators *Spm* and *Ac*. Carnegie Inst Washington Year Book 64: 527–534

McClintock B (1984) The significance of responses of the genome to challenge. Science 226: 792–801

Meyer C, Pouteau S, Rouzé P, Caboche M (1994) Isolation and molecular characterization of *dTnp*1, a mobile and defective tranposable element of *Nicotiana plumbaginifolia*. Mol Gen Genet 242: 194–200

Michel D, Salamini F, Motto M, Döring H-P (1994) An unstable allele at the maize Opaque2 locus is caused by the insertion of a double *Ac* element. Mol Gen Genet 243: 334–342

Miler SM, Schmitt R, Kirk DL (1993) *Jordan*, an active *Volvox* transposable element similar to higher-plant transposons. Plant Cell 5: 1125–1138

Moreno MA, Chen J, Greenblatt I, Dellaporta SL (1992) Reconstitutional mutagenesis of the maize *P* gene by short-range Ac transpositions. Genetics 131: 939–956

Muller-Neumann M, Yoder JI, Starlinger P (1984) The DNA sequence of the transposable element *Ac* of *Zea mays* L. Mol Gen Genet 198: 19–24

Murai N, Kawagoe Y, Hayashimoto A (1991) Transposition of the maize *Activator* element in transgenic rice plants. Nucleic Acids Res 19: 617–622

Nelsen-Salz B, Döring H-P (1990) Rare de novo methylation within the transposable element *Activator* (*Ac*) in transgeinc tobacco plants. Mol Gen genet 223: 87–96

O'Hare K, Rubin GM (1993) Structures of *P* transposable elements and their sites of insertion and excision in the *Drosophila melanogaster* genome. Cell 34: 25–35

Osborne BI, Corr CA, Prince JP, Hehl R, Tanksley SD, McCormick S, Baker B (1991) *Ac* transposition from a T-DNA can generate linked and unlinked clusters of insertions in the tomato genome. Genetics 129: 833–844

Osterman JC (1991) Transposition of *AC-2* in response to temperature. Maydica 36: 147–151

Ott T, Nelsen-Salz B, Döring H-P (1992) PCR-aided genomic sequencing of 5' subterminal sequences of the maize transposable element *Activator* (*Ac*) in transgenic tobacco plants. Plant J 2: 705–711

Peacock WJ, Dennis ES, Gerlach WL, Sachs MM, Schwartz D (1984) Insertion and excision of *Ds* controlling elements in maize. Cold Spring Harb Symp Quant Biol 49: 347–354

Peleman J, Cottyn B, Van Camp W, Van Montagu M, Inze D (1991) Transient occurrence of extrachromosomal DNA of an *Arabidopsis thaliana* transposon-like element, *Tat*1. Proc Natl Acad Sci USA 88: 3618–3622

Pereira A, Schwarz-Sommer Z, Gierl A, Bertram I, Peterson PA, Saedler H (1985) Genetic and molecular analysis of the *Enhancer* (*En*) transposable element system of *Zea mays*. EMBO J 4: 17–23

Pereira A, Cuypers H, Gierl A, Schwarz-Sommer Z, Saedler H (1986) Molecular analysis of the *En/Spm* transposable element system of *Zea mays*. EMBO J 5: 835–841
Peschke VM, Phillips RL, Gengenbach BG (1987) Discovery of transposable element activity among progeny of tissue culture-derived maize plants. Science 238: 804–807
Peterson T (1990) Intragenic transposition of *Ac* generates a new allele of the maize *P* gene. Genetics 126: 469–476
Pisabarro AG, Martin WF, Peterson PA, Saedler H, Gierl A (1991) Molecular analysis of the *Ubiquitous* (*Uq*) transposable element system of *Zea mays*. Mol Gen Genet 230: 201–208
Pohlman RF, Fedoroff NV, Messing J (1994a) The nucleotide sequence of the maize controlling element *Activator*. Cell 37: 635–643
Pohlman RF, Fedoroff NV, Messing J (1984b) Correction: nucleotide sequence of *Ac*. Cell 39: 417
Raleigh EA, Kleckner N (1986) Quantitation of insertion sequence IS10 transposase gene expression by a method generally applicable to any rarely expressed gene. Proc Natl Acad Sci USA 83: 1787–1791
Ralston EJ, English J, Dooner HK (1989) Chromosome-breaking structure in maize involved in a fractured *Ac* element. Proc Natl Acad Sci USA 86: 9451–9455
Rhoades MM, Dempsey E (1983) Further studies on the two-unit mutable systems found in our high-loss studies and on the specificity of interaction of responding and controlling elements. Maize Genet Coop Newsl 57: 14–17
Rhodes PR, Vodkin LO (1988) Organization of the *Tgm* family of transposable elements in *Soybean*. Genetics 120: 597–604
Robbins TP, Carpenter R, Coen ES (1989) A Chromosome rearrangement suggests that donor and recipient sites are associated during *Tam3* transposition in *Antirrhinum majus*. EMBO J 8: 5–13
Robbins TP, Jenkin M, Courtney-Gutterson N (1994) Enhanced frequency of transposition of the maize transposable element *Activator* following excision from T-DNA in *Petunia hybrida*. Mol Gen Genet 244: 491–500
Roberts D, Hoopes BC, McClure WR, Kleckner N (1985) IS10 transposition is regualted by DNA adenine methylation. Cell 43: 117–130
Roberts MR, Kumar A, Scott R, Draper J (1990) Excision of the maize transposable element *Ac* in flax callus. Plant Cell Rep 9: 406–409
Robertson HM (1993) The *mariner* transposable element is widespread in insects. Nature 362: 241–245
Rommens CM, van Haaren MJ, Buchel AS, Mol JN, van Tunen AJ, Nijkamp HJ, Hille J (1992) Transactivation of *Ds* by *Ac*-transposase gene fusions in tobacco. Mol Gen Genet 231: 433–441
Rommens CMT, Munyikwa TRI, Overduin B, Nijkamp HJI, Hille J (1993) Transposition pattern of a modified *Ds* element in tomato. Plant Mol Biol 21: 1109–1119
Rosenzweig B, Liao LW, Hirsh D (1983) Sequence of the *C. elegans* transposable element *Tc*1. Nucleic Acids Res 11: 4201–4209
Saedler H, Nevers P (1985) Transposition in plants: a molecular model. EMBO J 4: 585–590
Schiefelbein JW, Furtek DB, Dooner HK, Nelson OE jr (1988) Two mutations in a *bronze-1* allele caused by transposable elements of the *Ac-Ds* family alter the quantity and quality of the gene product. Genetics 120: 767–777
Schmidt R, Willmitzer L (1989) The maize autonomous element *Activator* (*Ac*) shows a minimal germinal excision frequency of 0.2%–0.5% in transgenic *Arabidopsis thaliana* plants. Mol Gen Genet 220: 17–24
Schmidt-Rogge T, Weber B, Börner T, Brandenburg E, Schieder O, Meixner M (1994) Transposition and behavior of the maize transposable element *Ac* in transgenic haploid *Datura innoxia* Mill. Plant Sci 99: 63–74
Schwartz D (1984) Analysis of the *Ac* transposable element dosage effect. Mol Gen Genet 196: 81–84
Schwartz D (1986) Analysis of the autonomous *wx-m7* transposable element mutant of maize. Maydica 31: 123–129
Schwartz D (1989a) Gene-controlled cytosine demethylation in the promoter region of the *Ac* transposable element in maize. Proc Natl Acad Sci USA 86: 2789–2793
Schwartz D (1989b) Pattern of *Ac* transposition in maize. Genetics 121: 125–128
Schwartz D, Dennis ES (1986) Transposase activity of the *Ac* controlling element in maize is regulated by its degree of methylation. Mol Gen Genet 205: 476–482
Scofield SR, Harrison K, Nurrish SH, Jones JDG (1992) Promoter fusions to the *Activator* transposase gene cause distinct patterns of *Dissociation* excision in tobacco Cotyledons. Plant Cell 4: 573–582
Scofield SR, English JJ, Jones JDG (1993) High level expression of the *Activator* (*Ac*) transposase gene inhibits the excision of *Dissociation* (*Ds*) in tobacco cotyledons. Cell 75: 507–517

Shen WH, Hohn B (1992) Excision of a transposable element from a viral vector introduced into maize plants by agroinfection. Plant J 2: 35–42

Shen WH, Das S, Hohn B (1992) Mechanism of Ds1 excision from the genome of maize streak virus. Mol Gen Genet 233: 388–394

Shirsat AH (1988) A transposon-like structure in the 5'flanking sequence of a legumin gene from Pisum sativum. Mol Gen Genet 212: 129–133

Streck RD, MacGaffey JE, Beckendorf SK (1986) The structure of *hobo* transposable elements and their insertion sites. EMBO J 5: 3615–3623

Sutton WD, Gerlach WL, Schwartz D, Peacock WJ (1984) Molecular analysis of *Ds* controlling element mutations at the *Adhl* locus of maize. Science 223: 1265–1268

Swinburne J, Balcells L, Scofield SJ, Jones JDG, Coupland G (1992) Elevated levels of *Activator* transposase mRNA are associated with high frequencies of *dissociation* excision in *Arabidopsis*. Plant Cell 4: 583–595

Theres N, Scheele T, Starlinger P (1987) Cloning of the *Bz2* locus of *Zea mays* using the transposable element *Ds* as a gene tag. Mol Gen Genet 209: 193–197

Tsay YF, Frank MJ, Page T, Dean C, Crawford NM (1993) Identification of a mobile endogenous transposon in *Arabidopsis thaliana*. Science 260: 342–344

Tudor M, Lobocka M, Goodell M, Pettitt J, O'Hare K (1992) The *pogo* transposable element family of *Drosophila melanogaster*. Mol Gen Genet 232: 126–134

Van Sluys MA, Tempe J, Fedoroff NV (1987) Studies on the introduction and mobility of the maize *Activator* element in *Arabidopsis thaliana* and *Daucus carota*. EMBO J 6: 3881–3889

Van Sluys MA, Scortecci KC, Tempe J (1993) DNA methylation associated to *Ac* element in *Arabidopsis*. Plant Physiol Biochem 31: 805–813

Varagona M, Wessler SR (1990) Implications for the *Cis*-requirements for *Ds* transposition based on the sequence of the *wxB4 Ds* element. Mol Gen Genet 220: 414–418

Walbot V (1992) Strategies for mutagenesis and gene cloning using transposon tagging and T-DNA insertional mutagenesis. Annu Rev Plant Physiol 43: 49–82

Weck E, Courage U, Döring H-P, Fedoroff NV, Starlinger P (1984) Analysis of *sh-m6233*, a mutation induced by the transposable element *Ds* in the sucrose synthase gene of *Zea mays*. EMBO J 3: 1713–1716

Weil CF, Wessler SR (1993) Molecular evidence that chromosome breakage by *Ds* elements is casued by aberrant transposition. Plant Cell 5: 512–522

Weil CF, Marillonnet S, Burr B, Wessler SR (1992) Changes in state of the *Wx-m5* allele of maize are due to intragenic transposition of *Ds*. Genetics 130: 175–185

Yang CH, Ellis JG, Michelmore RW (1993) Infrequent transposition of *Ac* in lettuce, *Lactuca sativa*. Plant Mol Biol 22: 793–805

Yoder JI (1990) Rapid proliferation of the maize transposable element *Activator* in transgenic tomato. Plant Cell 2: 723–730

Yoder JI, Palys J, Alpert K, Lassner M (1988) *Ac* transposition in transgenic tomato plants. Mol Gen Genet 213: 291–296

Zhou JH, Atherly AG (1990) In situ detection of transposition of the maize controlling element (*Ac*) in transgenic soybean tissues. Plant Cell Rep 8: 542–545

The *Mutator* Transposable Element System of Maize

J.L. BENNETZEN

1 Introduction

The *Mutator* trait was first identified by Robertson as a heritable high forward-mutation rate exhibited by lines derived from a single maize stock (ROBERTSON

Department of Biological Sciences, Purdue University, West Lafayette, IN 47907, USA

1978). Many of these de novo mutations exhibited somatic instability, primarily apparent reversion to wild type. This instability, or mutability, was suggestive of the transposable element-induced mutations that had first been carefully characterized by McClintock (1948, 1949) in maize. Subsequent molecular studies confirmed this similarity, with the eventual cloning of *Mu* transposable elements from mutant loci generated in *Mutator* stocks (Bennetzen et al. 1984; Taylor and Walbot 1987; Oishi and Freeling 1988; Fleenor et al. 1990; Hershberger et al. 1991).

Robertson originally studied *Mutator* as a quantitative (i.e., multigenic) trait. Due to high frequencies of *Mutator* amplification and high rates of (possibly self-induced) inactivation, often competing in the same stocks, *Mutator* was only recently reduced to a simply inherited phenomenon (Robertson and Stinard 1989). Soon thereafter, an autonomous *Mu* element that can drive the activities of the *Mutator* element family was cloned and characterized (Qin and Ellingboe 1990; Chomet et al. 1991; Hershberger et al. 1991; Qin et al. 1991). Several reviews have recently been published that focus on various aspects of *Mutator* (Walbot 1991; Chandler and Hardeman 1992; Bennetzen et al. 1993), reflecting the ever-increasing interest in this system.

The exceptional level of investigation that has been engendered by *Mutator* is partly due to its unequalled productivity in yielding "tagged" mutations in maize (catalogued in Bennetzen et al. 1993). Beyond this, the somewhat idiosynchratic analyses performed in the *Mutator* system prior to its reduction to a two-component system have also yielded observations and materials not available from any other plant transposable element system. The addition of autonomous element studies to the arsenal of *Mutator* characterization tools will now greatly enhance this broad, multi-tiered array of studies.

2 *Mutator* Activity

Mutator was first defined by its ability to generate a high forward-mutation rate for an array of seed and seedling traits. Most of these new mutations, detected in self-crossed progeny, were recessive, and many were somatically mutable (Robertson 1978). Since then, several other manifestations of the *Mutator* system have been identified, providing insights into the mechanisms and regulation of *Mutator* activity.

2.1 *Mutator*-associated Mutation

The first *Mutator*-derived mutations characterized were found to be due to the insertion of a *Mu* transposable element into the affected gene (Strommer et al. 1982; Bennetzen et al. 1984; Taylor et al. 1986), and this remains the most common type of mutation. Careful and quantitative investigations at a large number of well-dispersed loci indicate that *Mu* elements can insert into most or all genes, at detected frequencies standardly ranging form 10^{-3} to 10^{-5} (summarized in BENNETZEN et al. 1993). Insertions in exons disrupt reading frames and generally lead to complete gene inactivations (BROWN et al. 1989; MCCARTY et al. 1989; SCHNABLE et al. 1989; FLEENOR et al. 1990; NASH et al. 1990; BRITT and WALBOT 1991; HERSHBERGER et al. 1991; HAN et al. 1992; HARDEMAN and CHANDLER 1993). Insertions in introns often cause only partial loss of gene activity (STROMMER et al. 1982; BENNETZEN et al. 1984; BRITT and WALBOT 1991) due to altered processing of the derived transcript and/or transcriptional termination within the *Mu* element (VAYDA and FREELING 1986; TAYLOR and WALBOT 1987; STROMMER and ORTIZ 1989; LUEHRSEN and WALBOT 1990; ORTIZ and STROMMER 1990). Insertions are sometimes observed in promoters or other regulatory regions, and these can lead to altered gene regulation (CHEN et al. 1987; ORTIZ et al. 1988; BARKAN and MARTIENSSEN 1991; GREENE et al. 1994).

The true insertion frequency of *Mu* elements into genes appears to be significantly higher than that observed in most studies. A very high percentage of *Mutator*-induced mutations are phenotypically suppressed soon or immediately after mutant induction (MARTIENSSEN et al. 1989; BENNETZEN et al. 1993; MARTIENSSEN and BARON 1994), suggesting that many mutational events may be missed altogether. Moreover, as with any mutagenic process, alterations in a gene that do not significantly affect function would not be detected by standard phenotypic screening. Recently, using polymerase chain reaction (PCR) technology with one primer anchored in a *Mu* end and one primer anchored in a known gene sequence, screened *Mutator* populations have been found to yield independent *Mu* insertions in specific genes at frequencies two or more times higher than that expected for phenotypic screens (S. Briggs, personal communication; R. Martienssen, personal communication).

Other types of genome rearrangements are also associated with *Mutator* elements, including short adjacent deletions (TAYLOR and WALBOT 1985; LEVY and WALBOT 1991) and duplications (STINARD et al. 1993). Excision of a *Mu* element is usually accompanied by a net change in the target locus; all or part of the flanking target direct repeats are commonly left behind (SCHNABLE et al. 1989; BRITT and WALBOT 1991; DOSEFF et al. 1991; LEVY and WALBOT 1991; KLOECKENER-GRUISSEM et al. 1992), and additional nearby sequences can be incorporated at the excision site (DOSEFF et al. 1991).

Due to the inactivational nature of most insertion events, the majority of *Mutator*-induced mutations are recessive. There are exceptions to this, however, in the cases of dominant *Kn1*, *Ae1*, and *Les* mutations that have been derived from *Mutator* stocks (BENNETZEN et al. 1993; STINARD et al. 1993; GREENE et al. 1994; MARTIENSSEN and BARON 1994).

Robertson and co-workers have identified high levels of terminal and internal deletions on the short arm of chromosome 9 in some *Mutator* stocks (ROBERTSON and STINARD 1987; ROBERTSON et al. 1994). Although the high frequency of recovered deletions in these *Mutator* lines suggests that *Mu* activity is responsible, tests have not been performed to check whether similar deletions would be generated in *Mutator*-loss lines derived from these stocks. As Robertson and co-workers have proposed (ROBERTSON abd STINARD 1987; ROBERTSON et al. 1994), unequal recombination between *Mu* elements in these lines could be responsible for the deletions observed and, thereby, not require the presence of *Mutator* transposase. Such a model would also suggest a similarly high frequency of additional rearrangement classes, including translocations and inversions. However, ectopic recombination between any of the other classes of repetitive elements scattered throughout the maize genome (HAKE and WALBOT 1980; BENNETZEN et al. 1994; SPRINGER et al. 1994) could also account for these deletions and other rearrangements. Hence, the case for *Mutator*-promoted deletions of long chromosomal segments requires further support.

As with many other transposable element systems, *Mutator* elements appear to be highly self-mutagenic. This is particularly true of the autonomous *Mutator* element, *MuDR*, which undergoes frequent internal deletion during development (LISCH et al. 1995; P. Schnable, personal communication). Additional analyses will be required to see whether, as expected, this instability is dependent upon *Mutator* activity. The question is particularly pertinent for *MuDR*, since several labs have independently noted that the autonomous element is difficult to maintain in an intact form in *Escherichia coli*, including in recombination-defective *E. coli* strains.

2.2 Mutability and Excision

Many of the inactivational mutations caused by *Mu* element insertion exhibit somatic sectoring patterns associated with apparent reversion to wild type. The frequency of somatic reversion events, as indicated by sector number, is highly variable (BENNETZEN 1985a; ROBERTSON et al. 1985; WALBOT 1986). However, most of these "mutable" mutations show a distinctive size distribution of uniformly small sectors in the leaf and/or seed. This specific and rather narrow window for *Mutator*-driven somatic reversion does not vary greatly between most *Mutator* lines or between different *Mutator*-derived insertion alleles (BROWN et al. 1989; LEVY et al. 1989; MCCARTY et al. 1989). Recently, Walbot selected a *Mutator* stock in which somatic reversion events often occur early in endosperm development. This change appears to be in a transactivational component, since several *Mu* insertion reporter alleles show a similar large-seed-sector pattern (WALBOT 1992a). However, other studies that have shown major effects on the frequency of *Mutator* action due to *MuDR* chromosomal position or deletion status have not yielded observations of altered developmental timing of *Mu* action (LISCH et al. 1995).

When molecularly examined, most somatic reversion events have been associated with excision of the *Mu* element inserted into the mutable allele. Only in rare cases have the somatic sectors been large enough to make direct cloning and subsequent sequence analysis possible (MARTIENSSEN et al. 1989). With the standard small-sector patterns, investigators have relied on PCR amplification of DNA prepared from the tissue exhibiting instability. These experiments indicate a variety of excision events, only a subset of which would give rise to an apparent revertant phenotype. Other excisions, in addition to linked deletions, would be predicted to yield nonfunctional loci due to insertions/deletions/frameshifts generated in an imprecise or aborted excision process (BRITT and WALBOT 1991; DOSEFF et al. 1991).

With all *Mutator* lines and alleles, germinal revertants are significantly rarer than somatic revertant sectors. Rates of germinal excision for different *Mu* insertions are usually around 10^{-4} (BROWN et al. 1989; LEVY et al. 1989; WALBOT 1992a), much lower than the frequencies of germinal reversion associated with the excision of the *Ac* or *En(Spm)* transposable element families. Walbot's "big spot" *Mu* line, which exhibits somatic reversion unusually early in endosperm development, also gives an approximately tenfold increase in germinal reversion frequency (WALBOT 1992a). Reversion is five to ten times more frequent through the pollen than through the oocyte in this line (WALBOT 1992a), but a similar gametophytic bias of excision has not been observed (or carefully scored for) in standard *Mutator* lines.

As with somatic events, germinal mutability of some *Mu*-insertion alleles is not exclusively associated with element excision. Element-adjacent deletions and other rearrangements have been observed at frequencies as high as 1 in 40 with some insertion alleles (TAYLOR and WALBOT 1985; BENNETZEN et al. 1988; LEVY and WALBOT 1991). Moreover, Lowe and co-workers have determined that *Mu1* and *Mu8* element insertions in the duplicated *kn1–O* locus increase the rate of loss of one of the duplications by over 100-fold, but only in an active *Mutator* line. The loss of the duplication is dependent upon the *Kn1* allele on the other homologue and does not involve chromatid exchange, suggestive of gene conversion (LOWE et al. 1992). All *Mu* elements do not appear to enhance recombinational/conversion activities, however, since two analyzed *Mu1* insertions at *bz1* behaved in a manner similar to that of point mutations in intragenic recombination studies (DOONER and RALSTON 1990).

2.3 *Mu* Element Transposition and Amplification

Contrary to the relatively low rate of detected *Mu* element excision in germinal tissues, *Mu* elements transpose to new sites very frequently during gametophytic development. In many cases, transposition frequencies average more than once per element per plant generation (ALLEMAN and FREELING 1986; BENNETZEN et al. 1987; HARDEMAN and CHANDLER 1989). Transposition of *Mu* elements is usually associated with an amplification of their copy number, suggesting that germinal transposition does not usually involve excision of the *Mu* element (BENNETZEN 1984; ALLEMAN and FREELING 1986; BENNETZEN et al. 1987; WALBOT and WARREN 1988; CHOMET et al. 1991; LISCH et al. 1995). In test crosses of lines homozygous for a single *MuDR* element, LISCH et al. (1995) have directly demonstrated that whenever this element is found at a new site, the donor site still contains the *MuDR* element. Hence, this and other evidence has led to the model

(DOSEFF et al. 1991) that *Mutator* elements usually transpose in germinal tissues via a gap-repair mechanism like that first identified with the *P* elements of *Drosophila* (ENGELS et al. 1990).

Maize lines greatly differ in their *Mu* element content (BENNETZEN 1984; BARKER et al. 1984; ALLEMAN and FREELING 1986; CHANDLER et al. 1986; TALBERT et al. 1989; BENNETZEN et al. 1993). This is particularly true for the copy number of the very active *Mu1* elements, which are present in about 0–4 copies in most standard (non-*Mutator*) maize lines and often have more than 50 copies in active *Mutator* lines (reviewed in BENNETZEN et al. 1993). The frequent transposition and resultant amplification of these elements keeps them from being diluted in recurrent crosses to lines with no or few *Mu1* elements; however, once *Mutator* activity is lost, these elements are diluted with simple Mendelian kinetics of twofold per generation (BENNETZEN et al. 1987; WALBOT and WARREN 1988).

Transposition of *Mu* elements in somatic tissues has been more difficult to monitor, since gel-blot hybridization analysis of a sectored individual often yields bands of varying intensity that are difficult to distinguish from background. Nevertheless, genetic and molecular tests have demonstrated that *MuDR* elements do transpose duplicatively in developing somatic tissues (CHOMET et al. 1991; LISCH et al. 1995; MARTIENSSEN and BARON 1994). Some cases where very high *Mu* element copy numbers are seen via slot-blot analysis are likely to be due, in part, to the summation of multiple independent somatic transpositions (HERSHBERGER et al. 1991; WALBOT 1992b).

Free circular forms of *Mu* elements are detected in active *Mutator* lines, and these circles are missing from non-*Mutator* and *Mu*-loss lines (SUNDARESAN and FREELING 1987; SUNDARESAN 1988). None of these elements have been cloned, so their exact structure is not clear. Inverted repeat-containing transposable elements rarely generate circles, and a biological role has not been established for *Mu* circles. Although these circles could be intermediates in transposition and/or excision, they could also be outcomes of transposition of the element into itself, as seen with the Tn*10* element of *E. coli* (BENJAMIN and KLECKNER 1989).

2.4 Gene Regulation by *Mutator*

McClintock originally referred to transposable elements as "controlling elements" because of their ability to alter the expression patterns of loci that they have transposed into or near (MCCLINTOCK 1956). *Mutator* can also affect a target gene's expression pattern, at a number of levels and sometimes in a manner that is variable with the activity of *Mu* itself. As such, *Mutator* fits the definition of a controlling element system.

2.4.1 Alterations in Transcript Level or Tissue Specificity

The insertion of a *Mu3* element into the promoter of the *Adh1* locus led to an element-flanking duplication of the presumptive TATA sequence that positions the transcription initiation site of this gene. This was found to alter the tissue specificity of *Adh1*, reducing expression to about 10% of normal in the root and scutellum but leaving pollen expression virtually unaffected. These effects were directly correlated with parallel alterations in levels of mRNA accumulation and associated with alterations in the site of mRNA initiation in the anaerobic root (CHEN et al. 1987b; KLOECKNER-GRUISSEM et al. 1992).

A *Mu1* insertion 2 bp upstream of the *Sh1* transcription start site was found to decrease *Sh1* mRNA to about 4% of standard levels, in a tissue-nonspecific manner. Initiation of transcription at this allele was now found to occur primarily within the 5' terminal inverted repeat (TIR) of the insertion. The major effect of this mutation was apparently due to an approximately sevenfold decrease in initiation frequency, while deficiencies in RNA processing and/or stability were probably responsible for the remaining twofold decrease in expression (ORTIZ et al. 1988; STROMMER and ORTIZ 1989).

Mutator insertions into protein-encoding domains of exons lead to null phenotypes, although often with frequent somatic and occasional germinal reversions. Hence, it does not appear that any known *Mu* element behaves as a reasonably competent intron, as can maize elements of the *Ac/Ds* and *En(Spm)* families (KIM et al. 1987; WESSLER et al. 1987). *Mu* insertions in exons have been seen to affect splicing (TAYLOR and WALBOT 1987), including a case where one TIR of *Mu1* contributed a 5' splice site (ORTIZ and STROMMER 1990) but not leading to a functional gene product in either allele. In cases of insertion within the protein-encoding portions of genes, alterations in the general activity of the *Mutator* system have not been seen to alter the underlying null phenotype of the mutation, although mutability is generally affected.

Insertions of *Mu1* in the first intron of the *Adh1* gene of maize have been found to cause partial inactivation of the gene, yielding about 40% of wild-type message levels (STROMMER et al. 1982; BENNETZEN et al. 1984). This effect was found to be due primarily to termination of more than half of the transcripts within the *Mu1* element; the precursor mRNA molecules that were transcribed through and beyond the inserted *Mu1* element were apparently processed into normal mature mRNAs (VAYDA and FREELING 1986; LUEHRSEN and WALBOT 1990; ORTIZ and STROMMER 1990). Loss of *Mutator* activity by a line containing this *Adh1* allele led to stabilization of the mutation but was seen to not alter the effect this insertion has on *Adh1* expression (WALBOT and WARREN 1990).

2.4.2 Suppression

In contrast to the above cases, where the target locus phenotypes of insertions have not been seen to be altered by the general status of the *Mutator* system, *Mu* insertions in the promoters of *A1*, *hcf106*, *Kn1–O*, and *Vp1* provide examples of gene expression that comes under regulation by *Mutator* (MARTIENSSEN et al. 1990; BARKAN and MARTIENSSEN 1991; CHOMET et al. 1991; LOWE et al. 1992; MARTIENSSEN and BARON 1994). In each case, the inactivational phenotype (usually including somatic revertant sectors) is observed only when the *Mu* system is active. When *Mutator* activity is lost by the maize plant harboring the mutation, the phenotype of the mutant allele containing the *Mu* element insertion decreases in severity. The best-studied case of this *Mutator* suppression is the *hcf106::Mu1* allele, where a *Mu1* element has inserted near the transcription initiation site of the gene, leading to no accumulation of *hcf106* mRNA. When *Mutator* activity is lost, however, a new transcription-initiation site within the downstream *Mu1* TIR is utilized, and sufficient mRNA is produced to yield phenotypic suppression of the mutation (Barkan and Martienssen 1991). This *Mutator* control of *hcf106::Mu1* gene expression is presumably due to the loss of a *Mu* end-binding factor or factors (e.g., transposase) in the *Mu*-off line, thereby allowing access by standard transcription factors to the gene's promoter.

An especially interesting example of gene regulation by *Mutator* is exemplified by several dominant alleles of the *Kn1* locus (Greene et al. 1994). Nine independent *Mu* insertions, four of *Mu1* elements and five of *Mu8* elements, are all within a 310-bp region of the third intron of *Kn1*, and all impart ectopic expression patterns to the mutant alleles. This mutant *Kn1* phenotype is now suppressed by loss of *Mutator* activity in the line, and this loss of *Mutator* activity is associated with loss of ectopic *kn1* transcript production (Greene et al. 1994).

Given that many hundreds of independent *Mutator*-derived mutations have been observed at a wide variety of loci (reviewed in Bennetzen et al. 1993), the rare detection of *Mutator*-suppressible mutations suggests that this is an exceptional genetic outcome of *Mu*-element insertion. However, the majority of *Mu* insertions that have been traced to promoters or 5' leader regions have been found to exhibit the *Mu*-suppressible phenotype. Moreover, Robertson and all subsequent researchers have seen that many *Mutator*-induced mutations are not well transmitted, and this may be due to suppression of the phenotype in a subset of the progeny. Martienssen and Baron (1994) have directly demonstrated that poor transmission/penetrance of the suppresible *hcf106::Mu1* and *Les28* loci was due to sectorial suppression in the developing ear. Hence, suppressibility of *Mu*-induced mutations is probably a very common phenomenon, and may even be used as a diagnostic tool to confirm a *Mutator* origin of some mutations.

3 *Mutator* Elements

3.1 Subfamilies and Their Origins

The initial cloning of mutant alleles induced in *Mutator* backgrounds led to the discovery of a heterogeneous family of small (1-kb to 2.2-kb) transposable elements. All of these elements shared approximately 200-bp terminal inverted repeats (TIRs), and all generally created a 9-bp duplication of target DNA flanking the site of insertion (reviewed in Walbot 1991; Chandler and Hardeman 1992; Bennetzen et al. 1993). Subsequent studies uncovered the existence of the autonomous *Mutator* element (CHOMET et al. 1991; HERSHBERGER et al. 1991; QIN et al. 1991). Originally called *MuA2, MuR1*, or *Mu9*, the autonomous element is now designated *MuDR* in honor of Dr. Donald Robertson. The 4942-bp *MuDR* element has internal sequences related to the previously characterized *Mu5* (HERSHBERGER et al. 1991). The internal sequences of many of these elements differ greatly, and allowed the resolution of at least six subfamilies: *Mu1/Mu2, Mu3, Mu4, Mu6/Mu7, Mu8*, and *MuDR/Mu5* (Fig. 1) (BENNETZEN et al. 1993). Members of a single subfamily have related internal sequences (e.g., *Mu1* appears to be a single deletion derivative of *Mu2*, and *Mu5* differs from *MuDR* by at least two apparent deletions), while different subfamilies have no detected similarity at all in their internal sequences. Screening recombinant libraries containing maize genomic DNA with *Mu* TIR probes has led to the identification of many of the *Mu* subfamilies (CHANDLER et al. 1988; TALBERT et al. 1989), but it is likely that additional subfamilies could be found by a broad and rigorous search.

The internal variability within *Mu* elements is one of the exceptional characteristics of the *Mutator* system and has raised sig, ificant interest in the processes

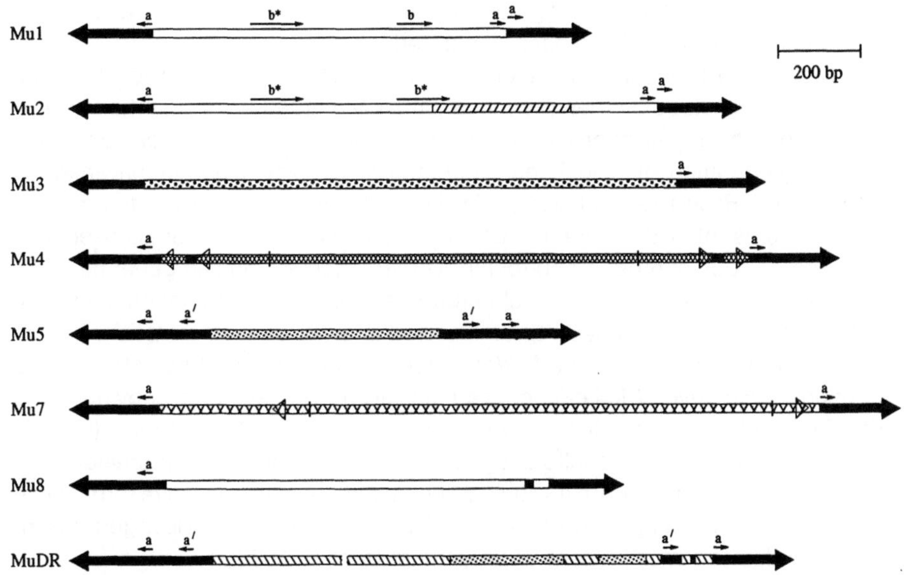

Fig. 1. Structures of sequenced *Mutator* transposable elements. *Arrows within bars* indicate terminal inverted repeat (TIR) sequences, with the *dark-filled arrows* representing homology to the 359-bp TIRs of *Mu5* (TALBERT et al. 1989). *Light arrows* ("*a*", "*a*", and "*b*") *above bars* indicate position and orientation of key repeats within *Mu*-element sequences. The *fill* of the *bars* indicates sequence homologies between genes. Repeats "a" and "a" are 24 bp in length and differ by several nucleotide changes (BENNETZEN and SPRINGER 1994). The element *Mu2*, a member of the *Mu1/Mu2* subfamily, was previously called *Mu1.7* (TAYLOR and WALBOT 1987), while the autonomous element *MuDR* was previously known as *Mu9, MuA2, MuR1*, or *Cy* (CHOMET et al. 1991; HERSHBERGER et al. 1991; JAMES et al. 1993; P. Schnable, personal communication) *Broken bar* in *MuDR* indicates sequences unique to this element that are omitted from the figure in order to save space. (From BENNETZEN and SPRINGER 1994)

by which new element subfamilies are generated. Cluster and principal component analyses of the most-terminal 172 bp (*Mu6's* left TIR and *Mu3's* right TIR) or 171 bp (all other sequenced *Mu* TIRs) of the termini shared by all elements (Fig. 2) indicate relationships between elements that could not have been predicted by their internal sequences. It seems likely that an autonomous *Mutator* element, much like *MuDR*, would have been the progenitor of the first defective *Mu* element subfamily. However, many of the intermediates in the generation of the known subfamilies may be missing, due either to their extinction from currently existing maize populations or to their not having been cloned and sequenced. The analysis in Fig. 2 suggests, for instance, that *Mu3* and *Mu4*, despite no internal homology, are closely related and that they have been derived from *MuDR (Mu9)* independently of the other *Mu* subfamilies. The *Mu8* subfamily appears to have also arisen directly from a *MuDR*-like element, probably more recently than any other element subfamily. In contrast, the *Mu1/Mu2* subfamily may have arisen from a *Mu5*-like derivative of *MuDR*. The relatively high dissimilarity between the TIRs of *MuDR* and other *Mu* TIRs may, as suggested by Benito and Walbot (1994), be due to selection for promoter function in the *MuDR* TIRs. Alternatively, it may be due to selection against such promoter activity in nonautonomous elements [particularly in *Mu5*-like elements, which might serve as inhibitors of *Mutator* activity (see below)] or to a long period of mutational drift since any of the nonautonomous members have been generated.

Members of the various *Mu*-element subfamilies vary a great deal in their activity, as measured by their presence within *Mutator*-derived insertion alleles, by their level of amplification (i.e., increase in copy number) in *Mutator* stocks, and by their generation of new bands detected in gel-blot hybridization analyses (reviewed in Bennetzen et al. 1993). At the extremes of this variability are *Mu1* (which is present in 33 of the first 61 *Mu*-insertion mutations characterized, can increase its copy number by 50-fold in *Mutator* stocks, and can generate more than one new band per parental element in a single plant generation) and *Mu4* (which has not yet been seen to be inserted in a *Mu*-derived mutation, has no apparent amplification in some *Mutator* stocks, and has rarely been observed to transpose). However, *Mu4* elements are usually present at higher copy numbers (2–15 per diploid genome) in non-*Mutator* stocks than are *Mu1* elements (0–4 per diploid genome). These results suggest that *Mu4* elements were created long ago and have accumulated in a widely dispersed but relatively inactive form, while the *Mu1* elements may have a relatively high level of activity and low germplasm dispersal due to a more recent origin (Bennetzen et al. 1993).

One unusual aspect of the different *Mu*-element subfamilies is the size (185 bp to 514 bp) and high sequence homology (>90%) of the TIRs within any single element. In all cases, the most terminal 200 bp of at least one element end is >80% identical to the most terminal 200 bp of *Mu1*. Apparently, in the cases of *Mu4* and *Mu5*, internal sequences have been duplicated at one internal end to increase the size of the TIR (Fig. 1). It has been proposed that gene conversion/repair, using either ectopic pairing between two double-stranded TIRs or cruciform pairing between single-stranded TIRs, accounts for this conservation of

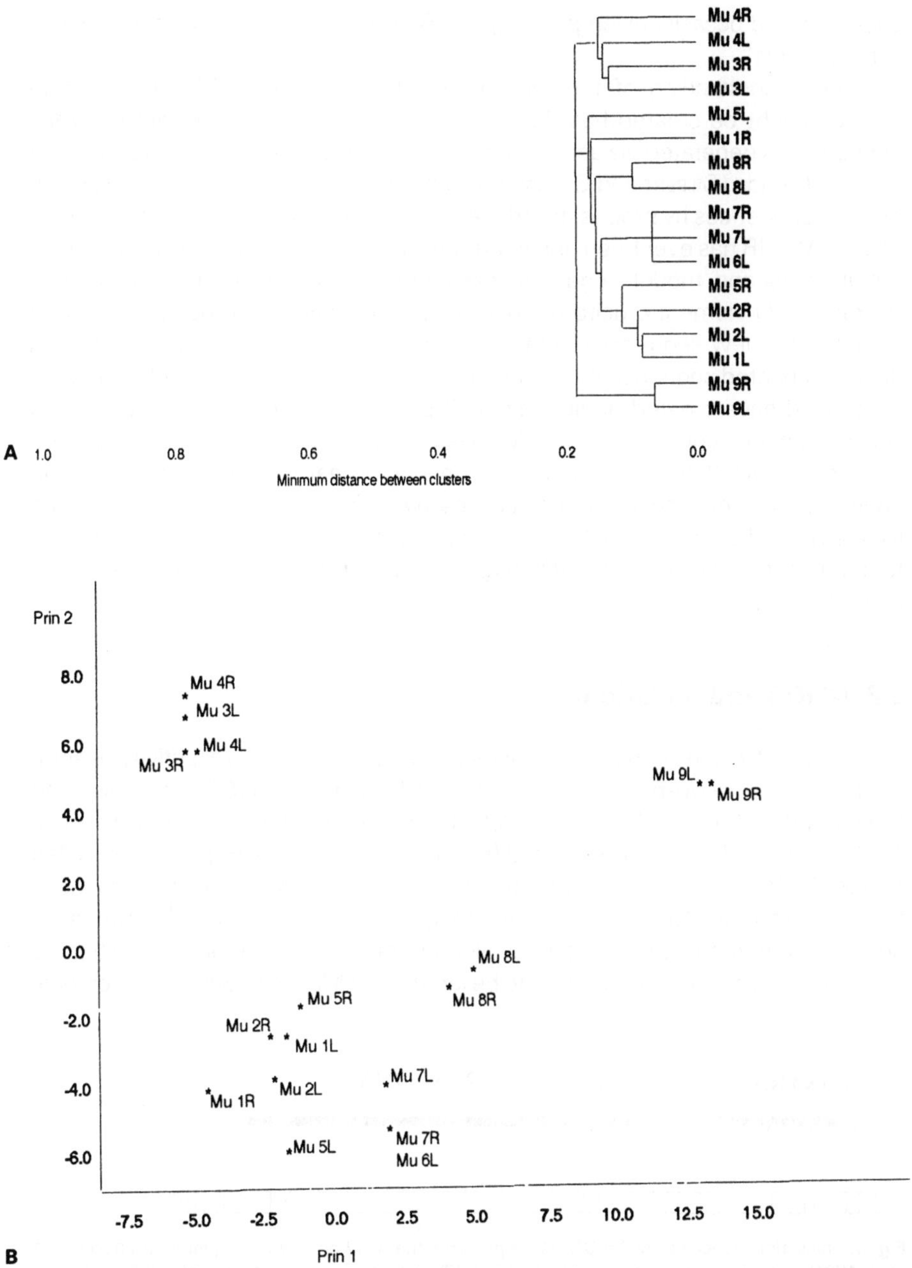

Fig. 2A,B. Evolutionary relationships of *Mu*-element TIRs, as determined by single-linkage cluster analysis **(A)** and principal component analysis **(B)** of DNA sequence similarity of the most-terminal 171 or 172 bp. Analyses were performed using the SAS system (SAS Institute, Cary, NC). "*L*" and "*R*" refer to the left and right TIRs, arbitrarily chosen, of each element. *MuDR* is referred to as *Mu9* in this figure. Sequence information is derived from previously published information (TAYLOR and WALBOT 1987; CHANDLER et al. 1988; OISHI and FREELING 1988; TALBERT et al. 1989; FLEENOR et al. 1990; HERSHBERGER et al. 1991), excepting the left TIR of *Mu6* (V. Chandler, personal communication)

sequence and occasional lengthening of TIRs (TALBERT et al. 1989; BENNETZEN and SPRINGER 1994).

Our understanding of the mechanisms by which new *Mu*-element subfamilies might be created has derived from the key observation that the *Mu2* element was generated via acquisition of normal maize genomic sequences by a pair of *Mu* ends (TALBERT and CHANDLER 1988). One model explains this acquisition of novel sequences by proposing independent transposition of *Mu* TIRs, although no solo *Mu* TIR has ever been observed (TALBERT and CHANDLER 1988; TALBERT et al. 1989). A second model suggests that ectopic strand invasion by a partially degraded *Mu* element cruciform could lead to a conversion/repair process that would insert new sequences within *Mu* ends (BENNETZEN and SPRINGER 1994). A directly repeated sequence ("a" in Fig. 1) at the internal end of most *Mu* elements TIRs could have created a slipped-strand pairing event that would generate a single-stranded sequence suitable for invasion of the genomic sequences picked up to create *Mu2*. It is interesting that this "a" repeat is routinely found at the boundary of sequence discontinuities between the *Mu*-element subfamilies (BENNETZEN and SPRINGER 1994) and also contains the transcriptional initiation sites for the transcripts unique to *MuDR* (Fig. 3) (BENITO and WALBOT 1994; HERSHBERGER et al. 1994).

3.2 *MuDR* and its Origin

Genetic and molecular analyses of lines segregating for *Mutator* activity eventually led to the discovery of an autonomous *Mu* element, *MuDR* (ROBERTSON and STINARD 1989; QIN and ELLINGBOE 1990; CHOMET et al. 1991; HERSHBERGER et al. 1991; QIN et al. 1991). The several different isolates of *MuDR*, previously called *MuA2, MuR1,* and *Mu9*, were found to have nearly identical sequences (HERSHBERGER et al. 1991; JAMES et al. 1994; M. Freeling, personal communication). In addition, the autonomous *Cy* element of the independently investigated *Cy/rcy* system has been confirmed to be identical to *MuDR* (P. Schnable, personal

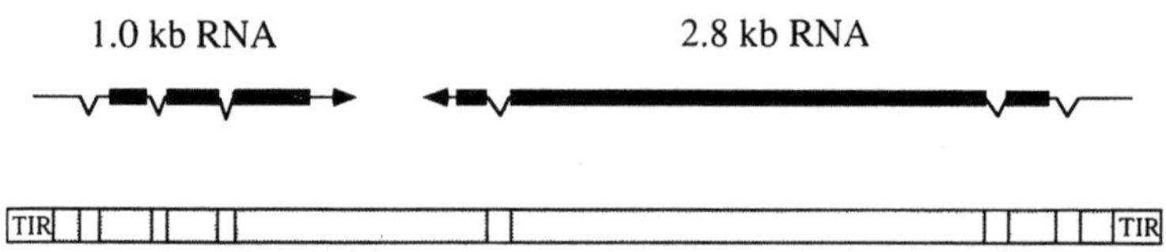

Fig. 3. Transcripts specified by *MuDR*. *Bar* represents the 4942-bp *MuDR* sequence (HERSHBERGER et al. 1991), with each terminal inverted repeat (*TIR*) and intron shown as an *unfilled box*. *Arrows above* the *bar* indicate the size, position, and orientation of two detected *MuDR* transcripts, with introns indicated as below-the-line "V". The third, most-3' intron in the smaller transcript is represented by a *lightly shaded fill* because it is very rarely removed (HERSHBERGER et al. 1994). This 120-base intron, when retained in the mature transcript, adds 40 amino acids to the putative protein encoded by this RNA and would make the transcript approximately 1.0kb. The *thick portions of the transcript arrows* indicate the long open reading frames that are predicted to encode the proteins specified by these RNAs (HERSHBERGER et al. 1994)

communication), as previously suggested by genetic and molecular studies (Schnable and Peterson 1989; Schnable et al. 1989). *MuDR* encodes two major transcripts, one of about 1.0 kb and the other of about 2.8 kb, and these are convergently transcribed from promoters in each TIR (Fig. 3) (Qin and Ellingboe 1990; Chomet et al. 1991; Qin et al. 1991; Benito and Walbot 1994; Hershberger et al. 1991, 1995; James et al. 1993). The 1.0-kb transcript is encoded by a gene containing three introns, but the last of these is rarely spliced out (Hershberger et al. 1995). The 2.8-kb RNA contains an open reading frame that would encode an 823-amino acid peptide. The 1.0-kb transcript would yield a protein of 207 amino acids. For each of these genes, antisense transcripts and transcripts with unspliced introns have been observed (Hershberger et al. 1995).

At least one *MuDR* element has been found to be very unstable, yielding deletion derivatives at a high frequency in somatic and germinal tissues (Lisch et al. 1995). Deletions that lead to loss of the 1.0-kb transcript still permit transposition and mutability conditioned by the *Mutator* system, but suppression activity appears to be lost. Deletions that remove the 2.8-kb transcript, while leaving the smaller transcription unit intact, cause loss of all *Mutator*-associated activities (D. Lisch and M. Freeling, personal communication). This frequent deletional inactivation of *MuDR* elements may partly explain their rarity in standard maize lines (Schnable and Peterson 1986; V. Walbot, personal communication). As with many other rapidly evolving repetitive DNAs (Bennetzen et al. 1994), *Mu*-element ends do not exhibit hybridizational signals upon gel-blot hybridization to DNA from species beyond the genus *Zea* and the closely related species *Tripsacum andersonii* (Talbert et al. 1990). However, both cDNA and genomic clone homologues to *MuDR* have been identified in rice (M. Freeling, personal communication; V. Walbot, personal communication). This suggests that *Mutator* has been present in plants for at least the 60–70 million years since the ancestors of maize and rice diverged.

4 *Mutator* Specificities

By most criteria, *Mutator* is among the least biased transposable elements in its mutational and transpositional properties. However, preferences are observed at a number of levels and suggest aspects of *Mu*-element regulation and transposition mechanisms.

4.1 Mutation

Robertson (1978) initially demonstrated *Mutator*-associated mutation at a large number of independent loci, and subsequent studies have detected mutations at essentially any locus for which mutations are screened (reviewed in Bennetzen et al. 1993). All of the confirmed mutations have been nuclear in origin, and most are recessive. The relative rarity of dominant mutations is true of essentially any mutagenic agent in maize or other eukaryotes. In fact, the few dominant mutations that have been associated with *Mutator* insertion (BENNETZEN et al. 1993; GREENE et al. 1994; MARTIENSSEN and BARON 1994) are numerous enough to suggest that *Mu*-element insertions are more likely to create dominant alleles than most mutagenesis processes.

The same *Mutator* stock has been shown to have up to an eightfold difference in the frequencies of mutation at different genes (ROBERTSON 1985), although this difference could be due to a smaller effective target size for some genes than for others. Very different frequencies of mutation at particular loci are often observed in independent screening programs (reviewed in BENNETZEN et al. 1993), but this is probably due primarily to differences in overall *Mutator* activity in the lines employed.

Among cloned mutations, the majority of insertion alleles have been found to contain elements of the *Mu1/Mu2* subfamily (reviewed in BENNETZEN et al. 1993). These elements are among the most active in a *Mutator* stock, showing both frequent transposition and high levels of amplification. *Mu*-induced mutations at some loci (e.g., *Bz1*; 25/26 so far) are nearly always caused by insertion of members of the *Mu1/Mu2* subfamily (TAYLOR et al. 1986; BROWN et al. 1989; HARDEMAN and CHANDLER 1989, 1993; SCHNABLE et al. 1989). Other genes (e.g., *Kn1*; 4 *Mu1*, 1 *Mu6/7*, 6 *Mu8*) show more frequent insertions by other *Mutator* elements (LOWE et al. 1992; HARDEMAN and CHANDLER 1993; GREENE et al. 1994). Part of this variation may be due to different phenotypic outcomes of insertion by different element types. In the case of *Kn1*, for instance, dominant mutations were mostly sought after; it may be that only some *Mu* elements can impart the ectopic expression patterns needed to detect this phenotype (GREENE et al. 1994).

Within genes, the majority of *Mutator* insertions have been associated with exons. This is probably because such insertions usually yield obvious, and generally nonsuppressible, phenotypes. Insertions in introns and promoters will be underrepresented in the *Mutator* mutant collection, due to an expected absence of detected insertional phenotype or rapid *Mutator* suppression after mutant induction (GREENE et al. 1994; MARTIENSSEN and BARON 1994). Given the small target size for promoter mutations and their suppressibility, it seems likely that *Mu* elements actually show preferential insertion in promoters, as observed for many prokaryotic transposons.

4.2 General Transpositional Preferences

Initial analyses of transposed *Mu1* bands by gel-blot hybridization indicated a preferential insertion of these elements into unmethylated DNA (BENNETZEN 1985b; BENNETZEN et al. 1988, 1994), and this phenomenon has also been observed for *Ac* and *En(Spm)* elements (CHEN et al. 1987a; CONE et al. 1988). Because unmethylated regions are distinctive to genes in maize (ANTEQUERA and

BIRD 1988; WALBOT and WARREN 1990; BENNETZEN et al. 1994), we were not surprised to find preferential insertion of *Mu1* and *Mu2* elements into single-copy, gene-like DNA (CRESSE 1992; CRESSE et al. 1995). Fractionation of total plant DNA over density gradients has also indicated that *Mu1* and other maize-transposable elements are preferentially associated with the low GC-content DNA often associated with genes (CAPEL et al. 1993).

Analyses of 19 randomly chosen *Mu1/Mu2* insertion sites detected a reasonable consensus sequence in the flanking direct repeats. This 9-base target consensus of 5'-G-T-T-G-G/C-A-G-G/A-G-3' is matched at three to eight sites in all 32 *Mu1*-related insertion sites that have been sequenced to date, with an average match of 5.4 (CRESSE et al. 1994). Random matches to this sequence should average 3.0. Insertions by other *Mu* element subfamilies average a match of only 4.0 to this consensus. This suggests that different subfamilies might have different insertion-site-sequence biases, and that this might account for all or some of the differences observed in the likelihood of each *Mu* subfamily inserting into a given gene (CRESSE et al. 1995).

Multiple independent *Mu*-element insertions within *Bz1* were seen to be scattered throughout the locus, but the majority of insertions were found to be near the 3' end of the lone *Bz1* intron (TAYLOR and WALBOT 1987; BROWN et al. 1989; HARDEMAN and CHANDLER 1989, 1993; SCHNABLE et al. 1989; BRITT and WALBOT 1991). Three insertions in *Bz1* (a *Mu1*, a partially deleted *Mu1*, and a *Mu7*) were all found at the same base pair (SCHNABLE et al. 1989; BRITT and WALBOT 1991; CHANDLER and HARDEMAN 1992). Hence, within-gene preferences for insertion are real, if not absolute. The most drastic case of within-gene insertion bias in the *Mutator* system is demonstrated at the *Kn1* locus, where nine independent insertions cluster within a 310-bp region of the third intron. This case is almost certainly an outcome of selection for an exceptional mutant phenotype, however, since these insertions all lead to the suppressible acquisition of transcriptional competence at additional times in development (GREENE et al. 1994).

Analysis of *Mu1/Mu2* segregation in crosses to lines lacking most such elements indicated that these elements are often found in clusters (INGELS et al. 1992). A similar clustering of *Mu* elements was suggested by the work of COOK (1988), showing that mutations yielding altered seedling fluorescence were commonly found on a single chromosome arm when *Mutator* was the mutational agent, but were more evenly distributed throughout the genome when chemical mutagens were employed. These results could both be outcomes of preferential transposition of *Mu* elements to linked sites, as has been well characterized for the *Ac/Ds* system (VAN SCHAIK and BRINK 1959; GREENBLATT 1984; DOONER and BELACHEW 1989). Genetic analysis of transposed *MuDR* elements also suggested a possible preference for transposition to linked sites (ROBERTSON and STINARD 1992). However, direct molecular mapping studies in a line with only one *Mu1* element have shown that six of six *Mu1* transpositions were to unlinked sites (LISCH et al. 1995). It appears that larger data sets will be needed before conclusions can be drawn regarding whether *Mu* elements preferentially transpose to linked sites.

5 Regulation of *Mutator*

The consistently late, relatively invariable timing of *Mutator*-derived somatic mutability has been a powerful reminder of the tight developmental regulation of this system. Regulation of *Mu*-element function, both positive and negative, has now also been associated with the autonomous *MuDR* element. As with other transposable element systems, our growing comprehension of the interplay between element-determined factors and necessary "host" functions is providing novel insights into both *Mutator* regulation and more general aspects of gene regulation.

5.1 Developmental Control of *Mu* Activities in Somatic Tissues

With some *Mu*-element insertion mutations, like *vp1-mum1*, the majority of apparent revertant sectors in the aleurone layer are observed to include only a single cell (McCarty et al. 1989). However, a more typical mutability pattern for a *Mu*-induced mutation is exhibited by *bz2::Mu1*, which yields purple (phenotypic revertant) sectors in the aleurone that range in size from 1 to 256 cells, with a median at 11–20 cells (Levy et al. 1989). Similarly small sectors are also observed with *Mu*-derived mutations that exhibit phenotypic reversion in the leaf. This late timing of mutability is not usually affected by genetic background, the dosage of *MuDR* or other *Mu* elements, or the relative activity of the *Mutator* system (Bennetzen 1985a, 1994; Robertson et al. 1985; Walbot 1986; Cresse 1992).

Walbot (1992a) recently identified the first heritable case of a change in timing for *Mutator*-derived mutability. The *bz1::Mu1* mutation, caused by the insertion of a *Mu1* element near the 3' boundary of the *Bz1* intron (Britt and Walbot 1991), usually yields phenotypic revertant (purple) sectors primarily containing 8–64 cells (Levy and Walbot 1990). Some outcross ears from a particular *bz1::Mu1* stock were observed to generate multiple sectors in the aleurone with 128 or more purple cells. Systematic selection and advancement of seed with larger sectors produced subsequent progeny from this lineage with many half kernel or larger purple sectors. This "big spot" line also yielded an at least 40-fold enhanced frequency of germinal reversion at *bz1::Mu1* to about 5×10^{-3} (Walbot 1992a). The "big spot" *Mu* line conditioned earlier somatic reversion, and more frequent germinal reversion, in other *Mu* reporter alleles (*bz2::Mu1*, *c2::Mu1*, and *bz2::MuDR*), indicating that the alteration in this line was in the time of action of a *trans*-acting factor needed for *Mu*-element excision (Walbot 1992a). It seems likely that this change has been in some previously unidentified host function, because various changes (especially deletions and position effects) in *MuDR* element structure and activity have not been found to alter the developmental timing of *Mu*-driven mutability (Lisch et al. 1995). Given that this new phenotype arose in a *Mutator* stock, it is possible that a mutation in this *trans*-acting factor was induced/tagged by a *Mu*-element insertion.

Other activities of *Mutator* elements also appear to be under developmental regulation in somatic tissues, but with very different timing patterns than that observed for excision. The sectorial acquisition of the suppression phenotype, correlated with deletions inside *MuDR* elements (LISCH et al. 1995), occurs early in plant development to yield large sectors in individual leaves, sectors that cover parts of more than one leaf, or multiseed sectors (MARTIENSSEN et al. 1990; MARTIENSSEN and BARON 1994). Preliminary analyses of the somatic transposition of *MuDR* have indicated that this element can move at developmental times similar to those seen for acquisition of suppression (LISCH et al. 1995; MARTIENSSEN and BARON 1994). The sizes and placements of these suppression and transposition sectors suggest that these activities occur during embryogenesis in cells that will contribute to the shoot apical meristem (MARTIENSSEN et al. 1990; MARTIENSSEN and BARON 1994).

5.2 Germinal *Mutator* Regulation

In most *Mutator* lines, with the notable exception of the "big spot" *Mu* stocks, germinal revertants are both rare (frequencies averaging around 10^{-4}) and confined mostly to single-seed events in the female lineage (BROWN et al. 1989; LEVY et al. 1989; WALBOT 1992a). The detected excision rate is two to four orders of magnitude less than that for *Mu*-element transposition; this disparity was one of the early bases for arguing that transposition must normally be via a replicative mechanism (BENNETZEN 1985a, b; ALLEMAN and FREELING 1986; BENNETZEN et al. 1987; HARDEMAN and CHANDLER 1989; LISCH et al. 1995). Subsequent studies with *MuDR* have proven that the autonomous element usually transposes without excision (LISCH et al. 1995). The occasional addition of nearby sequences to excision sites (DOSEFF et al. 1991) suggests that transposition is by a gap-repair mechanism like that seen for *P* elements in *Drosophila* (ENGELS et al. 1990) and Tc1 of *Caenorhabditis* (PLASTERK and GROENEN 1992).

In careful fate-mapping studies, ROBERTSON (1980, 1981) determined that about 80% of new mutations arising in the female gametophyte were represented by only a single seed, although sectors as large as 11 seeds were observed. By analysis of discordant kernels (where the embryo and endosperm possess different genotypes due to postmeiotic events), ROBERTSON and STINARD (1993) have shown that mutations can also be induced after meiosis in both the male and female lineages. Taken in their entirety, these studies show a narrow range of germinally significant *Mutator* mutagenesis, from the last few mitoses before meiosis, during meiosis (the most frequent time of mutation), to the first mitotic division of pregametic nuclei after meiosis (ROBERTSON and STINARD 1993). Hence, *Mutator*-driven mutation and *Mu*-element excision (BROWN et al. 1989; LEVY et al. 1989; WALBOT 1992a) show very similar, and late, developmental timing in these gametophytic tissues.

This concurrence of late excision and mutagenesis timing in germinal tissues contrasts with the earlier somatic timing for detected transposition of *MuDR*

(Lisch et al. 1995; Martienssen and Baron 1994). In theory, this disparity could be due to the fact that most *Mutator*-induced mutations are associated with the insertion of *Mu* elements other than *MuDR* (reviewed in Bennetzen et al. 1993), implying that *MuDR* might have a different timing for its activity. However, standard late timing for somatic excision is observed at *bz2::Mu9* (Walbot 1992a). Probably, frequent transposition of *MuDR* and other *Mu* elements also occurs very late in development, but these individual events are not easily visualized in genetic or molecular assays of somatic tissues. The very high copy numbers and degree of amplification of *MuDR* elements determined by slot-blot analyses may, in fact, be due to a summing of many separate amplifications that have occurred in small clones of somatic cells (Hershberger et al. 1991; Walbot 1992b).

Mutator-driven mutagenesis, *Mu*-associated mutability, *Mutator* loss, and reactivation of *Mutator* loss have all been seen to be more frequent in the progeny of test crosses where the *Mutator* parent was the female (Walbot 1986; Martienssen et al. 1990; Brown and Sundaresan 1992; Bennetzen 1994; Lisch et al. 1995; Martienssen and Baron 1994). This greater female contribution is observed only in a subset of the crosses performed and is not associated with any change in the developmental timing of the *Mutator* activities monitored in the progeny. Often, resultant mutability of a *Mu*-insertion allele is assayed in the F1 kernel, where the 2N female contribution to the triploid endosperm could account for some or all of the detected difference in gamete contribution.

Recent studies (Lisch et al. 1995; Martienssen and Baron 1994) indicate that these differential gametic contributions to *Mutator* traits may involve at least three different preferences. Early duplication/transposition of *MuDR* elements appears to occur more frequently in the tissues giving rise to the ear, while later (gametophytic) transposition/duplication of *MuDR* is more frequent in the tassel lineage (Lisch et al. 1995). Inactivation of *MuDR* elements and suppression appear to arise with the same developmental timing and male bias observed for early transposition events (Lisch et al. 1995; Martienssen and Baron 1994). Since only a subset of progeny will undergo a *MuDR* transposition, inactivation, and/or excision event in any generation, the relative rates of each of these phenomena will lead to different frequencies of outcomes in a given cross but will not yield absolute biases.

Taken in their entirety, studies of *Mutator* activity in the developing plant indicate that this transposable-element system is tightly regulated by "host" factors that are active at different times in development. Some transposition and most *MuDR* deletion/suppression occur primarily during embryogenesis in the cells that will make up the apical shoot meristem. Most excisions and most transpositions occur during a specific narrow range of cell division late in the development of both somatic and gametophytic tissues, giving rise to the majority of heritable mutations. Because the *Mutator* element array is fairly constant in the different tissues of a single *Mutator* plant, tissue-specific and stage-specific cellular factors must regulate these tissue-specific and stage-specific *Mutator* activities. Since the two different times of *Mutator* action seem to have different preferred outcomes (*MuDR* deletion/suppression/transposition early versus rare excision/frequent transposition late in development), it is likely that these cellular factors also strongly influence the likelihood of a given outcome of any interaction between the *Mu* transposase and a *Mu* element.

We have very little information regarding the molecular nature of *Mutator* regulation during development. ZHAO and SUNDARESAN (1991) have identified protein factors that bind *Mu* TIRs, and their presence in non-*Mutator* stocks suggests that they are "host" factors. The presence of different levels and variable processing of *MuDR* transcripts in various tissues of a *Mutator* plant (HERSHBERGER et al. 1995) suggests transcriptional and post-transcriptional control of *Mutator* function by tissue-specific factors. Further study of these possible modes of regulation is warranted, particularly using mutants that are altered in developmental timing of *Mutator* action (WALBOT 1992a).

5.3 Positive Regulatory Factors: *MuDR*

The discovery of *Mutator* lines that segregated 1:1 for *Mu* activity in outcrosses to non-*Mutator* lines allowed the eventual discovery of *MuDR* (ROBERTSON and STINARD 1989; QIN and ELLINGBOE 1990; CHOMET et al. 1991; HERSHBERGER et al. 1991; QIN et al. 1991). Transactivation of *Mu*-derived mutability co-segregates with *MuDR*, and the element is itself able to transpose at high rates to generate lines that have transactivational capacity encoded at a new chromosomal position (CHOMET et al. 1991; ROBERTSON and STINARD 1992; LISCH et al. 1995). Hence, *MuDR* fulfills all of the requirements to be considered an autonomous transposable element.

MuDR encodes two primary transcripts, one of 2.8 kb and one of 1.0 kb, that are transcribed convergently from initiation sites in the *MuDR* TIRs (Fig. 3) (QIN and ELLINGBOE 1990; CHOMET et al. 1991; HERSHBERGER et al. 1991, 1995; BENITO and WALBOT 1994). The transcripts are polyadenylated and terminate at nonoverlapping sites that contain five different sets of direct repeats which are 11 to 27 bp in size and are repeated three to five times. Each gene has three introns. The first intron of the 2.8-kb RNA is in the nontranslated leader and has two possible splice sites, yielding cDNAs with either 92 or 96 bp removed but no predicted difference in protein coding capacity. The third intron of the smaller transcript is rarely processed. The first introns of the transcripts, which are within the noncoding leaders, are almost always processed out, but the other introns of each gene are left unspliced approximately 20% of the time. Some tissue-specific differences are observed in the relative efficiencies of splicing. RNA from each of these genes is observed in all tissues examined by gel-blot hybridization analysis, although at high levels primarily in meristematic tissues (HERSHBERGER et al. 1995).

Both of the proteins that have been seen to tightly bind *Mu1* TIRs in vitro associate specifically with sequences that serve as promoters in the *MuDR* TIRs. One of these protein factors is found in *Mutator* lines, non-*Mutator* lines, and *Mutator*-loss lines at similar levels, but the second is present only in *Mutator* stocks (at higher levels) and non-*Mutator* stocks (at low levels) (ZHAO and SUNDARESAN 1991). Hence, these both appear to be "host" factors that may be involved in the regulation of *MuDR* transcription.

An antibody has been prepared that is specific to the 23-kD protein predicted to be encoded by the 1.0-kb transcript. This antibody detects a protein of about 30 kD that is found only in maize lines with an active *MuDR* element. The protein is found in cellular extracts of root, seedling leaf, mature leaf, tassel, and ear tissues; highest levels are found in the tassel, ear, and seedling leaf (M. Donlin and M. Freeling, personal communication). This protein level study agrees with the work of HERSHBERGER et al. (1995), which indicated that both transcripts are present in most or all tissues but are most abundant in meristematic tissues.

Deletions of *MuDR* often occur during plant development (LISCH et al. 1995; P. Schnable, personal communication). Deletions that severely alter the large transcription unit lead to loss of *Mutator* function, while deletions that remove only the smaller transcript and the protein it encodes have little detected effect on the transposition and transactivational competence of *MuDR* (D. Lisch, M. Donlin, and M. Freeling, personal communication). Hence, the larger transcript appears to encode the *Mu*-specific transposase. Deleted *MuDR* elements that cannot produce the smaller transcript show constitutive suppression at *Mu*-suppressible loci and permit only a low frequency of mutability at reporter alleles (D. Lisch and M. Freeling, personal communication). This suggests that the smaller transcript is an enhancer of transposase action. Further experiments should investigate whether the protein encoded by the smaller transcript affects the level of transcription, translation, or *Mu* binding of the *MuDR* transposase.

The position of a *MuDR* element in the genome also affects its transpositional and transactivational contribution to the *Mutator* trait. LISCH et al. (1995) have seen that a weak *MuDR* element (one that rarely transposes and transactivates relatively low frequencies of phenotypic reversion at reporter loci) can become a stronger *MuDR* element when it transposes to a new location. Thus, the *MuDR* element can apparently "cycle" (McCLINTOCK 1958) between levels of activity through changes in its genomic position that do not involve alterations of its internal sequences. The weak *MuDR* elements exhibit a positive dosage effect for the frequency (but not timing) of reporter allele mutability that they condition (LISCH et al. 1995). This observation supports a model indicating that transposition of *Mu* elements requires a certain threshold level of *MuDR*-derived factors (e.g., transposase), and that the tissue specificities observed for this system are due to that threshold being reached only in tissues that permit high levels of *MuDR* expression.

5.4 Negative Regulation of *Mutator* Functions

Initial studies by ROBERTSON (1978, 1983) indicated that *Mutator* function (as measured by mutagenic activity) could be lost both in outcrosses to non-*Mutator* stocks and in intercrosses between *Mutator* lines. Loss of *Mutator* activity in outcrosses occurred in a subset (averaging about 10%) of progeny in each serial outcross, and this *Mutator*-loss state was fully reversible by crossing to an active *Mutator* line (ROBERTSON 1986; BENNETZEN 1994). Although many of these outcross-loss lines were probably due to segregational loss of a *MuDR* element, molecular studies showed that some still conditioned a low level of *Mu* element transposition (BENNETZEN et al. 1987). Because the mutagenicity assay required the presence of both a *MuDR* element and many nonautonomous *Mu* elements to induce a high frequency of mutation, genetic identification of the autonomous element was possible only when *Mutator*-derived mutability was used as the assay (ROBERTSON and STINARD 1989).

5.4.1 *Mu*-Element Modification

In intercrosses between vigorous *Mutator* stocks, inbreeding for the *Mutator* trait eventually led to complete inactivation of *Mu* activity in all progeny (ROBERTSON 1983). This intercross-loss was found to be dominant for *Mutator* inactivation (ROBERTSON 1983, 1986) and was associated with 5-methylation of cytosines at 5'-CG-3" and 5'-CNG-3' sequences inside reporter *Mu1* elements (BENNETZEN 1985a, 1987; BENNETZEN et al. 1988; BROWN et al. 1994). In addition, both intercross and outcross losses of *Mutator*-derived mutability were associated with modification of *Mu* elements at various *Mu1* internal and TIR sites (WALBOT et al. 1985; CHANDLER and WALBOT 1986; BROWN and SUNDARESAN 1992). Similar observations of modification-associated inactivation were also made in the *Ac/Ds* and *En(Spm)* systems (DELLAPORTA and CHOMET 1985; SCHWARTZ and DENNIS 1986; CHOMET et al. 1987; BANKS et al. 1988).

In many eukaryotes, including plants, the lack of methylation at cytosines in symmetric 5'-CG-3' sequences is associated with gene activity (ANTEQUERA and BIRD 1988; WALBOT and WARREN 1990; BENNETZEN et al. 1994). The majority of the DNA in some higher eukaryotes, including maize, is repetitive, much of it transposons and retroelements. Modification of these sequences, and their expected condensation into heterochromatin, appears to be a generic mechanism by which most of these sequences are kept in an inactive state. Activation of such "quiescent" elements can occur under conditions of genomic shock (reviewed in MCCLINTOCK 1984), such as chromosome breakage, that probably lead to element demethylation via a hyperactive repair process that runs ahead of maintenance DNA methylases.

It is not clear whether element modification is a cause or an outcome of *Mu* inactivation. Once modified, a *Mutator* element could be maintained in a modified state by standard eukaryotic DNA methylases that will rapidly methylate cytosines in hemimethylated 5'-CG-3' and 5'-CNG-3' sites. However, observations that some of the modifications of *Ac* and *MuDR* elements are at sites other than those containing 5'-CG-3' or 5'-CNG-3' sequences (SCHWARTZ 1989; MARTIENSSEN and BARON 1994) suggest that a currently uncharacterized mechanism may be at least partly responsible for the establishment and maintenance of transposable element modification.

It is likely that modification is involved in keeping *Mu* elements in an inactive state; ultraviolet or gamma ray reactivation of *Mu* activity in previously modified *Mutator*-loss lines is correlated with *Mu* element demethylation (WALBOT 1988,

1992b). These reactivation studies also indicated that some *Mutator*-loss lines still contain a potentially functional *MuDR* that has been epigenetically inactivated.

Unlike the *Ac* and *En(Spm)* systems, where specific element modification events are associated with element inactivation (Schwartz and Dennis 1986; Banks et al. 1988; Schwartz 1989), clear relationships between specific *Mu*-element modifications and either *cis* or *trans* components of *Mutator* activity have not been identified. Part of this problem is due to the fact that most modification studies have focused on a *Mu1* reporter element, and not on *MuDR* (Martienssen and Baron 1994). It is clear that modifications are largely limited to the elements themselves, with modification rarely detected in flanking target DNA (Bennetzen et al. 1988; Martienssen et al. 1990; Walbot and Warren 1990; Walbot 1991). Variations in *Mu1* element modification have been seen in lines lacking *Mutator* activity, including elements with detected modifications only in their TIRs, or in both internal and TIR sites, or only in internal regions (reviewed in Bennetzen et al. 1993). Particular levels and sites of modification are often attained by the entire population of *Mu1*-related elements in a *Mutator* plant within a single plant generation (Bennetzen et al. 1988; Brown et al. 1994), associated with the simultaneous loss of mutability at multiple reporter loci (Bennetzen 1994; Martienssen and Baron 1994).

5.4.2 *MuDR* Effects on *Mu* Element Modification

All molecular studies show at least a transient demethylation of *Mu* elements in crosses between modified *Mutator*-loss lines and active *Mutator* lines (Bennetzen 1987; Bennetzen et al. 1988; Brown and Sundaresan 1992; Brown et al. 1994). This would not necessarily require an active demethylation process; if, for instance, the binding of *Mu* transposase to a modified *Mu* element blocked methylation of the hemimethylated DNA replication product, then the *Mu* element would be unmodified by the second cell generation.

Segregational loss of a *MuDR* element has been associated with the modification of *Hin*f 1 sites in the TIRs of reporter *Mu1* elements (Chomet et al. 1991; Lisch et al. 1995), and these same sites were modified in a *Mu1* element found in a non-*Mutator* line (Chandler et al. 1988). Hence, *MuDR* activity is required to prevent modification as well as remove it, perhaps by the same mechanism. As originally proposed by Chandler et al. (1988), these results imply that modification of *Mu* elements is a default host function that somehow recognizes and acts on *Mu* elements at their insertion sites.

Studies of the absolute loss of *Mutator* activity that eventually arose in all intercrossed *Mu* lines suggested that inactivation was caused by a high level of *Mu* activity (Robertson 1983). Observations that this intercross loss was associated both with initially high element copy numbers and with internal modification of reporter *Mu1* elements suggested that high levels of *Mu* activity elicited the modification process (Bennetzen 1985a, 1987; Bennetzen et al. 1987). Because most of these intercross-loss progeny must contain *MuDR* elements, it seemed likely that the default modification associated with segregational loss of *MuDR* was unrelated to the modification associated with *Mutator* inbreeding (BENNETZEN et al.1993). However, recent studies of the *MuDR* element suggest that a common mechanism may link these apparently different phenomena (LISCH et al. 1995; MARTIENSSEN and BARON 1994).

5.4.3 The Case for Inhibitory *Mu* Elements

In studies of "minimal" *Mutator* lines, mostly containing only one *MuDR* element and one *Mu1* reporter element, LISCH et al. (1995) observed that all somatic sectors showing loss of *Mutator* activity contained both modified *Mu1* elements and a de novo deletion of the *MuDR* element present. *Mu*-element modification is also seen in sectors that lose *Mutator* activity in lines with multiple *MuDR* elements (MARTIENSSEN and BARON 1994). There are one or more deleted *MuDR* elements in each of these somatic inactivations in multiple-*MuDR* lines, suggesting that these defective *MuDR* derivatives could serve as dominant inhibitors of intact *MuDR* elements in the same nucleus (MARTIENSSEN and BARON 1994). Such dominant inhibitory elements, all derived from deletion of an autonomous element, have also been observed in the *En(Spm)* system of maize (CUYPERS et al. 1988) and the *P*-element system of *Drosophila* (JACKSON et al. 1988; ROBERTSON and ENGELS 1989; MISRA and RIO 1990).

A defective *MuDR* element (*dMuDR)* could act as an inhibitor of standard *MuDR* elements in a number of mutually nonexclusive ways. Internal deletions that remove the transcription stop/RNA processing site in the center of the *MuDR* element would lead to the production of antisense transcripts that should decrease transposase levels. Low-abundance antisense transcripts that hybridize to *MuDR* have been seen (HERSHBERGER et al. 1995). Deletions that lead to the production of truncated proteins could also produce an inhibitory effect, particularly if the transposase acts as a multimer. Regardless of the mode of action, the production of inhibitory *MuDR* derivatives via internal deletion would explain the basic biology of both outcross-loss and intercross-loss of *Mutator* activity.

One problem with a model employing deletions of *MuDR* as the source of *Mutator* inhibition is that genetic programs with high rates of loss of *Mutator* activity often do not see many (or sometimes any) *dMuDR* elements. Only one cDNA has been found that looks like a transcription product of a *dMuDR*, representing a 2.2-kb RNA coming from the 2.8-kb transcript (HERSHBERGER et al. 1995). Moreover, the chief antisense transcript seen to be homologous to *MuDR* was found to be about 4.9 kb in size, suggesting that it is the product of an intact *MuDR* element that has been transcribed through the central transcription termination/RNA processing site (V. Walbot, personal communication). It may be that the very frequent generation of deletion derivatives of *MuDR* seen by LISCH et al. (1995) is an exceptional outcome of this element's chromosomal position (and low activity) or low copy number. Hence, the alternate processing of *MuDR* transcripts and/or its synthesis of antisense RNAs (perhaps influenced by genomic position of the element) are also possible origins of negative/inhibitory *Mutator* regulation by a *MuDR* element (HERSHBERGER et al. 1995).

5.4.4 Co-suppression as a Process for Element Modification

If the absence of effective transposase leads to specific *Mu*-element modification, then we still do not know what factors are directly responsible for the specificity in *Mu*-element modification (Chandler et al. 1988). One model for the coordinate modification and inactivation of multiple *Mu* elements in a plant can be derived by analogy to the phenomena termed paramutation and co-suppression (Brink et al. 1968; Matzke et al. 1989; Napoli et al. 1990; Linn et al. 1990; van der Krol et al. 1990). In each process, homologous sequences at different genomic locations interact to epigenetically alter (usually, decrease) gene expression by a process associated with DNA methylation in the affected gene(s). Co-suppression in transgenic plants is particularly comparable to the concomitant inactivation of *Mu* transposable elements; in each case, an expressed DNA is found at an "abnormal" site (either by random integration of a transgene or by transposition), and the modification of this gene then leads to the modification of genes with homologous sequences. One possible mechanism for this coordinate inactivation would involve ectopic pairing of modified and unmodified sequences, leading to rapid methylation of the hemimethylated product generated by reciprocal strand exchange (Bennetzen et al. 1993). Recent studies of co-suppression have shown that some inactivations are not associated with altered transcription or detected DNA methylation. In these cases, it appears that high-level expression of a transgene leads to rapid turnover of RNA from all genes with a similar sequence (reviewed in Flavell 1994). The mechanism of this RNA degradation is not known, but the recognition of a high-level RNA expressed in the wrong "region" (for instance) of the folded chromatin within an interphase nucleus could bring about its specific turnover. This specific and previously undiscovered RNA-level mechanism for inactivation of a family of related sequences has all the hallmarks of a process used by the "host" plant to provide resistance to the detrimental activities of viruses and transposons.

Mutator-element modification by a default process could be fully explained by an ectopic pairing model. The integration of a *Mu* element into a site that acquires methylation (e.g., via "position effect") is then perpetuated through the entire population of related *Mu* elements by ectopic pairing. This would explain the relatively sharp boundaries of the methylation events as being due to the homology requirement, and would also propose that the pairing either is TIR mediated or would require a modified member of each subfamily. This coordinate process would be prevented in a line containing an active *MuDR* element, perhaps via active demethylation of *Mu* elements by the transposase or (more likely) via prevention of ectopic pairing and/or maintenance methylation of *Mu* elements by the presence of a bound transposase. It is also possible that RNA-level repression, seen in some co-suppression events, could also contribute to the recognition and modification of *Mu* elements by some unknown mechanism.

5.5 The Inheritance of *Mutator* Activity

For many investigators, the most confusing aspect of the *Mutator* system has been its complex patterns of inheritance. Although ROBERTSON (1978) reported fairly consistent 90% retention of *Mutator* activity in outcrosses, this was probably due to a combination of recurrent selection of most-active *Mutator* lines and the averaging of results from many separate crosses (reviewed in BENNETZEN et al. 1993). Other laboratories, following *Mutator*-derived somatic mutability at a reporter allele, have seen essentially every possible frequency of transmission from 0% to 100% in an outcross (BENNETZEN 1985a, 1994; WALBOT 1986; ROBERTSON and STINARD 1989; CHOMET et al. 1991; BROWN and SUNDARESAN 1992; CRESSE 1992). Certain patterns were observed in these segregational results, indicating that a particular *Mutator* stock was "conditioned" (BENNETZEN et al. 1993; BENNETZEN 1994) for a particular inheritance pattern in subsequent generations. The genetic background into which *Mutator* activity was crossed had no detectable effect on subsequent *Mutator* inheritance patterns (CRESSE 1992), indicating that the nature of the founding *Mu*-element population was the most significant factor in determining *Mutator* transmission (BENNETZEN 1985a; BENNETZEN et al. 1993). Recent data on the activities of autonomous elements, their transcription products, and their relationships to *Mu*-element modification (CHOMET et al. 1991; HERSHBERGER et al. 1995; LISCH et al. 1995; MARTIENSSEN and BARON 1994) now allow explanation of these complex inheritance patterns as the outcomes of competing biological processes acting on the population of *Mu* elements in a particular *Mutator* lineage.

Probably the most important factor influencing the retention of *Mutator* activity in a crossing program will be the number and activity of functioning *MuDR* elements and inhibitory *Mu* elements in the plant. *Mutator* lineages with few *MuDR* elements are likely to lose activity at Mendelian frequencies in crosses to non-*Mutator* stocks (ROBERTSON and STINARD 1989; CHOMET et al. 1991; LISCH et al. 1995). In crosses to *Mutator*-loss lines with multiple inhibitory *Mu* elements, these minimal *MuDR* lines are likely to be rapidly overcome, perhaps after an initial phase in which they bring about demodification of *Mu* elements (BENNETZEN 1987; BENNETZEN et al. 1988; BROWN et al. 1994). Hence, one would see dominance of the *Mutator*-loss phenotype in such a line. Due to their high concentrations of *MuDR* elements, intercross-loss lines would be expected to have generated multiple *MuDR* derivatives with inhibitory functions, perhaps by deletion and/or by position effects upon transcription and processing. This expected high level of inhibitory *Mu* elements would explain the consistent dominance of intercross *Mu*-loss lines (ROBERTSON 1983, 1986; BENNETZEN 1987, 1994; BENNETZEN et al. 1988; BROWN et al. 1994).

In lines with many active *MuDR* elements and/or few inhibitory *Mu* elements, activity would be maintained in most outcrosses and would rarely be suppressed during somatic development. Transposition of *MuDR*, associated with amplification of the element, would partially overcome dilution of the autonomous element in serial outcrosses. If the ratio of inhibitory *Mu* elements to active *MuDR* elements edged over a threshold leading to suppression and poor transmission of

Mutator activity, then crossing to an active *Mutator* line would be likely to lead to complete reactivation of *Mutator* activity. Hence, lines that have barely lost *Mu* activity (perhaps with a complete level of *Mutator*-element methylation but containing few inhibitory *Mu* elements), would be fully recessive for inactivation in crosses to active *Mu* lines (MARTIENSSEN et al. 1990; BROWN and SUNDARESAN 1992; BENNETZEN 1994).

The position of a *MuDR* element can also quantitatively affect its expression (LISCH et al. 1994). It seems likely that a "weak" *MuDR*, due either to genome position or internal mutation, would not be as effective in overcoming the action of an inhibitory *Mu* element as would a fully active *MuDR* element. Of course, different inhibitory *Mu* elements are also likely to be of varying strength, due either to genome position effects or to their different internal rearrangement structures.

Hence, the *Mu*-element founder effect, which I have referred to as the "conditioning" of a *Mutator* lineage (BENNETZEN et al. 1993; BENNETZEN 1994), would amount to a combination of the ratio of intact *MuDR* elements to inhibitory *Mu* elements and the relative activity of each of these elements. Activity, expressed as transcriptional strength and/or ectopic pairing efficiency, would vary due to elements' chromosomal positions, internal structures (e.g., vis-a-vis deletions and other mutations), and level of acquired modification. Differential segregation of these factors would tend to differentiate *Mutator* transmission frequencies in the branches within a lineage. However, amplification of autonomous elements via transposition and cooperativity in both the demodification and modification of *Mu* elements would minimize these differences, accounting for the consistent patterns observed within a given lineage (WALBOT 1986; BENNETZEN 1994).

6 Uses of the *Mutator* System

6.1 Mutagenesis

Mutator is an unusually active agent for the creation of de novo mutations, and the low specificity for *Mu* insertion makes it an effective mutagen for the generation of a wide variety of mutant alleles (reviewed in BENNETZEN et al. 1993). As for other transposable element systems, *Mutator* has the advantage of generating mutations with phenotypes unlike those seen for other mutagenic activities. Dominant and regulatory mutations, normally rare, are fairly common with the *Mutator* system (CHEN et al. 1987b; BENNETZEN et al. 1993; STINARD et al. 1993; GREENE et al. 1994; MARTIENSSEN and BARON 1994). Some genes, for instance those with an essential function or belonging to a multigene family, may be detected only genetically as dominant mutations. In addition, mutations induced by *Mutator* are often mutable and/or suppressible, and this can be used as a useful tool in fate mapping of developing tissues (MARTIENSSEN and BARON 1994).

Often, molecular biologists will identify a particular mRNA (and/or its protein product) as potentially significant to a particular biological process merely by its tissue-specific or inducible/repressible expression pattern. With higher eukaryotes, a common problem in the study of a gene that encodes such an mRNA is determination of the significance of its function. A recent technical innovation in maize (S. Briggs, personal communication) should allow one to convert sequence information for a cloned gene into a phenotype, if any, conditioned by mutation at that gene. Briggs has generated a population of some 25 000 self-crossed *Mutator* lines and individually stored DNAs made from pools of the progeny of each cross. One can now PCR screen these 25 000 pools with a *Mu* TIR primer and a primer homologous to the gene of interest. Amplification in a given DNA preparation will indicate that the self-cross progeny contain a *Mu* element insertion in the targetted locus, and the size of the amplification product will indicate the position of the insertion. One can then grow the seed from the self-cross(es) that yielded the positive result, and determine the phenotype(s) of the mutation(s). In preliminary experiments, Briggs (personal communication) has found several independent *Mu* insertions for each sequence tested and has identified the expected phenotypes in a subset of the segregating progeny. Hence, as is also true of the standard mutagenesis described in the preceding paragraph, *Mutator* mutagenesis can both allow the discovery of new genes and assist in understanding the function and regulation of tagged genes.

6.2 Tagging

The most extensive use of the *Mutator* system, by numerous laboratories, has been in the generation of mutations in maize that can be cloned via the technique called transposon tagging. This approach has been very successful (reviewed in Bennetzen et al. 1993), and additional genes have been isolated by *Mu* tagging recently (Stinard et al. 1993; Schnable and Wise 1994; P. Schnable, personal communication).

Programs are also underway to transfer *Mutator* into other plant species, where a tagging system with *Mu*'s high activity and low specificity would be desirable. Unfortunately, the instability of *MuDR* elements in *E.coli*, particularly that portion encoding the 2.8-kb transposase transcript (Benito and Walbot 1994), has slowed this process greatly.

6.3 Genetic Engineering

The generation of transgenic plants by naked DNA uptake is a very inefficient process. In *Drosophila*, *P* elements have been used as vectors to increase the efficiency of transgene integration in injected oocytes (Rubin and Spradling 1982). *Mutator*'s frequent transposition activity, and multiple other properties in common with the *P* system of *Drosophila*, suggest that it might also be a useful vector to enhance maize transformation efficiencies. At the moment, this would require transformation of a *Mutator* line, because the instability of *MuDR* in *E. coli* currently makes co-injection of a *MuDR* element and the transformation vector impossible. *Mutator* activity has now been entered into a number of maize lines (Cresse 1992), so this should not limit the approach greatly. However, for attempts to improve transformation frequencies in a full range of maize lines and other plant species, a mechanism to generate an intact *Mu* transposase gene will need to be developed. One possible approach to this problem would be to prepare complementary portions of the transposase gene in two different plasmids, then directly transform with the reaction mix generated in the ligation of these segments. Alternatively, if expression of the *Mu* transposase gene in *E. coli* is the cause of its instability, then regulatory sequences will need to be attached to the gene that allow no expression in the bacterium but still fully promote expression in the plant.

The proposed mechanism of *Mu* element transposition, via gap repair of an excised element, suggests that these elements could also be used to directly target transgenes to specific regions of the genome. This approach has been successful with the *P* elements of *Drosophila* (Gloor et al. 1991) and the Tc1 elements of *C. elegans* (Plasterk and Groenen 1992).

7 Future Directions

The most exciting recent discovery in the *Mutator* system has been the isolation and initial characterization of the autonomous element *MuDR*. Much future work will focus on determination of the mechanism of action of this element, including its involvement in the regulation of *Mu* activity. The roles of the two *MuDR* transcripts need to be more fully analyzed and will be particularly productive in "minimal" *Mutator* lines where the interactions between different *MuDRs* and their derivatives can be minimized (Chomet et al. 1991; Lisch et al. 1995). In particular, it will be interesting to note whether different *MuDR* elements or different *MuDR* positions generate different ratios of sense, antisense, and variably spliced transcripts (Hershberger et al. 1994). Transgenic systems should also provide a good background for *MuDR* studies, when and if constructs can be prepared for transfer to other plants.

A central goal of *Mutator* studies should be a convincing test of whether some *MuDR* derivatives can have a dominant inhibitory effect (Martienssen and Baron 1994). The high frequency of somatic and germinal deletion of *MuDR* sequences in maize (Lisch et al. 1995) will allow the rapid generation of a variety of biological materials suited to both transcript functional analysis and d*MuDR* studies. Given that the correlations between *Mu* inactivation and co-suppression are so extensive, it will be interesting to determine whether any effects of an inhibitory *dMuDR* require transcription of the element or are affected by its genomic integration site and methylation level.

The proposed "conditioning" of *Mutator* lines, a function of the nature, number, modification, and genomic location of *Mu* elements in a line (BENNETZEN 1985a, 1994; BENNETZEN et al. 1993), can now be tested in lines containing known numbers of *MuDR* and *dMuDR* elements. The proposed significance of the ratio of *dMuDR* elements to *MuDR* elements can be determined in crosses that test the dominance of inactivation or reactivation with different ratios of these presumably key elements. Such crossing programs and analyses could also allow determination of whether *MuDR* deletion is dependent upon *Mutator* activity. For instance, if *MuDR* elements are truly self-mutagenic, then a modified and quiescent *MuDR* in a *Mu*-off line would not undergo deletion at the same rate as would a *MuDR* element in an active *Mutator* line.

The coordinate modification of multiple *Mu* elements in a single plant, apparently by a default mechanism independent of *Mutator* function, suggests the existence of a plant process for inactivation of potentially detrimental mobile nucleic acids. An analogous, perhaps identical, process acts on transgenes (in the phenomenon called co-suppression) and may be involved in limiting some viral virulences. Study of the cooperative inactivation of *Mutator* may provide insights into this interesting process. Especially interesting will be studies that determine whether coordinate inactivation of *Mutator* activity (for instance, in intercross-loss lines) is associated with rapid turnover of *MuDR* transcripts, suggesting RNA-level effects like those seen in some co-suppression phenomena.

At the moment, the mechanism of *Mu* element transposition has not been directly demonstrated, but it is known to be by a replicative process that has much in common with the gap-repair system used by some animal transposable elements. Direct demonstration of this process should now be feasible and may even be useful for directing transgenes to specific genomic locations (GLOOR et al. 1991; PLASTERK and GROENEN 1992).

The interaction of *Mu* elements with "host" factors is responsible for the tight developmental control of *Mutator* activity and may also be involved in their concomitant methylation/inactivation. Further efforts to identify these host activities is needed, at both the biochemical level (ZHAO and SUNDARESAN 1991) and by genetic characterization (WALBOT 1992a). The lack of correlation between excision and transposition events is a particularly fascinating characteristic of the *Mutator* system. Because these different outcomes of *Mu*-element interaction with *Mu* transposase appear to show different tissue and timing specificities, they may be especially interesting to characterize vis-a-vis binding factors or altered *Mutator* regulation. The ability of transposable elements to induce regulatory mutations and the phenomenon of *Mu*-element suppression (BARKAN and MARTIENSSEN 1991) provide ample demonstration of the uses that can be made of *Mutator* as a sensitive probe for standard nuclear factors that regulate maize genes.

The origin of the *Mutator* system and of *Mu* element subfamilies deserves further investigation. The creation of the *Mu1/Mu2* subfamily by acquisition of standard genomic DNA within *Mu* TIRs suggests a similar origin of many of the other subfamilies. A search for other sequences similar to *Mu*-element internal regions, but lacking *Mu* TIRs, is warranted. This ability to acquire and amplify "host" DNA is a common property of retroviruses and has now been seen with one maize retrotransposon (JIN and BENNETZEN 1994) but is only rarely observed with elements that carry TIRs. Identification of the source of these internal sequences and their comparison to the current *Mu* elements would provide insights into the nature of the acquisition process and may assist us in understanding the mechanisms of *Mu* element transposition, excision, and self-mutagenesis. Transposable elements constitute such a large part of the genomes of maize and other higher eukaryotes, including major contributions to promoter construction (BUREAU and WESSLER 1994), that further understanding of their organization, source, inactivations, and reactivations will provide key information regarding genome function.

The *Mutator* system will continue to be a powerful tool for isolating maize genes by transposon tagging and can now even be used to determine the phenotypic contribution of any maize gene with a known sequence (S. Briggs, personal communication). Progress toward transferring *Mutator* to other plant species is needed. This will allow a wide array of basic *Mu* studies and should provide these species with a tagging system that has a high frequency and low specificity of insertion.

A great deal of progress has been made in understanding the *Mutator* system since its discovery by ROBERTSON (1978). Robertson's extensive genetic studies have laid the foundation for all subsequent discoveries, including the use of *Mu* for tagging and cloning plant genes. We are now poised to employ *Mutator* in a wider variety of uses and a broader range of species. In addition, the molecular mechanisms of *Mu* action and regulation are beginning to be uncovered. Hence, *Mutator* research has a bright future and should yield more unexpected insights into transposable element activity and into plant genome structure and function.

Acknowledgments. I thank S. Briggs, V. Chandler, M. Donlin, M. Freeling, S. Hake, J. Hershberger, D. Lisch, R. Martienssen, P. Schnable, and V. Walbot for providing information prior to publication. I am especially grateful to M. Freeling, D. Lisch, R. Martienssen, and V. Walbot for their helpful discussions, and to A. Oliveira for his analyses of *Mu* TIR homologies. The preparation of this manuscript was supported by USDA grant 91-37300-6498.

References

Alleman M, Freeling M (1986) The *Mu* transposable elements of maize: evidence for transposition and copy number regulation during development. Genetics 112: 107–119

Antequera F, Bird AP (1988) Unmethylated CpG islands associated with genes in higher plant DNA. EMBO J 7: 2295–2299

Banks JA, Masson P, Fedoroff N (1988) Molecular mechanisms in the developmental regulation of the maize suppressor-mutator transposable element. Genes Dev 2: 1364–1380

Barkan A, Martienssen RA (1991) Inactivation of maize transposon *Mu* suppresses a mutant phenotype by activating an outward-reading promoter near the end of *Mu1*. Proc Natl Acad Sci USA 88: 3502–3506

Barker RF, Thompson DV, Talbot DR, Swanson J, Bennetzen JL (1984) Nucleotide sequence of the maize transposable element *Mu1*. Nucleic Acids Res 12: 5955–5967

Benito M-I, Walbot V (1994) The terminal inverted repeat sequences of *MuDR* are functionally active promoters in maize cells. Maydica 39: 255–264

Benjamin HW, Kleckner N (1989) Intramolecular transposition by Tn*10*. Cell 59: 373–383

Bennetzen JL (1984) Transposable element *Mu1* is found in multiple copies only in Robertson's *Mutator* maize lines. J Mol Appl Gen 2: 519–524

Bennetzen JL (1985a) The regulation of *Mutator* function and *Mu1* transposition. In: Freeling M (ed) Plant genetics. Liss, New York, pp 343–353

Bennetzen JL (1985b) The mechanism, rate, regulation, and specificity of *Mu1* transposition in maize. J Cell Biochem 9: 211

Bennetzen JL (1987) Covalent DNA modification and the regulation of *Mutator* element transposition in maize. Mol Gen Genet 208: 45–51

Bennetzen JL (1994) Inactivation and reactivation of mutability at a *Mutator*-derived *Bronze-1* allele in maize. Maydica 39: 309–317

Bennetzen JL, Springer PS (1994) The generation of *Mutator* transposable element subfamilies in maize. Theor Appl Genet 87: 657–667

Bennetzen JL, Swanson J, Taylor WC, Freeling M (1984) DNA insertion in the first intron of maize *Adh1* affects message levels: cloning of progenitor and mutant alleles. Proc Natl Acad Sci USA 81: 4125–4128

Bennetzen JL, Fracasso RP, Morris DW, Robertson DS, Skogen-Hagenson MJ (1987) Concomitant regulation of *Mu1* transposition and *Mutator* acitivity in maize. Mol Gen Genet 208: 57–62

Bennetzen JL, Brown WE, Springer PS (1988) The state of DNA modification within and flanking maize transposable elements. In: Nelson OE (ed) Plant transposable elements. Plenum, New York, pp 237–250

Bennetzen JL, Springer PS, Cresse AD, Hendrickx M (1993) Specificity and regulation of the *Mutator* transposable element system of maize. Crit Rev Plant Sci 12: 57–95

Bennetzen JL, Schrick K, Springer PS, Brown WE, SanMiguel P (1994) Active maize genes are unmodified and flanked by diverse classes of modified, highly repetitive DNA. Genome 37: 565–576

Brink RA, Styles ED, Axtell JD (1968) Paramutation: directed genetic change. Science 159: 161–170

Britt AB, Walbot V (1991) Germinal and somatic products of *Mu1* excision from the *bronze-1* gene of *Zea mays*. Mol Gen Genet 227: 267–276

Brown J, Sundaresan V (1992) Genetic study of the loss and restoration of *Mutator* transposon activity in maize: evidence against dominant-negative regulator associated with loss of activity. Genetics 130: 889–898

Brown WE, Robertson DS, Bennetzen JL (1989) Molecular analysis of multiple *Mutator*-derived alleles of the *bronze* locus of maize. Genetics 122: 439–445

Brown WE, Springer PS, Bennetzen JL (1994) Progressive modification of *Mu* transposable elements during development. Maydica 39: 119–126

Bureau TE, Wessler SR (1994) *Stowaway*: a new family of inverted repeat elements associated with the genes of both monocotyledenous and dicotyledenous plants. Plant Cell 6: 907–916

Capel J, Montero LM, Martinez-Zapater JM, Salinas J (1993) Nonrandom distribution of transposable elements in the nuclear genome of plants. Nucleic Acids Res 21: 2369–2373

Chandler VL, Hardeman KJ (1992) The *Mu* elements of *Zea mays*. Adv Genet 30: 77–122

Chandler VL, Walbot V (1986) DNA modification of a maize transposable element correlates with loss of activity. Proc Natl Acad Sci USA 83: 1767–1771

Chandler V, Rivin C, Walbot V (1986) Stable non-*Mutator* stocks of maize have sequences homologous to the *Mu1* transposable element. Genetics 114: 1007–1021

Chandler VL, Talbert LE, Raymond F (1988) Sequence, genomic distribution and DNA modification of a *Mu1* element from non-*Mutator* maize stocks. Genetics 119: 951–958

Chandler VL, Talbert LE, Mann L, Faber C (1989) Structure and DNA modification of endogenous *Mu* elements. In: Nelson OE (ed) Plant transposable elements. Plenum, New York, pp 339–350

Chen J, Greenblatt IM, Dellaporta SL (1987a) Transposition of *Ac* from the *P* locus of maize into unreplicated chromosomal sites. Genetics 117: 109–116

Chen C-H, Oishi KK, Kloeckener-Gruissem B, Freeling M (1987b) Organ-specific expression of maize *Adh1* is altered after a *Mu* transposon insertion. Genetics 116: 469–477

Chomet PS, Wessler SR, Dellaporta SL (1987) Inactivation of the maize transposable element *Activator* *(Ac)* is associated with its DNA modification. EMBO J 6: 295–302

Chomet P, Lisch D, Hardeman KJ, Chandler VL, Freeling M (1991) Identification of a regulatory transposon that controls the *Mutator* transposable element system in maize. Genetics 129: 261–270

Cone KC, Schmidt RJ, Burr B, Burr FA (1988) Advantages and limitations of using *Spm* as a transposon tag. In: Nelson OE (ed) Plant transposable elements. Plenum, New York, pp 149–159
Cook WB (1988) Isolation and characterization of photosynthetic mutants from a Robertson's *Mutator* line of maize (*Zea mays*), PhD thesis, University of Missouri
Cresse AD (1992) An investigation of insertion specificity and genetic background effects in the *Mutator* transposable element system of maize. PhD thesis, Purdue University
Cresse AD, Hulbert SH, Brown WE, Lucas JR, Bennetzen JL (1995) *Mu1*-related transposable elements of maize preferentially insert into low copy number DNA. Genetics 140: 315–324
Cuypers H, Dash S, Peterson PA, Saedler H, Gierl A (1988) The defective *En*-1102 element encodes a product reducing the mutability of the *En/Spm* transposable element system of *Zea mays*. EMBO J 7: 2953–2960
Dellaporta SL, Chomet PS (1985) The action of maize controlling elements. In: Hohn B, Dennis ES (eds) Plant gene research: genetic flux in plants. Springer, Berlin Heidelberg New York, pp 169–216
Dooner HK, Belachew A (1989) Transposition pattern of the maize element *Ac* from the *bz-m2 (Ac)* allele. Genetics 122: 447–457
Dooner HK, Ralston EJ (1990) Effect of the *Mu1* insertion on intragenic recombination at the *bz* locus in maize. Maydica 35: 333–337
Doseff A, Martienssen R, Sundaresan V (1991) Somatic excision of the *Mu1* transposable element of maize. Nucleic Acids Res 19: 579–584
Engels WR, Johnson-Schlitz DM, Eggleston WB, Sved J (1990) High-frequency *P* element loss in *Drosophila* is homolog dependent. Cell 62: 515–525
Flavell R (1994) Inactivation of gene expression in plants as a consequence of specific sequence duplication. Proc Natl Acad Sci USA 91: 3490–3496
Fleenor D, Spell M, Robertson D, Wessler S (1990) Nucleotide sequence of the maize *Mutator* element, *Mu8*. Nucleic Acids Res 18: 6725
Gloor GB, Nassif NA, Johnson-Schlitz DM, Preston CR, Engels WR (1991) Targeted gene replacement in *Drosophila* via *P* element-induced gap repair. Science 253: 1110–1117
Greenblatt IM (1984) A chromosome replication pattern deduced from pericarp phenotypes resulting from movements of the transposable element, *Modulator*, in maize. Genetics 108: 471–485
Greene B, Walko R, Hake S (1994) *Mutator* insertions in an intron of the maize *knotted-1* gene result in dominant suppressible mutations. Genetics 138: 1275–1285
Hake S, Walbot V (1980) The genome of *Zea mays*, its organization and homology to related grasses. Chromosoma 79: 369–373
Han C, Coe EH jr, Martienssen RA (1992) Molecular cloning and characterization of *lojap* (*ij*), a pattern striping gene of maize. EMBO J 11: 4037–4046
Hardeman KJ, Chandler VL (1989) Characterization of *bz1* mutants isolated from *Mutator* stocks with high and low numbers of *Mu1* elements. Dev Genet 10: 460–472
Hardeman KJ, Chandler VL (1993) Two maize genes are each targeted predominantly by distinct classes of *Mu* elements. Genetics 135: 1141–1150
Hershberger RJ, Warren CA, Walbot V (1991) *Mutator* activity in maize correlates with the presence and expression of the *Mu* transposable element *Mu9*. Proc Natl Acad Sci USA 88: 10198–10202
Hershberger RJ, Benito M-I, Hardeman KJ, Warren CA, Chandler VL, Walbot V (1995) Characterization of the major transcripts encoded by the regulatory *MuDR* transposable element of maize. Genes Dev (in press)
Ingels SC, Bennetzen JL, Hulbert SH, Qin M, Ellingboe AH (1992) *Mutator* transposable elements that occur in clusters in the maize genome. J Hered 83: 114–118
Jackson MS, Black DM, Dover GA (1988) Amplification of *KP* elements associated with the repression of hybrid dysgenesis in *Drosophila melanogaster*. Genetics 120: 1003–1013
James MG, Scanlon MJ, Qin M-M, Robertson D, Myers AM (1993) DNA sequence and transcript analysis of transposon *MuA2*, a regulator of *Mutator* transposable element activity in maize. Plant Mol Biol 21: 1181–1185
Jin Y-K, Bennetzen JL (1994) Integration and nonrandom mutation of a plasma membrane proton ATPase gene fragment within the *Bs1* retroelement of maize. Plant Cell 6: 1177–1186
Kim H-Y, Schiefelbein JW, Raboy V, Furtek DB, Nelson OE jr (1987) RNA splicing permits expression of a maize gene with a defective Suppressor-mutator transposable element insertion in an exon. Proc Natl Acad Sci USA 84: 5863–5867
Kloeckener-Gruissem B, Vogel JM, Freeling M (1992) The TATA box promoter region of maize *Adh1* affects its organ-specific expression. EMBO J 11: 157–166
Levy AA, Walbot V (1990) Regulation of the timing of transposable element excision during maize development. Science 248: 1534–1537

Levy AA, Walbot V (1991) Molecular analysis of the loss of somatic instability in the *bz2: mu1*allele of maize. Mol Gen Genet 229: 147–151

Levy AA, Britt AB, Luehrsen KR, Chandler VL, Warren C, Walbot V (1989) Developmental and genetic aspects of *Mutator* excision in maize. Dev Genet 10: 520–531

Linn F, Heidmann I, Saedler H, Meyer P (1990) Epigenetic changes in the expression of the maize *A1* gene in *Petunia hybrida*: role of numbers of integrated gene copies and state of methylation. Mol Gen Genet 222: 329–336

Lisch D, Chomet P, Freeling M (1995) Genetic characterization of the *Mutator* system in maize: behavior and regulation of *Mu* transposons in a minimal line. Genetics 139: 1777–1796

Lowe B, Mathern J, Hake S (1992) Active *Mutator* elements suppress the *knotted* phenotype and increase recombination at the *kn1–O* tandem duplication. Genetics 132: 813–822

Luehrsen KR, Walbot V (1990) Insertion of *Mu1* elements in the first intron of the *Adh1-S* gene of maize results in novel RNA processing events. Plant Cell 2: 1225–1238

Martienssen R, Baron A (1994) Coordinate expression of mutations caused by Robertson's *Mutator* transposons in maize. Genetics 136: 1157–1170

Martienssen RA, Barkan A, Freeling M, Taylor WC (1989) Molecular cloning of a maize gene involved in photosynthetic membrane organization that is regulated by Robertson's *Mutator*. EMBO J 8: 1633–1639

Martienssen RA, Barkan A, Taylor WC, Freeling M (1990) Somatically heritable switches in the DNA modification of *Mu* transposable elements monitored with a suppressible mutant in maize. Genes Dev 4: 331–343

Matzke MA, Primig M, Trnovsky J, Matzke AJM (1989) Reversible methylation and inactivation of marker genes in sequentially transformed tobacco plants. EMBO J 8: 643–649

McCarty DR, Carson CB, Lazar M, Simonds SC (1989a) Transposable element-induced mutations of the *viviparous-1* gene in maize. Dev Genet 10: 473–481

McCarty DR, Carson CB, Stinard PS, Robertson DS (1989b) Molecular analysis of *viviparous-1*: an abscisic acid insensitive mutant of maize. Plant Cell 1: 523–532

McClintock B (1948) Mutable loci in maize. Carnegie Inst Washington Yearbook 47: 155–169

McClintock B (1949) Mutable loci in maize. Carnegie Inst Washington Yearbook 48: 142–154

McClintock B (1956) Controlling elements and the gene. Cold Spring Harb Symp Quant Biol 21: 197–216

McClintock B (1958) The suppressor-mutator system of control of gene action in maize. Carnegie Inst Washington Year Book 57: 415–429

McClintock B (1984) The significance of responses of the genome to challenge. Science 226: 792–801

Misra S, Rio DC (1990) Cytotype control of *Drosophila P* element transposition: the 66-kD protein is a repressor of transposase activity. Cell 62: 269–284

Napoli C, Lemieux C, Jorgensen R (1990) Introduction of a chimeric chalcone synthase gene into petunia results in reversible co-suppression of homologous genes in trans. Plant Cell 2: 279–289

Nash J, Luehrsen KR, Walbot V (1990) *Bronze-2* gene of maize: reconstruction of a wild-type allele and analysis of transcription and splicing. Plant Cell 2: 1039–1049

Oishi KK, Freeling M (I1988) A new *Mu* element from a Robertson's *Mutator* line. In: Nelson OE (ed) Plant transposable elements. Plenum, New York, pp 289–291

Ortiz DF, Strommer JN (1990) The *Mu1* maize transposable element induces tissue-specific aberrant splicing and polyadenylation in two *adh1* mutants. Mol Cell Biol 10: 2090–2095

Ortiz DF, Rowland LJ, Gregerson RG, Strommer JN (1988) Insertion of *Mu* into the *Shrunken1* gene of maize affects transcriptional and post-transcriptional regulation of *Sh1* RNA. Mol Gen Genet 214: 135–141

Plasterk RHA, Groenen TM (1992) Targeted alterations of the *Caenorhabditis elegans* genome by transgene instructed DNA double-strand break repair following Tc1 excision. EMBO J 11: 287–290

Qin M, Ellingboe AH (1990) A transcript identified by *MuA* of maize is associated with *Mutator* activity. Mol Gen Genet 224: 357–363

Qin M, Robertson DS, Ellingboe AH (1991) Cloning of the *Mutator* transposable element *MuA2*, a putative regulator of somatic mutability of the *a1-Mum2* allele in maize. Genetics 129: 845–854

Robertson DS (1978) Characterization of a mutator system in maize. Mutat Res 51: 21–28

Robertson DS (1980) The timing of *Mu* activity in maize. Genetics 94: 969–978

Robertson DS (1981) *Mutator* activity in maize: timing of its activation in ontogeny. Science 213: 1515–1517

Robertson DS (1983) A possible dose-dependent inactivation of *Mutator (Mu)* in maize. Mol Gen Genet 191: 86–90

Robertson DS (1985) Differential activity of the maize mutator *Mu* at different loci and in different cell lineages. Mol Gen Genet 200: 9–13
Robertson DS (1986) Genetic studies on the loss of *Mu* mutator activity in maize. Genetics 113: 765–773
Robertson DS, Stinard PS (1987) Genetic evidence of *Mutator*-induced deletions in the short arm of chromosome 9 of maize. Genetics 115: 353–361
Robertson DS, Stinard PS (1989) Genetic analyses of putative two element systems regulating somatic mutability in *Mutator*-induced aleurone mutants in maize. Dev Genet 10: 482–506
Robertson DS, Stinard PS (1992) Genetic regulation of somatic mutability of two *Mu*-induced *a1* mutants of maize. Theor Appl Genet 84: 225–236
Robertson DS, Stinard PS (1993) Evidence for *Mutator* activity in the male and female gametophytes of maize. Maydica 38: 145–150
Robertson DS, Stinard PS, Wheeler JG, Morris DW (1985) Genetic and molecular studies on germinal and somatic instability in *Mutator*-induced aleurone mutants of maize. In: Freeling M (ed) Plant genetics. Liss, New York, pp 317–332
Robertson DS, Stinard PS, Maguire MP (1994) Genetic evidence of *Mutator*-induced deletions in the short arm of chromosome 9 of maize. II. *wd* deletions. Genetics 136: 1143–1149
Robertson HM, Engels WR (1989) Modified *P* elements that mimic the P cytotype in *Drosophila melanogaster*. Genetics 123: 815–824
Rubin GM, Spradling AC (1982) Genetic transformation of *Drosophila* with transposable element vectors. Science 218: 348–353
Schnable PS, Peterson PA (1986) Distribution of genetically active *Cy* transposable elements among diverse maize lines. Maydica 31: 59–81
Schnable PS, Peterson PA (1989) Genetic evidence of a relationship between two maize transposable element systems: *Cy* and *Mutator*. Mol Gen Genet 215: 317–321
Schnable PS, Wise RP (1994) Recovery of heritable, transposon-induced, mutant alleles of the *rf2* nuclear restorer of T-cytoplasm maize. Genetics 136: 1171–1185
Schnable PS, Peterson PA, Saedler H (1989) The *bz-rcy* allele of the *Cy* transposable element system of *Zea mays* contains a *Mu*-like element insertion. Mol Gen Genet 217: 459–463
Schwartz D (1989) Gene controlled cytosine demethylation in the promoter region of the *Ac* transposable element in maize. Proc Natl Acad Sci USA 86: 2789–2793
Schwartz D, Dennis ES (1986) Transposase activity of the *Ac* controlling element in maize is regulated by its degree of methylation. Mol Gen Genet 205: 476–482
Springer PS, Edwards KJ, Bennetzen JL (1994) DNA class organization on maize *Adh1* yeast artificial chromosomes. Proc Natl Acad Sci USA 91: 863–867
Stinard PS, Robertson DS, Schnable PS (1993) Genetic isolation, cloning, and analysis of a *Mutator*-induced, dominant antimorph of the maize *amylose extender1* locus. Plant Cell 5: 1555–1566
Strommer JN, Ortiz D (1989) *Mu1*-induced mutant alleles of maize exhibit background-dependent changes in expression and RNA processing. Dev Genet 10: 452–459
Strommer JN, Hake S, Bennetzen JL, Taylor WC, Freeling M (1982) Regulatory mutants of the maize *Adh1* gene caused by DNA insertions. Nature 300: 542–544
Sundaresan V (1988) Extrachromsomal *Mu*. In: Nelson OE (ed) Plant transposable elements. Plenum, New York , pp 251–259
Sundaresan V, Freeling M (1987) An extrachromsomal form of the *Mu* transposons of maize. Proc Natl Acad Sci USA 84: 4924–4928
Talbert LE, Chandler VL (1988) Characterization of a highly conserved sequence related to *Mutator* transposable elements in maize. Mol Biol Evol 5: 519–529
Talbert LE, Patterson GI, Chandler VL (1989) *Mu* transposable elements are structurally diverse and distributed throughout the genus *Zea*. J Mol Evol 29: 28–39
Talbert LE, Doebley JF, Larson S, Chandler VL (1990) *Tripsacum andersonii* is a natural hybrid involving *Zea* and *Tripsacum*: molecular evidence. Am J Bot 77: 722–726
Taylor LP, Walbot V (1985) A deletion adjacent to the maize transposable element *Mu1* accompanies loss of *Adh1* expression. EMBO J 4: 869–876
Taylor LP, Walbot V (1987) Isolation and characterization of a 1.7-kb transposable element from a *Mutator* line of maize. Genetics 117: 297–307
Taylor LP, Chandler VL, Walbot V (1986) Insertion of 1.4-kb and 1.7-kb *Mu* elements into the *Bronze1* gene of *Zea mays* L. Maydica 31: 31–45
van der Krol AR, Mur LA, Beld M, Mol JNM, Stuitje AR (1990) Flavonoid genes in petunia: addition of a limited number of gene copies may lead to a suppression of gene expression. Plant Cell 2: 291–299

Van Schaik NW, Brink RA (1959) Transpositions of *Modulator*, a component of the variegated pericarp allele in maize. Genetics 44: 725–738

Vayda ME, Freeling M (1986) Insertion of the *Mu1* transposable element into the first intron of maize *Adh1* interferes with transcript elongation but does not disrupt chromatin structure. Plant Mol Biol 6: 441–454

Walbot V (1986) Inheritance of *Mutator* activity in *Zea mays* as assayed by somatic instability of the *bz2-mu1* allele. Genetics 114: 1293–1312

Walbot V (1988) Reactivation of the *Mutator* transposable element system following gamma irradiation of seed. Mol Gen Genet 212: 259–264

Walbot V (1991) The *Mutator* transposable element family of maize. In: Setlow JK (ed) Genetic engineering, vol 13. Plenum, New York, pp 1–37

Walbot V (1992a) Developmental regulation of excision timing of *Mutator* transposons of maize: comparison of standard lines and an early excision *bz1:: Mu1* line. Dev Genet 13: 376–386

Walbot V (1992b) Reactivation of *Mutator* transposable elements of maize by ultraviolet light. Mol Gen Genet 234: 353–360

Walbot V, Warren C (1988) Regulation of *Mu* element copy number in maize lines with an active or inactive *Mutator* transposable element system. Mol Gen Genet 211: 27–34

Walbot V, Warren C (1990) DNA methylation in the *Alcohol dehydrogenase-1* gene of maize. Plant Mol Biol 15: 121–125

Walbot V, Chandler V, Taylor L (1985) Alterations in the *Mutator* transposable element family of *Zea mays*. In: Freeling M (ed) Plant genetics. Liss, New York, pp 333–342

Walbot V, Briggs CP, Chandler VL (1986) Properties of mutable alleles recovered from mutator stocks of *Zea mays* L. In: Gustafson JP, Stebbins GL, Ayala FJ (eds) Genetics, development and evolution. Plenum, New York, pp 115–142

Wessler SR, Baran G, Varagona M (1987) The maize transposable element *Ds* is spliced from RNA. Science 237: 916–918

Zhao Z-Y, Sundaresan V (1991) Binding sites for maize nuclear proteins in the terminal inverted repeats of the *Mu1* transposable element. Mol Gen Genet 229: 17–26

Subject Index

Current Topics in Microbiology and Immunology

Volumes published since 1989 (and still available)

Vol. 161: **Racaniello, Vincent R. (Ed.):** Picornaviruses. 1990. 12 figs. X, 194 pp. ISBN 3-540-52429-0

Vol. 162: **Roy, Polly; Gorman, Barry M. (Eds.):** Bluetongue Viruses. 1990. 37 figs. X, 200 pp. ISBN 3-540-51922-X

Vol. 163: **Turner, Peter C.; Moyer, Richard W. (Eds.):** Poxviruses. 1990. 23 figs. X, 210 pp. ISBN 3-540-52430-4

Vol. 164: **Bækkeskov, Steinnun; Hansen, Bruno (Eds.):** Human Diabetes. 1990. 9 figs. X, 198 pp. ISBN 3-540-52652-8

Vol. 165: **Bothwell, Mark (Ed.):** Neuronal Growth Factors. 1991. 14 figs. IX, 173 pp. ISBN 3-540-52654-4

Vol. 166: **Potter, Michael; Melchers, Fritz (Eds.):** Mechanisms in B-Cell Neoplasia. 1990. 143 figs. XIX, 380 pp. ISBN 3-540-52886-5

Vol. 167: **Kaufmann, Stefan H. E. (Ed.):** Heat Shock Proteins and Immune Response. 1991. 18 figs. IX, 214 pp. ISBN 3-540-52857-1

Vol. 168: **Mason, William S.; Seeger, Christoph (Eds.):** Hepadnaviruses. Molecular Biology and Pathogenesis. 1991. 21 figs. X, 206 pp. ISBN 3-540-53060-6

Vol. 169: **Kolakofsky, Daniel (Ed.):** Bunyaviridae. 1991. 34 figs. X, 256 pp. ISBN 3-540-53061-4

Vol. 170: **Compans, Richard W. (Ed.):** Protein Traffic in Eukaryotic Cells. Selected Reviews. 1991. 14 figs. X, 186 pp. ISBN 3-540-53631-0

Vol. 171: **Kung, Hsing-Jien; Vogt, Peter K. (Eds.):** Retroviral Insertion and Oncogene Activation. 1991. 18 figs. X, 179 pp. ISBN 3-540-53857-7

Vol. 172: **Chesebro, Bruce W. (Ed.):** Transmissible Spongiform Encephalopathies. 1991. 48 figs. X, 288 pp. ISBN 3-540-53883-6

Vol. 173: **Pfeffer, Klaus; Heeg, Klaus; Wagner, Hermann; Riethmüller, Gert (Eds.):** Function and Specificity of γ/δ TCells. 1991. 41 figs. XII, 296 pp. ISBN 3-540-53781-3

Vol. 174: **Fleischer, Bernhard; Sjögren, Hans Olov (Eds.):** Superantigens. 1991. 13 figs. IX, 137 pp. ISBN 3-540-54205-1

Vol. 175: **Aktories, Klaus (Ed.):** ADP-Ribosylating Toxins. 1992. 23 figs. IX, 148 pp. ISBN 3-540-54598-0

Vol. 176: **Holland, John J. (Ed.):** Genetic Diversity of RNA Viruses. 1992. 34 figs. IX, 226 pp. ISBN 3-540-54652-9

Vol. 177: **Müller-Sieburg, Christa; Torok-Storb, Beverly; Visser, Jan; Storb, Rainer (Eds.):** Hematopoietic Stem Cells. 1992. 18 figs. XIII, 143 pp. ISBN 3-540-54531-X

Vol. 178: **Parker, Charles J. (Ed.):** Membrane Defenses Against Attack by Complement and Perforins. 1992. 26 figs. VIII, 188 pp. ISBN 3-540-54653-7

Vol. 179: **Rouse, Barry T. (Ed.):** Herpes Simplex Virus. 1992. 9 figs. X, 180 pp. ISBN 3-540-55066-6

Vol. 180: **Sansonetti, P. J. (Ed.):** Pathogenesis of Shigellosis. 1992. 15 figs. X, 143 pp. ISBN 3-540-55058-5

Vol. 181: **Russell, Stephen W.; Gordon, Siamon (Eds.):** Macrophage Biology and Activation. 1992. 42 figs. IX, 299 pp. ISBN 3-540-55293-6

Vol. 182: **Potter, Michael; Melchers, Fritz (Eds.):** Mechanisms in B-Cell Neoplasia. 1992. 188 figs. XX, 499 pp. ISBN 3-540-55658-3

Vol. 183: **Dimmock, Nigel J.:** Neutralization of Animal Viruses. 1993. 10 figs. VII, 149 pp. ISBN 3-540-56030-0

Vol. 184: **Dunon, Dominique; Mackay, Charles R.; Imhof, Beat A. (Eds.):** Adhesion in Leukocyte Homing and Differentiation. 1993. 37 figs. IX, 260 pp. ISBN 3-540-56756-9

Vol. 185: **Ramig, Robert F. (Ed.):** Rotaviruses. 1994. 37 figs. X, 380 pp. ISBN 3-540-56761-5

Vol. 186: **zur Hausen, Harald (Ed.):** Human Pathogenic Papillomaviruses. 1994. 37 figs. XIII, 274 pp. ISBN 3-540-57193-0

Vol. 187: **Rupprecht, Charles E.; Dietzschold, Bernhard; Koprowski, Hilary (Eds.):** Lyssaviruses. 1994. 50 figs. IX, 352 pp. ISBN 3-540-57194-9

Vol. 188: **Letvin, Norman L.; Desrosiers, Ronald C. (Eds.):** Simian Immunodeficiency Virus. 1994. 37 figs. X, 240 pp. ISBN 3-540-57274-0

Vol. 189: **Oldstone, Michael B. A. (Ed.):** Cytotoxic T-Lymphocytes in Human Viral and Malaria Infections. 1994. 37 figs. IX, 210 pp. ISBN 3-540-57259-7

Vol. 190: **Koprowski, Hilary; Lipkin, W. Ian (Eds.):** Borna Disease. 1995. 33 figs. IX, 134 pp. ISBN 3-540-57388-7

Vol. 191: **ter Meulen, Volker; Billeter, Martin A. (Eds.):** Measles Virus. 1995. 23 figs. IX, 196 pp. ISBN 3-540-57389-5

Vol. 192: **Dangl, Jeffrey L. (Ed.):** Bacterial Pathogenesis of Plants and Animals. 1994. 41 figs. IX, 343 pp. ISBN 3-540-57391-7

Vol. 193: **Chen, Irvin S. Y.; Koprowski, Hilary; Srinivasan, Alagarsamy; Vogt, Peter K. (Eds.):** Transacting Functions of Human Retroviruses. 1995. 49 figs. IX, 240 pp. ISBN 3-540-57901-X

Vol. 194: **Potter, Michael; Melchers, Fritz (Eds.):** Mechanisms in B-cell Neoplasia. 1995. 152 figs. XXV, 458 pp. ISBN 3-540-58447-1

Vol. 195: **Montecucco, Cesare (Ed.):** Clostridial Neurotoxins. 1995. 28 figs. XI., 278 pp. ISBN 3-540-58452-8

Vol. 196: **Koprowski, Hilary; Maeda, Hiroshi (Eds.):** The Role of Nitric Oxide in Physiology and Pathophysiology. 1995. 21 figs. IX, 90 pp. ISBN 3-540-58214-2

Vol. 197: **Meyer, Peter (Ed.):** Gene Silencing in Higher Plants and Related Phenomena in Other Eukaryotes. 1995. 17 figs. IX, 232 pp. ISBN 3-540-58236-3

Vol. 198: **Griffiths, Gillian M.; Tschopp, Jürg (Eds.):** Pathways for Cytolysis. 1995. 45 figs. IX, 224 pp. ISBN 3-540-58725-X

Vol. 199/I: **Doerfler, Walter; Böhm, Petra (Eds.):** The Molecular Repertoire of Adenoviruses I. 1995. 51 figs. XIII, 280 pp. ISBN 3-540-58828-0

Vol. 199/II: **Doerfler, Walter; Böhm, Petra (Eds.):** The Molecular Repertoire of Adenoviruses II. 1995. 36 figs. XIII, 278 pp. ISBN 3-540-58829-9

Vol. 199/III: **Doerfler, Walter; Böhm, Petra (Eds.):** The Molecular Repertoire of Adenoviruses III. 1995. 51 figs. XIII, 310 pp. ISBN 3-540-58987-2

Vol. 200: **Kroemer, Guido; Martinez-A., Carlos (Eds.):** Apoptosis in Immunology. 1995. 14 figs. XI, 242 pp. ISBN 3-540-58756-X

Vol. 201: **Kosco-Vilbois, Marie H. (Ed.):** An Antigen Depository of the Immune System: Follicular Dendritic Cells. 1995. 39 figs. IX, 209 pp. ISBN 3-540-59013-7

Vol. 202: **Oldstone, Michael B. A.; Vitković, Ljubiša (Eds.):** HIV and Dementia. 1995. 40 figs. XIII, 279 pp. ISBN 3-540-59117-6

Vol. 203: **Sarnow, Peter (Ed.):** Cap-Independent Translation. 1995. 31 figs. XI, 183 pp. ISBN 3-540-59121-4

Springer-Verlag and the Environment

We at Springer-Verlag firmly believe that an international science publisher has a special obligation to the environment, and our corporate policies consistently reflect this conviction.

We also expect our business partners – paper mills, printers, packaging manufacturers, etc. – to commit themselves to using environmentally friendly materials and production processes.

The paper in this book is made from low- or no-chlorine pulp and is acid free, in conformance with international standards for paper permanency.

MIX
Papier aus verantwortungsvollen Quellen
Paper from responsible sources
FSC® C105338
FSC
www.fsc.org

If you have any concerns about our products,
you can contact us on
ProductSafety@springernature.com

In case Publisher is established outside the EU,
the EU authorized representative is:
Springer Nature Customer Service Center GmbH
Europaplatz 3, 69115 Heidelberg, Germany

Printed by Libri Plureos GmbH
in Hamburg, Germany